Complex Variables

HOLDEN-DAY SERIES IN MATHEMATICS

Earl A. Coddington and Andrew Gleason, Editors

Complex Variables

Norman Levinson
Massachusetts Institute
of Technology

Raymond M. Redheffer
University of California,
Los Angeles

Holden-Day, Inc.
San Francisco, Cambridge,
London, Amsterdam

McGraw Hill Education (India) Private Limited
NEW DELHI

McGraw Hill Education Offices
New Delhi New York St Louis San Francisco Auckland Bogotá Caracas
Kuala Lumpur Lisbon London Madrid Mexico City Milan Montreal
San Juan Santiago Singapore Sydney Tokyo Toronto

McGraw Hill Education (India) Private Limited

McGraw Hill Education (India) Edition 2015

RXLCRRZVRQAQQ

Reprinted in India by arrangement with McGraw-Hill Global Education Holdings, LLC, New York

Sales territories: India, Pakistan, Nepal, Bangladesh, Sri Lanka and Bhutan

Library of Congress Catalog Card No. 76-113833

ISBN-13: 978-93-392-2157-7
ISBN-10: 93-392-2157-5

Published by McGraw Hill Education (India) Private Limited,
P-24, Green Park Extension New Delhi 110 016, and printed at
Pashupati Printers, Pvt. Ltd., Delhi 110 095

Visit us at: www.mheducation.co.in

US Edition Price: $72.67

ACKNOWLEDGMENT

It is a pleasure to express our appreciation to

FREDERICK H. MURPHY,

president of Holden-Day, whose creativity as a publisher brought this collaboration into being.

Preface

This book is addressed to students with diverse backgrounds who are interested in complex analysis mainly for its usefulness to them. The content represents what we consider to be the minimum content that is indispensible alike for mathematicians, mathematical physicists, and scientific engineers.

Since there is little unanimous agreement concerning the foundations of mathematics, in analysis one usually starts with a set of postulates which are by no means universally accepted, and one then proceeds with the further development. We have assumed many facts of real analysis here. In view of the general upgrading of scientific training that has taken place in major institutions throughout the country, the needed subjects are included not only in courses of advanced calculus and real analysis, but in undergraduate courses addressed to engineers and physicists. The student not familiar with some of the results referred to can simply accept the statements which are based on these results, and proceed with his study starting from there. In this sense the logical development is self-contained.

The goal in the study of mathematics should be not only the passive learning and aesthetic enjoyment of the subject, but also the ability to use the material in other areas of mathematics, in the creation of new mathematics and in the application of mathematics to other disciplines. We have tried to give a glimpse of the breadth of complex analysis as well as its depth.

Norman Levinson
Raymond M. Redheffer

To the instructor

For some objectives, sections marked * need not be studied in detail but can be used for occasional reference. One reason for starring a section is that the contents parallel real analysis so closely that many readers would be willing to take the results on faith. Sections presenting important but isolated applications are also starred. Occasionally, a starred section includes essential new ideas, but with a degree of precision or generality that may seem unnecessary to those whose primary interest is in applications. The content is then summarized informally in the next unstarred section. According to our experience, by omission of starred sections, the instructor can get well into complex integration before the end of the fourth week. This rapidity is a source of satisfaction to students of applied mathematics whose time is limited.

To save space, many problems consist of unnumbered parts which are set across the page. In general, the technical difficulty increases horizontally while conceptual difficulty increases vertically; that is, with increasing problem number. The problems fall into three broad categories: (1) Those that promote the assimilation of material in the preceding section, (2) those that prepare the student for concepts introduced later, and (3) those that give new theorems and help to lay the foundations for original research. Some advanced problems are included for enrichment when the contents of this book are presented to graduate students, as has been done by both authors. In such cases, the prerequisites are clearly stated.

Since both of us have done much of our research in functions of a complex variable, the subject is to us a very vital and dynamic one. To bring the student quickly to the point where the theory falls into perspective and becomes alive to him, we have cut some corners. Analytic developments are sometimes simplified by reference to the geometry of the plane, and there are a few compromises in notation. The presentation is logical, but we place a high value on intuition for insight and motivation. With limited goals, we have tried to instill the attitude of the active doer rather than that of the passive spectator in this fundamental and elegant area of mathematics.

Norman Levinson
Raymond M. Redheffer

Suggestions for use of this book

The contents of this book are arranged in fifty lessons. Five of these, at the ends of chapters, are very short and provide an opportunity for review.

One-semester course for mathematicians

Chapter 1: 1,2,3,4,5,6
Chapter 2: 1,2,3,4,7
Chapter 3: 1,2,3,4,5,6,7,8,9,10
Chapter 4: 1,2,3,6,9
Chapter 5: 1,2,6,7,8
Chapter 6: 1,2,3

One-semester course for engineers

Chapter 1: 1,2,3,4
Chapter 2: 1,2,3,5,6
Chapter 5: 1,2,3,4,5
Chapter 3: 3,5,6,7,8,9
Chapter 4: 2,3,4,5,6,7,10
Chapter 6: 1,2,3,4

One-semester course for physicists

Chapter 1: 1,2,3,4
Chapter 2: 1,2,3,4,5
Chapter 3: 3,5,6,7,8,9
Chapter 4: 2,3,4,5,6,8,10
Chapter 5: 1,2,3,4
Chapter 6: 1,2,3,5,6,7,8,9

Two-semester course

First semester: Chapters 1,2,3
Second semester: Chapters 4,5,6

Two-semester course for advanced students

First semester: Most of this book.
Second semester: Supplementary topics from advanced texts and from original sources.

Contents

Complex Variables

Chapter 1

Complex numbers and functions

This chapter reviews familiar results of real analysis concerning polynomials, functions, regions, and limits in the setting of complex analysis. Although there are many points of similarity, it is found that complex numbers have a two-dimensional character which gives a distinctive flavor to the geometric description. For example, a real function of a real variable generally maps an interval of a line onto another interval, while a complex function of a complex variable generally maps a region of the plane onto another region. The two-dimensional character of complex numbers accounts, in part, for the prominence of plane topology in complex analysis, and also for the effectiveness of complex analysis in two-dimensional problems of mathematical physics.

1. Complex numbers. Since the square of every real number is positive or zero, the equation $x^2 = -1$ cannot be solved by real numbers. Complex numbers provide an extension of the real number system to a larger system in which this equation, and others like it, have solutions. The impossibility of solving equations such as $x^2 + 1 = 0$ by real numbers was long known to mathematicians. Nevertheless, development of the theory of complex numbers came centuries after this difficulty was apparent—a fact which indicates that the concept of complex numbers is by no means obvious.

At one time complex numbers were developed by adjoining a symbol $\sqrt{-1}$ to the real number system, which satisfies the equation $x^2 + 1 = 0$ by definition. This notation, however, is not very satisfactory. The difficulty with the symbol $\sqrt{-1}$ is the same as could arise in attempting to represent 1 by $\sqrt{1}$. Since -1 is also a square root of 1 it is possible to get into apparently paradoxical situations by loose procedures with $\sqrt{1}$. In the case of $\sqrt{-1}$, a common paradox arising from loose usage of terms is the following: $(\sqrt{-1})^2 = -1$ and, on the other hand, $(\sqrt{-1})^2 = \sqrt{-1}\sqrt{-1} = \sqrt{(-1)(-1)} = \sqrt{1} = 1$.

The notation i for $\sqrt{-1}$, introduced in 1779 by the Swiss mathematician Leonhard Euler, avoids this pitfall. The basic property $i^2 = -1$ is retained, but there is little danger of confusing $-i$ with i. The symbol i is sometimes called the imaginary unit, just as 1 is the real unit.

Using Euler's notation, we summarize the algebra of complex numbers here. It is assumed that the reader has already had experience with complex numbers and may therefore find some intellectual interest in a purely algebraic treatment. Familiarity with the real number system is presupposed, but for several pages the discussion makes no reference to geometry.

A complex number α is defined in terms of two real numbers a and b and is designated by

$$\alpha = a + ib.$$

The number a is called the real part of α and b is the imaginary part. Whenever we write an equation such as $\alpha = a + ib$ without further explanation, it is understood that a and b are real. Thus

$$a = \text{Re } \alpha, \qquad b = \text{Im } \alpha$$

where the symbols Re and Im denote real and imaginary parts, respectively.

If $\alpha = a + ib$ and $\beta = c + id$, then the complex numbers α and β are equal if and only if $a = c$ and $b = d$. The sum and product of α and β are

$$\begin{aligned} \alpha + \beta &= (a + c) + i(b + d) \\ \alpha\beta &= (ac - bd) + i(ad + bc). \end{aligned} \tag{1.1}$$

These are the results one would get using the ordinary rules of arithmetic with the added relation $i^2 = -1$. It is readily verified that (1.1) implies the commutative, associative and distributive laws:

$$\begin{aligned} \alpha + \beta &= \beta + \alpha & \alpha\beta &= \beta\alpha \\ \alpha + (\beta + \gamma) &= (\alpha + \beta) + \gamma & \alpha(\beta\gamma) &= (\alpha\beta)\gamma \\ & \alpha(\beta + \gamma) = \alpha\beta + \alpha\gamma. \end{aligned} \tag{1.2}$$

Setting $b = d = 0$ in (1.1), we see that

$$\begin{aligned}(a + i0) + (c + i0) &= (a + c) + i0 \\ (a + i0)(c + i0) &= (ac) + i0.\end{aligned} \tag{1.3}$$

Equations (1.3) indicate that the complex number $x + i0$ is equivalent in every respect, except notation, to the real number x. Hence we can, and do, consider that

$$x = x + i0.$$

This expresses the familiar fact that the complex numbers contain the reals as a special case.

By (1.1) it is seen that $0 + i0$ and $1 + i0$ are the zero and unit, respectively. That is,

$$\alpha + (0 + i0) = \alpha, \qquad \alpha(1 + i0) = \alpha \tag{1.4}$$

for all α. Since the above-mentioned correspondence of real and complex numbers gives $0 = 0 + i0$ and $1 = 1 + i0$, the equations (1.4) take the more familiar form $\alpha + 0 = \alpha$, $\alpha 1 = \alpha$.

By $-\alpha$ is meant the product $(-1)(\alpha)$. Clearly,

$$\alpha + (-\alpha) = (1)(\alpha) + (-1)(\alpha) = [(1) + (-1)](\alpha) = 0.$$

For any complex number α, the equation

$$\alpha + z = \beta \tag{1.5}$$

has a unique solution z, which is called the *difference* and is written $\beta - \alpha$. Indeed, if $-\alpha$ is added to both sides of (1.5), we get

$$z = \beta + (-\alpha)$$

and conversely, this z satisfies (1.5). Thus $\beta - \alpha = \beta + (-\alpha)$.

The *conjugate* of α is denoted by $\bar{\alpha}$ (read "alpha bar") and is defined by

$$\bar{\alpha} = a - ib.$$

The conjugate of $\bar{\alpha}$ is α. By (1.1),

$$\alpha\bar{\alpha} = a^2 + b^2$$

and so $\alpha\bar{\alpha} > 0$ unless $\alpha = 0$.

If $\alpha \neq 0$ and $z = x + iy$, then the equation

$$\alpha z = \beta \tag{1.6}$$

has a unique solution z which is called the *quotient* and is written β/α. To show this, the real and imaginary parts of both sides of (1.6) could be equated to yield a pair of linear equations for x and y with determinant $a^2 + b^2$. But a more convenient procedure is to multiply both sides of (1.6) by $\bar{\alpha}$ to give

$$\bar{\alpha}\alpha z = \bar{\alpha}\beta.$$

Since $\bar{\alpha}\alpha > 0$, the above can be multiplied by $1/(\bar{\alpha}\alpha)$ to give

$$z = \frac{1}{\bar{\alpha}\alpha}\bar{\alpha}\alpha z = \frac{1}{\bar{\alpha}\alpha}\bar{\alpha}\beta = \frac{1}{a^2 + b^2}(a - ib)(c + id). \tag{1.7}$$

Conversely, z in (1.7) satisfies (1.6).

This discussion shows that laws pertaining to addition, subtraction, multiplication and division of complex numbers are similar to those for real numbers, and hence familiar definitions based on these operations can be carried over to the complex case. For example, as the reader will recall from real analysis, an expression of the form

$$P(z) = a_n z^n + \cdots + a_1 z + a_0$$

is a polynomial in z with coefficients $a_0, a_1, \ldots, a_n$. If $a_n \neq 0$, the degree of the polynomial is n. The same definitions apply in the complex case, except that the coefficients and z may be complex. If all coefficients are real, the polynomial is said to be a real polynomial.

Construction of a polynomial $P(z)$ requires only addition and multiplication. If division is also allowed, the result can be represented as a quotient of two polynomials $P(z)/Q(z)$ and, as in real analysis, it is called a rational function. When both polynomials P and Q are real, the rational function is said to be real.

Extension of this terminology to analogous situations is obvious; for example, a real rational function of $\alpha, \beta, \ldots, \sigma$ is generated from $\alpha, \beta, \ldots, \sigma$ and from the set of real numbers by repeated addition, multiplication and division.

From the definition of conjugates, it follows that

$$\overline{\alpha + \beta} = \bar{\alpha} + \bar{\beta}, \qquad \overline{\alpha\beta} = \bar{\alpha}\bar{\beta}.$$

This can be extended by induction to any finite number of terms or factors to give

$$\overline{\alpha + \beta + \cdots + \sigma} = \bar{\alpha} + \bar{\beta} + \cdots + \bar{\sigma}$$
$$\overline{\alpha\beta \cdots \sigma} = \bar{\alpha}\bar{\beta} \cdots \bar{\sigma}.$$

We say, briefly, that the conjugate of a sum equals the sum of the conjugates, and the conjugate of a product equals the product of the conjugates.

If $\alpha z = \beta$, then $\bar{\alpha}\bar{z} = \bar{\beta}$ and, since $z = \beta/\alpha$, we get

$$\overline{\left(\frac{\beta}{\alpha}\right)} = \frac{\bar{\beta}}{\bar{\alpha}} \qquad (\alpha \neq 0).$$

Also, $\bar{c} = c$ if c is real, and so $\overline{c\alpha} = c\bar{\alpha}$ in that case. Combining these results it follows that if F denotes any real rational function of $\alpha, \beta, \ldots, \sigma$, then

$$\overline{F(\alpha, \beta, \ldots, \sigma)} = F(\bar{\alpha}, \bar{\beta}, \ldots, \bar{\sigma}).$$

Example 1.1. For any complex number α show that

$$\operatorname{Re} \alpha = \frac{\alpha + \bar{\alpha}}{2}, \qquad \operatorname{Im} \alpha = \frac{\alpha - \bar{\alpha}}{2i}.$$

If $\alpha = a + ib$, then $\alpha + \bar{\alpha} = (a + ib) + (a - ib) = 2a = 2\operatorname{Re} \alpha$, which gives the first equation. Proof of the second is similar.

Example 1.2. As in the text, solve $(1 + i)z = 2 + i$. Multiplying by $1 - i$, the conjugate of the coefficient of z, we get

$$(1 + i)(1 - i)z = (1 - i)(2 + i) = 2 - i - i^2,$$

or $2z = 3 - i$. Division by 2 now gives $z = 3/2 - (1/2)i$.

Example 1.3. Write $(2 + i)/(1 + i)$ in the form $a + ib$. Multiplying numerator and denominator by $1 - i$, the conjugate of the denominator, we get

$$\frac{2+i}{1+i} = \frac{(2+i)(1-i)}{(1+i)(1-i)} = \frac{3-i}{2} = \frac{3}{2} - \frac{1}{2}i.$$

We used the fact that $\beta/\alpha = \gamma\beta/\gamma\alpha$ if α and γ are not zero. This follows from the definition of quotient, according to which β/α and $\gamma\beta/\gamma\alpha$ are the unique solutions of

$$\alpha z = \beta \qquad \text{and} \qquad \gamma\alpha z = \gamma\beta,$$

respectively. The equations are clearly equivalent, and so their solutions coincide. Other operations with fractions are extended with equal ease from the real to the complex case and should be used without hesitation. A similar remark applies to operations with integral exponents, as

$$\alpha^m\alpha^n = \alpha^{m+n}, \qquad (\alpha^m)^n = \alpha^{mn}, \qquad (\alpha\beta)^n = \alpha^n\beta^n.$$

Example 1.4. If P is a polynomial, a real or complex number α, such that $P(\alpha) = 0$, is called a *zero* of the polynomial. Show that the complex zeros of a real polynomial occur in conjugate pairs.

If $P(\alpha) = 0$, it is to be proved that $P(\bar{\alpha}) = 0$. But

$$P(\bar{\alpha}) = \overline{P(\alpha)} = \bar{0} = 0,$$

since P is real, and this completes the proof.

Example 1.5. For each positive integer n show that

$$1 + z + z^2 + \cdots + z^{n-1} = \frac{1-z^n}{1-z}, \qquad z \neq 1.$$

If the left side is denoted by s, then s and sz are, respectively,

$$s = 1 + z + z^2 + \cdots + z^{n-2} + z^{n-1}$$
$$sz = z + z^2 + \cdots + z^{n-1} + z^n.$$

By subtraction, $s(1-z) = 1 - z^n$, and division by $1 - z$ gives s.

Example 1.6. If n is a positive integer, prove that there are positive real numbers $c_0, c_1, \ldots, c_n$ such that $c_0 = 1$, $c_1 = n$ and

$$(1+z)^n = c_0 + c_1 z + c_2 z^2 + \cdots + c_n z^n$$

for all complex z. The proof is by induction. The result is obvious for

$n = 1$, and if it holds for n, multiplication by $1 + z$ gives

$$(1 + z)^{n+1} = c_0 + (c_1 + c_0)z + \cdots + (c_n + c_{n-1})z^n + c_n z^{n+1}.$$

By inspection, the new coefficients are positive if the old ones were, the constant term $c_0 = 1$ is unchanged, and the new coefficient of z is $n + 1$ if the original coefficient c_1 was n.

This is the binomial expansion familiar from real analysis, and the coefficients c_k are the binomial coefficients. The proof shows that the expansion is valid for complex, as well as real, z.

Problems

1. Compute Re $(1+i)$, Im $(3-2i)$, $\overline{3-2i}$, $(1+i)+(3-2i)$, $(1+i)(3-2i)$.
2. Verify $(\alpha\beta)\gamma = \alpha(\beta\gamma)$ with $\alpha = 1 + 3i$, $\beta = 2 + 2i$, $\gamma = 3 + i$.
3. Solve as in Example 1.2 and check by substitution:
$$(1 + i)z = 1, \qquad (5 + i)z = 6 - i, \qquad (1 + ib)z = 1 - ib.$$
4. Write in the form $a + ib$ if $z = x + iy$, $z \neq 0$, $z \neq i$:
$$(1 + i)^2, \qquad (1 + i)^{11}, \qquad \frac{3 + 4i}{1 - 2i}, \qquad z^3, \qquad \bar{z}z, \qquad \frac{\bar{z}}{z}, \qquad \frac{z - i}{1 - \bar{z}i}.$$
5. Prove that Re $(1/z) > 0$ if and only if Re $z > 0$.
6. (a) In Example 1.6 let n be a multiple of 4. By setting $z = i$ and equating imaginary parts, show that the binomial coefficients, in this case, satisfy
$$c_1 - c_3 + c_5 - c_7 + \cdots = 0.$$
(b) Get another identity by equating real parts. (c) Consider other values of n.
7. In Example 1.6 set $z = \beta/\alpha$ and multiply by α^n to get
$$(\alpha + \beta)^n = c_0\alpha^n + c_1\alpha^{n-1}\beta + c_2\alpha^{n-2}\beta^2 + \cdots + c_n\beta^n.$$
8. In Example 1.5 set $z = \beta/\alpha$ and multiply by α^{n-1} to get
$$\alpha^{n-1} + \alpha^{n-2}\beta + \cdots + \alpha\beta^{n-2} + \beta^{n-1} = \frac{\alpha^n - \beta^n}{\alpha - \beta}, \qquad \alpha \neq \beta.$$
9. If $\alpha = \beta$ and $\gamma = \delta$, prove $\alpha + \gamma = \beta + \delta$ and $\alpha\gamma = \beta\delta$.
10. If $\alpha\beta = 0$, prove $\alpha = 0$ or $\beta = 0$. (When $\alpha \neq 0$, α^{-1} exists.)
11. If $\beta \neq 0$ and $\delta \neq 0$, then $(\alpha/\beta)(\gamma/\delta) = (\alpha\gamma)/(\beta\delta)$. (Show that each side satisfies $(\beta\delta)z = \alpha\gamma$.)

Polynomials

12. Let z_0 be a given complex number. A polynomial $P(z)$ is *divisible* by $z - z_0$ if $P(z) = (z - z_0)Q(z)$ where $Q(z)$ is a polynomial. For $n = 1, 2, 3, \ldots$ and c constant, show that $c(z^n - z_0^n)$ is divisible by $z - z_0$. (Set $\alpha = z$ and $\beta = z_0$ in Problem 8.)
13. If $P(z) = a_0 + a_1 z + \cdots + a_m z^m$ is an arbitrary polynomial, prove that
$$P(z) - P(z_0)$$
is divisible by $z - z_0$. (Express $P(z) - P(z_0)$ as a sum of terms like those in Problem 12.)
14. From Problem 13 deduce that $P(z)$ is divisible by $z - z_0$ if $P(z_0) = 0$.
15. Let $P(z)$ be a polynomial of degree $\leq n$ which vanishes at n distinct values $z_1, z_2, \ldots, z_n$. Prove that
$$P(z) = c(z - z_1)(z - z_2) \cdots (z - z_n)$$
where c is a constant. (By Problem 14, $P(z) = (z - z_n)Q(z)$ and $Q(z) = 0$ at $z_1, z_2, \ldots, z_{n-1}$. Use induction.)
16. If the polynomial $P(z)$ has degree $\leq n$ and vanishes at $n + 1$ distinct values z_k, show that $P(z) = 0$. (Setting $z = z_{n+1}$ in Problem 15 gives $c = 0$.)
17. If $P(z)$ and $Q(z)$ are polynomials of degree $\leq n$ which agree for $n + 1$ distinct values z_k, then $P(z) = Q(z)$. (Apply Problem 16 to $P(z) - Q(z)$.)

2. Absolute values. The subject of analysis is in large part based on inequalities. There is no relation $<$ between complex numbers that has the familiar properties of this relation as applied to real numbers.[1] However, inequalities involving complex numbers are easily expressed by the symbol for absolute value.

If $\alpha = a + ib$, the *absolute value* of α is designated by $|\alpha|$ and is defined to be the nonnegative square root of $a^2 + b^2$. Thus,

$$|\alpha|^2 = a^2 + b^2 = \alpha\bar{\alpha} \qquad \text{and} \qquad |\alpha| \geq 0.$$

Evidently $|\alpha| > 0$ unless $\alpha = 0$, and $|c| = c$ if c is positive. The expression $|\alpha|$ is often called the *modulus* of α.

From $|\alpha|^2 = \alpha\bar{\alpha}$, it follows that $|\bar{\alpha}| = |\alpha|$. Also

$$|\alpha\beta|^2 = \alpha\beta\,\overline{\alpha\beta} = \alpha\beta\,\bar{\alpha}\bar{\beta} = \alpha\bar{\alpha}\,\beta\bar{\beta} = |\alpha|^2\,|\beta|^2,$$

[1] Hence an inequality such as $a < b$ or $0 \leq t \leq 1$ automatically implies that the numbers are real.

and taking the square root,

$$|\alpha\beta| = |\alpha||\beta|. \tag{2.1}$$

This can be extended by induction to a product with any number of factors to give

$$|\alpha\beta \cdots \sigma| = |\alpha||\beta| \cdots |\sigma|,$$

which states that the absolute value of a product equals the product of the absolute values of the factors.

If $\alpha z = \beta$, then $|\alpha||z| = |\beta|$ and, since $z = \beta/\alpha$ for $\alpha \neq 0$,

$$\left|\frac{\beta}{\alpha}\right| = \frac{|\beta|}{|\alpha|} \qquad (\alpha \neq 0).$$

This states that the absolute value of a quotient equals the quotient of the absolute values.

Since $a^2 \leq a^2 + b^2$ and $b^2 \leq a^2 + b^2$ we see, on taking the square root, that $|a| \leq |\alpha|$ and $|b| \leq |\alpha|$, or

$$|\text{Re } \alpha| \leq |\alpha|, \qquad |\text{Im } \alpha| \leq |\alpha|.$$

If $\gamma = \alpha\bar{\beta}$, then $\alpha\bar{\beta} + \bar{\alpha}\beta = \gamma + \bar{\gamma} = 2\text{Re } \gamma$, and hence the formula

$$(\alpha + \beta)(\bar{\alpha} + \bar{\beta}) = \alpha\bar{\alpha} + \alpha\bar{\beta} + \bar{\alpha}\beta + \beta\bar{\beta}$$

gives

$$|\alpha + \beta|^2 = |\alpha|^2 + |\beta|^2 + 2\text{Re } (\alpha\bar{\beta}). \tag{2.2}$$

Since

$$|\text{Re } (\alpha\bar{\beta})| \leq |\alpha\bar{\beta}| = |\alpha||\beta|,$$

there results

$$|\alpha + \beta|^2 \leq |\alpha|^2 + |\beta|^2 + 2|\alpha||\beta| = (|\alpha| + |\beta|)^2$$

or

$$|\alpha + \beta| \leq |\alpha| + |\beta|. \tag{2.3}$$

This extends to any finite sum by induction, so that

$$|\alpha + \beta + \cdots + \sigma| \leq |\alpha| + |\beta| + \cdots + |\sigma|. \tag{2.4}$$

Hence the absolute value of a sum of terms cannot exceed the sum of the absolute values of the terms.

The equality sign holds in (2.3) if and only if Re $(\alpha\bar{\beta}) = |\alpha\bar{\beta}|$ in (2.2). This in turn holds if and only if $\alpha\bar{\beta} \geq 0$. Since $\alpha\bar{\beta} = (\alpha/\beta)|\beta|^2$ for $\beta \neq 0$, it follows that equality in (2.3) holds if and only if $\alpha/\beta \geq 0$ or $\beta = 0$.

We conclude this discussion of absolute values by mentioning two interpretations, the first in terms of vectors, the second in terms of the cartesian plane.

If the complex number $\alpha = a + ib$ is represented by a vector with horizontal coordinate a and vertical coordinate b, as shown in Figure 2-1, the addition rule

$$(a + ib) + (c + id) = (a + c) + i(b + d)$$

shows that complex numbers add vectorially (Figure 2-2). It is also clear that $|\alpha| = (a^2 + b^2)^{1/2}$ is the length of the vector representing α. Hence (2.3) states the obvious fact that the sum of two sides of a triangle is greater than or equal to the third side. For this reason (2.3) is sometimes called the triangle inequality. Similarly, in geometric terms, (2.4) states that a straight line is the shortest distance between two points, as illustrated in

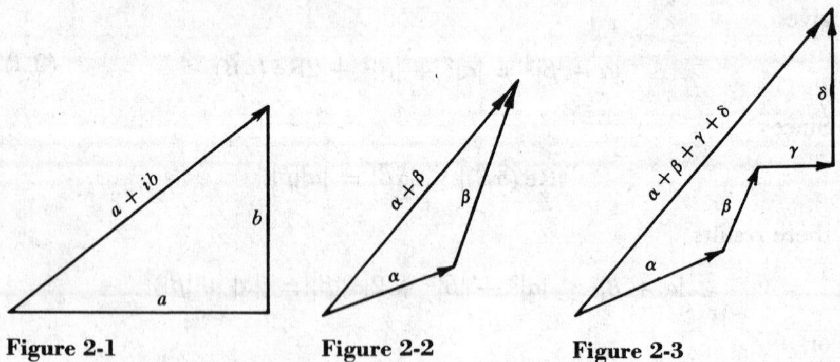

Figure 2-1 Figure 2-2 Figure 2-3

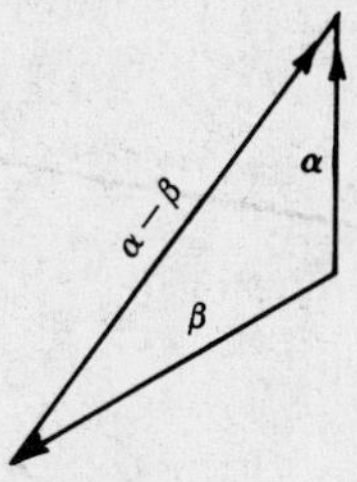

Figure 2-4

Figure 2-3. Subtraction of vectors is illustrated geometrically in Figure 2-4, where

$$\alpha = \beta + (\alpha - \beta). \tag{2.5}$$

Since the sum of the absolute values majorizes the absolute value of the sum, from (2.5) there follows $|\alpha| \leq |\beta| + |\alpha - \beta|$ or, subtracting $|\beta|$ from both sides,

$$|\alpha| - |\beta| \leq |\alpha - \beta|.$$

Interchanging α and β gives a similar inequality for $|\beta| - |\alpha|$, so that

$$\big||\alpha| - |\beta|\big| \leq |\alpha - \beta|. \tag{2.6}$$

In terms of Figure 2-4 this states that the difference of two sides of a triangle cannot exceed the third side.

The representation of a complex number by a vector as just described is useful in dynamics, in the theory of alternating current, and elsewhere. However the representation by a point in the plane as described below is more fundamental.

To each complex number $z = x + iy$ there corresponds a unique point (x,y) in the cartesian plane and conversely to each point (x,y) there corresponds the unique complex number $x + iy$. Thus, with no confusion, complex numbers can be referred to as points in the plane, and we speak indifferently of the point z or the complex number z.

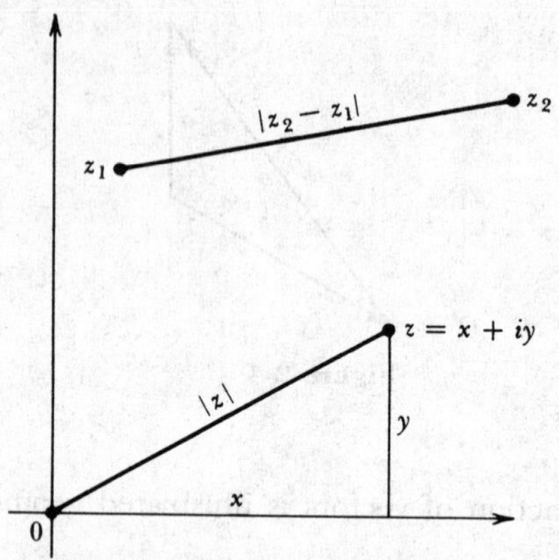

Figure 2-5

Evidently $|z| = (x^2 + y^2)^{1/2}$ represents the distance from z to the origin (Figure 2-5). The distance between two points

$$z_1 = x_1 + iy_1 \qquad \text{and} \qquad z_2 = x_2 + iy_2$$

is

$$[(x_2 - x_1)^2 + (y_2 - y_1)^2]^{1/2}$$

by a familiar formula of analytic geometry, and this agrees with $|z_2 - z_1|$ by the definition of absolute value. Hence $|z_2 - z_1|$ is interpreted geometrically as the distance from z_1 to z_2.

If we replace z by $z' = z + \alpha$, where $\alpha = a + ib$, the new coordinates are

$$x' = x + a, \qquad y' = y + b,$$

and so the transformation $z' = z + \alpha$ represents a translation. Since

$$|z_2' - z_1'| = |(z_2 + \alpha) - (z_1 + \alpha)| = |z_2 - z_1|,$$

the translation preserves distances, as one would expect geometrically.

If we replace z by $z' = \bar{z}$, the new coordinates are $(x,-y)$ and this transformation therefore represents reflection in the x axis (Figure 2-6). The reflection does not change distance, since

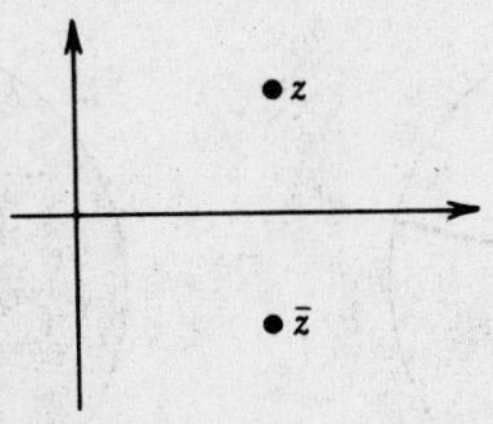

Figure 2-6

$$|z_2' - z_1'| = |\bar{z}_2 - \bar{z}_1| = |\overline{z_2 - z_1}| = |z_2 - z_1|.$$

If we replace z by $z' = \alpha z$, where α is any complex number of modulus 1, then

$$|z_2' - z_1'| = |\alpha z_2 - \alpha z_1| = |\alpha||z_2 - z_1| = |z_2 - z_1|.$$

Hence this transformation also preserves distances. As seen in the next section, it represents a rotation.

Example 2.1. If $|\alpha| < 1$, show that $|z| \le 1$ is equivalent to

$$\left|\frac{z - \alpha}{1 - \bar{\alpha}z}\right| \le 1.$$

Multiplication by $|1 - \bar{\alpha}z|$ and squaring shows that the inequality is equivalent to

$$|z - \alpha|^2 \le |1 - \bar{\alpha}z|^2,$$

and this in turn is equivalent to

$$(z - \alpha)(\bar{z} - \bar{\alpha}) \le (1 - \bar{\alpha}z)(1 - \alpha\bar{z}).$$

Canceling such terms as can be canceled, we find that the latter reduces to

$$(1 - |\alpha|^2)(1 - |z|^2) \ge 0,$$

which clearly holds if and only if $|z| \le 1$. The proof shows that equality holds for $|z| = 1$ and only then.

Example 2.2. A collection of points in the z plane is called a point set, or

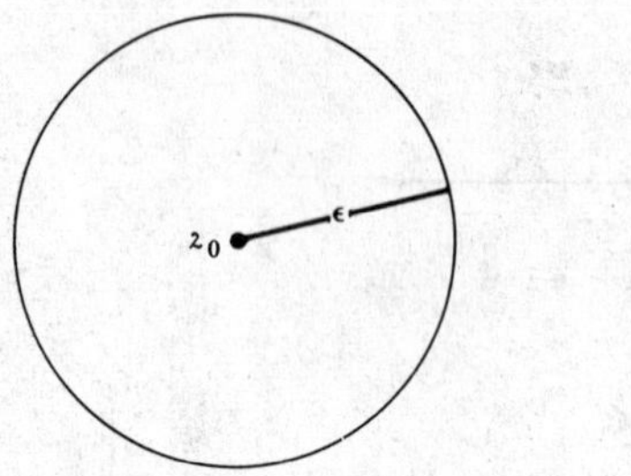

Figure 2-7 **Figure 2-8**

set of points. If ϵ is a given positive number, describe the set of points z such that $|z - z_0| = \epsilon$ and also the set such that $|z - z_0| < \epsilon$. Since $|z - z_0|$ is the distance from z to z_0, the first set represents a circle with center z_0 and radius ϵ as shown in Figure 2-7. The second set is the interior of this circle as shown by the circular disk in Figure 2-8.

In analysis, sets are often described by just giving the defining inequality, omitting the phrase "the set of all z such that . . . ." Thus one can speak of the circle $|z| = 1$, the disk $|z| < 1$, and so on.

Example 2.3. If $\beta \neq 0$, prove that

$$\left|\frac{(\alpha + \beta)^n - \alpha^n}{\beta} - n\alpha^{n-1}\right| \leq \frac{(|\alpha| + |\beta|)^n - |\alpha|^n}{|\beta|} - n|\alpha|^{n-1}. \quad (2.7)$$

By the binomial theorem (Example 1.6),

$$(\alpha + \beta)^n - \alpha^n = c_1\alpha^{n-1}\beta + c_2\alpha^{n-2}\beta^2 + \cdots + c_n\beta^n.$$

Since $c_1 = n$, the left-hand side of (2.7) reduces to $|F(\alpha,\beta)|$ where

$$F(\alpha,\beta) = c_2\alpha^{n-2}\beta + \cdots + c_n\beta^{n-1}.$$

The binomial coefficients are positive and so (2.4) followed by (2.1) gives

$$|F(\alpha,\beta)| \leq F(|\alpha|,|\beta|).$$

This is (2.7).

Example 2.4. Find $\sqrt{\alpha}$. With $\alpha = a + ib$, $z = x + iy$, and $z^2 = \alpha$ the equations $|z|^2 = |\alpha|$ and $\operatorname{Re} z^2 = \operatorname{Re} \alpha$ give, respectively,

$$x^2 + y^2 = |\alpha|, \qquad x^2 - y^2 = a.$$

Thus $x^2 = \frac{1}{2}(|\alpha| + a)$ and $y^2 = \frac{1}{2}(|\alpha| - a)$. If the signs of x and y are chosen so that $2xy = b$, it is easily checked that $z^2 = \alpha$. In the following section this equation is solved by another method which also applies to $z^n = \alpha$. Note that if $\alpha \neq 0$ there are two square roots of α.

Problems

1. If $z = x + iy$ and the denominators do not vanish, compute:
$$\left|\frac{(3+4i)(2+i)}{(1+2i)(3-4i)}\right|, \qquad |z-1|^2, \qquad |z|^4, \qquad \left|\frac{z+1}{z-1}\right|, \qquad \left|\frac{z+i}{1-iz}\right|.$$
2. (a) Find $\sqrt{i}$, $\sqrt{-i}$, $\sqrt{1+i}$, $\sqrt{2i+3}$, and check by squaring. (b) Show that a quadratic equation $z^2 + az + b = 0$ can be put into the form $w^2 - \alpha = 0$ by setting $z = w + h$ and choosing h appropriately.
3. Describe the set of points z such that

   (a) $\text{Re } z = 0$, $\text{Re } z > 0$, $|z| = 1$, $|z| > 1$, $\text{Im } z = 1$, $\text{Im } z < 1$, $1 < |z| < 2$.
   (b) $|z - 1| = 2$, $|z - 1| < 0.01$, $|z - 1| = |z + 1|$, $|\text{Re } z| + |\text{Im } z| = 1$.
4. Describe the following transformations of the plane:
$$z' = z + 1, \qquad z' = z + 1 + 2i, \qquad z' = 2z, \qquad z' = 2z + i, \qquad z' = \bar{z} + 1.$$
5. Given that $13 = 2^2 + 3^2$ and $74 = 5^2 + 7^2$, express $(13)(74) = 962$ as a sum of two squares. (Let $\alpha = 2 + 3i$, $\beta = 5 + 7i$ and use $|\alpha|^2|\beta|^2 = |\alpha\beta|^2$.)
6. If c is real and $\alpha \neq 0$, show that $\alpha z + \bar{\alpha}\bar{z} + c = 0$ is the equation of a straight line in the (x,y) plane and find its slope.
7. What does $z\bar{z} + \alpha z + \bar{\alpha}\bar{z} + c = 0$ represent if $|\alpha|^2 \geq c$?
8. If $\alpha \neq 0$, prove that α can be written uniquely as $\alpha = hv$ where $h > 0$ and $|v| = 1$. (Equating absolute values gives $h = |\alpha|$.)
9. The vectors α and β are parallel if $\alpha = t\beta$ or $\beta = t\alpha$ where t is real. Deduce $\text{Im } \alpha\bar{\beta} = 0$ as the condition for parallelism.
10. (a) Show that $\alpha \perp \beta$ if and only if $\text{Re } \alpha\bar{\beta} = 0$. (Perpendicularity is equivalent to the Pythagorean equation $|\alpha - \beta|^2 = |\alpha|^2 + |\beta|^2$.) (b) Show that $i\alpha \perp \alpha$ and $|i\alpha| = |\alpha|$, so that multiplication by i represents rotation of α through 90°.
11. (a) Referring to Problem 10, explain why $z' = iz$ represents rotation of the plane through 90°. Illustrate by plotting z and z' with $z = 1$, i, $1 + i$. (b) Multiplying by i again gives $z'' = i(iz) = -z$. Does $z'' = -z$ represent a rotation of the plane through 180°?
12. By (2.2) with β and $-\beta$, prove the *parallelogram equation*

$$|\alpha + \beta|^2 + |\alpha - \beta|^2 = 2(|\alpha|^2 + |\beta|^2).$$

Interpret by drawing α and β as vectors emanating from a common point.

13. By squaring, prove $2^{-1/2}(|a| + |b|) \leq |\alpha| \leq |a| + |b|$.

14. From (2.4) and Section 1, Problem 8, deduce

$$\left|\frac{\alpha^n - \beta^n}{\alpha - \beta}\right| \leq \frac{|\alpha|^n - |\beta|^n}{|\alpha| - |\beta|} \qquad |\alpha| \neq |\beta|,\ n = 1, 2, 3, \ldots.$$

15. Show that the angle θ between two nonzero vectors α and β satisfies

$$|\alpha||\beta| \cos\theta = \operatorname{Re} \alpha\bar{\beta}, \qquad \pm|\alpha||\beta| \sin\theta = \operatorname{Im} \alpha\bar{\beta}$$

and hence the area of the triangle formed by α, β, $\beta - \alpha$ is $\frac{1}{2}|\operatorname{Im} \alpha\bar{\beta}|$. (By the law of cosines, $|\alpha - \beta|^2 = |\alpha|^2 + |\beta|^2 - 2|\alpha||\beta| \cos\theta$.)

16. This problem requires knowledge of three-dimensional vectors. Associate $\alpha = a + ib$ with the vector $\mathbf{A} = a\mathbf{i} + b\mathbf{j} + 0\mathbf{k}$ and similarly β is associated with $\mathbf{B}$. Show that

$$\mathbf{A} \cdot \mathbf{B} = (\operatorname{Re} \alpha\bar{\beta}), \qquad \mathbf{A} \times \mathbf{B} = (\operatorname{Im} \alpha\bar{\beta})\mathbf{k}$$

and obtain the results of Problems 9, 10, and 15 from familiar results of vector algebra.

3. Multiplication and the complex plane. When the complex number z is associated with the point (x,y) as in Section 2, the (x,y) plane is called the z plane or the complex plane. The main difference between the complex plane and the cartesian plane of real analysis is that complex numbers can be multiplied. A geometric interpretation of complex multiplication is now obtained by means of polar coordinates, (r,θ).

The familiar relations $x = r\cos\theta$, $y = r\sin\theta$ suggested by Figure 3-1 lead to a corresponding representation for $z = x + iy$, namely,

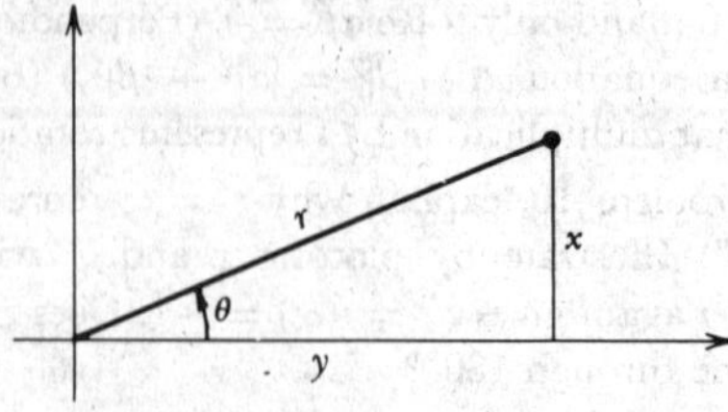

Figure 3-1

$$z = r(\cos\theta + i\sin\theta).$$

Analytically, r is given by the formula $r = (x^2 + y^2)^{1/2}$ and hence $r = |z|$. The coordinate θ is undefined for $z = 0$, but if $z \neq 0$, θ is any real number such that

$$\cos\theta = \frac{x}{|z|}, \qquad \sin\theta = \frac{y}{|z|}. \tag{3.1}$$

Of course, because of the periodicity of the sine and cosine, θ is determined only to within some multiple of 2π. The coordinate θ is known as the *argument* of z and is designated by arg z. The fact that arg z is determined only within a multiple of 2π has important consequences, as seen subsequently. The choice of arg z such that $-\pi < \arg z \leq \pi$ is referred to as the *principal value* of the argument and is denoted by Arg z.

Since arg z has infinitely many values it is necessary to establish a convention for equations involving this symbol. An equation such as

$$\arg z = \theta + 2\pi k$$

shall mean: Every choice of the integer k leads to one of the values of arg z, and every value of arg z is given by some choice of the integer k. The same relationship is also expressed by writing

$$\arg z = \theta \qquad \text{mod } 2\pi.$$

Thus $\arg z = \operatorname{Arg} z + 2\pi k$ for $z \neq 0$ or $\arg z = \operatorname{Arg} z \bmod 2\pi$. The condition $z \neq 0$ is always understood in equations involving arg z or Arg z, since (3.1) is meaningless for $z = 0$.

If θ_1 is a value of arg z_1 and θ_2 is a value of arg z_2, then

$$\begin{aligned} z_1z_2 &= r_1(\cos\theta_1 + i\sin\theta_1)r_2(\cos\theta_2 + i\sin\theta_2) \\ &= r_1r_2[(\cos\theta_1\cos\theta_2 - \sin\theta_1\sin\theta_2) + i(\sin\theta_1\cos\theta_2 + \cos\theta_1\sin\theta_2)] \\ &= |z_1||z_2|[\cos(\theta_1 + \theta_2) + i\sin(\theta_1 + \theta_2)]. \end{aligned}$$

This shows that $\theta_1 + \theta_2$ is one of the values of arg z_1z_2 and hence

$$\arg z_1z_2 = \arg z_1 + \arg z_2 + 2\pi k. \tag{3.2}$$

In multiplying complex numbers, their arguments add mod 2π. and as has already been shown in Section 2, their moduli multiply.

If $zz_1 = z_2$, then $\arg z + \arg z_1 = \arg z_2 + 2\pi k$ and since $z = z_2/z_1$, we get

$$\arg \frac{z_2}{z_1} = \arg z_2 - \arg z_1 + 2\pi k. \tag{3.3}$$

As a special case,

$$\arg \frac{1}{z} = -\arg z + 2\pi k. \tag{3.4}$$

Equation (3.2) gives, by induction,

$$\arg(z_1 z_2 \cdots z_n) = \arg z_1 + \arg z_2 + \cdots + \arg z_n + 2\pi k$$

and hence, taking all $z_k = z$,

$$\arg z^n = n \arg z + 2\pi k. \tag{3.5}$$

Since $z^{-n} = (1/z)^n$, (3.4) shows that this also holds for negative n. The result can be used to compute fractional powers of complex numbers, as shown next.

If p and q are integers, the equation $z = \alpha^{p/q}$ means that $z^q = \alpha^p$, and all solutions z of this equation are allowed. We assume that p and q have no common divisor; an exponent 4/6 would be written as 2/3, for example. The case $\alpha = 0$ being trivial, $\alpha \neq 0$ is also assumed.

From $z^q = \alpha^p$, it follows that

$$|z|^q = |\alpha|^p \qquad \text{and} \qquad q \arg z = p \arg \alpha + 2\pi k$$

by the results of Section 2 and (3.5), respectively. Hence

$$|z| = |\alpha|^{p/q} \qquad \text{and} \qquad \arg z = \frac{p}{q} \arg \alpha + \frac{2\pi k}{q} \tag{3.6}$$

where $|z|^{p/q}$ denotes the real positive root, supposed available from real analysis.

It is readily verified that with $k = 0, 1, \ldots, q-1$, (3.6) determines q distinct points in the complex plane on the circle $|z| = |\alpha|^{p/q}$ spaced at equal angles $2\pi n/q$. Any other choice of the integer k in (3.6) gives no new points. It is also clear that any of the q determinations of z by (3.6) satisfies $z^q = \alpha^p$. These q points are called the qth roots of α^p or the p/qth powers of α and are designated by $\sqrt[q]{\alpha^p}$ as well as by $\alpha^{p/q}$. It has been shown above that *there are exactly q such roots equally spaced around a circle with center at the origin.* If $\alpha > 0$, the symbol $\alpha^{p/q}$ will, however, be reserved for the root with argument zero, unless otherwise stated.

It follows from the above that a complex number α has two square roots spaced 180° apart, three cube roots spaced 120° apart, and so forth (Figures 3-2 and 3-3). In particular, the equation

$$z^n - \alpha = 0$$

has at least one root for each positive integer n and every complex α. This is a special case of the fundamental theorem of algebra, which asserts that not only $z^n - \alpha$ but every nonconstant polynomial $P(z)$ has a zero. In Problem 5.3 at the end of this chapter it is seen that the general theorem follows from the special case.

We conclude this discussion of the complex plane with some remarks on the relation of geometry to analysis. Historically, the development of complex numbers was greatly facilitated by geometric representation. The same is true of the development of functions of a complex variable. The complex plane provides so much motivation and intuitive insight that it seems to us that to develop function theory without it would handicap most students severely, and we refer to the complex plane whenever such reference makes the analysis easier to follow.

Nevertheless, it should not be thought that analysis is logically dependent on geometry. The logical dependence is just the other way around—analysis is used to give precise definitions of geometric concepts. For example, the circle of radius r_0 centered at the origin is the collection of all points z such that

$$|z| = r_0.$$

Although "circle" is a geometric concept, this is an analytic definition. The statement that z is in the upper half-plane has an immediate translation into the analytic statement $\text{Im}\, z > 0$, and so on. Such definitions are possible because the algebraic development of Sections 1 and 2 was independent of geometry.

Assuming knowledge of the trigonometric functions, the analytic definition of θ given in (3.1) can be justified by consideration of

$$\tan\theta = \frac{y}{x}, \qquad \cot\theta = \frac{x}{y}. \tag{3.7}$$

These equations also provide an effective means of computation. A definition which does not assume knowledge of trigonometry is outlined in the problems following Chapter 2, Section 4. This discussion hinges on the fact that $\mathrm{Tan}^{-1}t$ can be defined by an integral, as can the constant π. Other definitions are based on power series (Chapter 6, Section 3) or on differential equations.

Whichever method is used it is found that $\sin\theta$, $\cos\theta$ and θ have all the properties suggested by plane geometry, and these properties should be used without hesitation.

Example 3.1. Find all the values of $\sqrt[3]{-i}$. As suggested by Figure 3-2, the angle associated with $-i$ is $\mathrm{Arg}(-i) = -90°$, and so one of the values of $\arg(-i)^{1/3}$ is $(⅓)(-90°)$ or $-30°$. All other values are obtained by adding multiples of $(⅓)(360°)$, and hence

$$\arg(-i)^{1/3} = -30°, 90°, 210°, \ldots.$$

Since $|-i| = 1$, corresponding values of $(-i)^{1/3}$ are

$$\cos(-30°) + i\sin(-30°), \qquad \cos(90°) + i\sin(90°), \ldots,$$

which reduces to

$$\frac{1}{2}\sqrt{3} - \frac{1}{2}i, \qquad i, \qquad -\frac{1}{2}\sqrt{3} - \frac{1}{2}i,$$

as shown in Figure 3-2. The reader should check that the foregoing analysis agrees with (3.6) when $\alpha = -i$, $p = 1$ and $q = 3$.

Example 3.2. Which of the values of $(3 - 4i)^{-3/8}$ lie closest to the imaginary axis? Here $\mathrm{Arg}(3 - 4i)$ is found from the equation

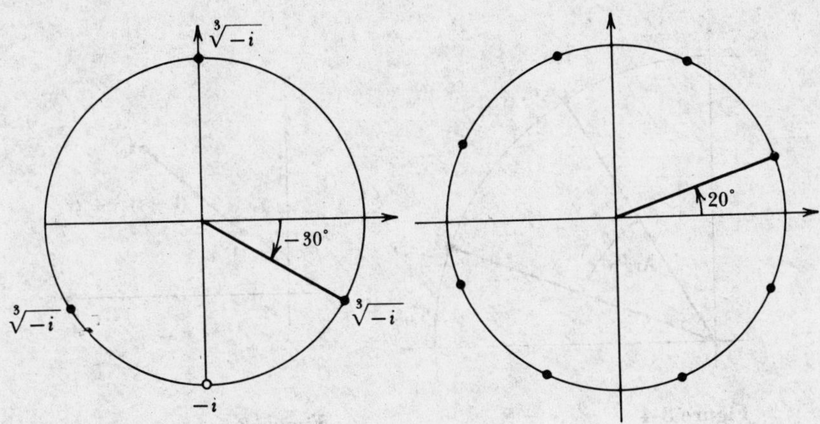

Figure 3-2

Figure 3-3

$$\cot\theta = \frac{x}{y} = -\frac{3}{4}.$$

Since $\sin\theta = y = -4/5$ is negative, θ is negative, and trigonometric tables give[1] $\theta \doteq -53°$. Since $(-3/8)(-53) \doteq 20$ we get

$$\arg(3-4i)^{-3/8} \doteq 20°, 20° + 45°, 20° + 90°, \ldots,$$

which leads to points as shown in Figure 3-3. The radius of the circle is

$$|(3-4i)^{-3/8}| = |3-4i|^{-3/8} = 5^{-3/8},$$

but is not needed for this problem. Indeed, regardless of the radius, the roots in question are the ones whose arguments are closest to 110° and $-70°$. These two roots are equidistant from the y axis.

Example 3.3. If $|\alpha| = 1$, interpret the transformation $z' = \alpha z$.

Since $|z'| = |\alpha||z| = |z|$, the distance to the origin is unchanged. The equation $\arg z' = \operatorname{Arg} z + \operatorname{Arg}\alpha \bmod 2\pi$ shows that the transformation represents a rotation through an angle $\operatorname{Arg}\alpha$ about an axis passing through the origin (Figure 3-4).

[1] We use $\doteq$ to denote approximate equality.

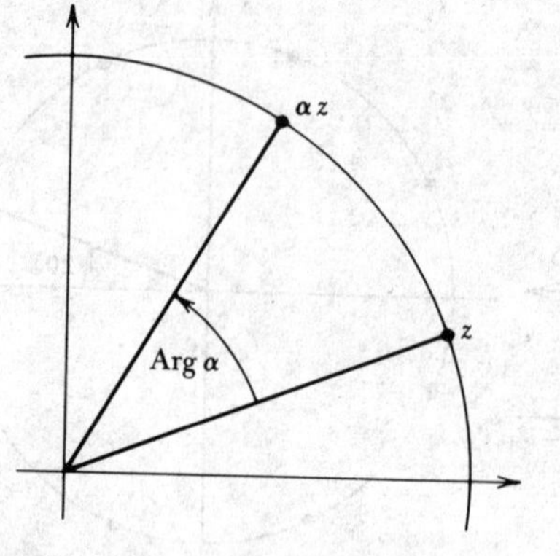

Figure 3-4

Figure 3-5

Example 3.4. If z_1 and z_2 are given complex numbers and $z_1 \neq z_2$, describe the locus of all points z satisfying

$$z = (1 - t)z_1 + tz_2, \qquad 0 \leq t \leq 1. \quad (3.8)$$

Since $t = 0$ gives $z = z_1$ and $t = 1$ gives $z = z_2$, the locus contains the points z_1 and z_2. Equating real and imaginary parts gives, in an obvious notation,

$$x = (1 - t)x_1 + tx_2, \qquad y = (1 - t)y_1 + ty_2,$$

which are the parametric equations of a straight line. The restriction of t to the interval $0 \leq t \leq 1$ indicates that the locus is the line segment joining z_1 and z_2, as shown in Figure 3-5.

In keeping with the spirit of the foregoing remarks, (3.8) is used to define the line segment. The geometric discussion is needed to make it clear that this definition is appropriate.

Problems

1. (a) Find all cube roots of $1, -8, i$. (b) Solve $(z^4 - 16)(z^3 + 1) = 0$.
2. Compute all roots and plot them in the complex plane:

$$(-1)^{1/4}, \quad (1 + i)^{1/2}, \quad (1 - i)^{-1/2}, \quad 1^{1/6}, \quad (-4)^{3/4}, \quad (-1 + i)^{8/3}.$$

3. Interpret the transformation $z' = \alpha z$ by writing in polar form when

$$\alpha = -i, \qquad \alpha = 2^{-1/2}(1 - i), \qquad \alpha = (1/2)(\sqrt{3} + i).$$

4. From results of the text, deduce $z^n = |z|^n(\cos n\theta + i \sin n\theta)$ if n is an integer, and taking $|z| = 1$ get *de Moivre's theorem*,

$$(\cos\theta + i\sin\theta)^n = \cos n\theta + i \sin n\theta.$$

5. (a) From Problem 4, get $\cos 3\theta = \cos^3\theta - 3\cos\theta\sin^2\theta$ and a similar formula for $\sin 3\theta$. (b) Generalize by the binomial theorem.

6. The notation cis θ, from c(os+)is(in), is used for $\cos\theta + i\sin\theta$. If $z = r \operatorname{cis}\theta$ and n is an integer, show that $z^{1/n}$ is given by

$$r^{1/n} \operatorname{cis}\left(\frac{\theta}{n} + \frac{2\pi k}{n}\right), \qquad k = 0, 1, 2, \ldots, n-1.$$

7. Let z_1, z_2, z_3 be three distinct points, and also z_1', z_2', z_3'. (a) Explain why the triangles (z_1,z_2,z_3) and (z_1', z_2', z_3') are similar, with this correspondence of vertices, if and only if

$$\frac{z_1 - z_2}{z_1' - z_2'} = \frac{z_2 - z_3}{z_2' - z_3'} = \frac{z_3 - z_1}{z_3' - z_1'}.$$

(b) Show that the above condition is equivalent to

$$\begin{vmatrix} 1 & 1 & 1 \\ z_1 & z_2 & z_3 \\ z_1' & z_2' & z_3' \end{vmatrix} = 0.$$

(Reduce to 2-by-2 determinants by subtracting columns.)

8. Show that the three distinct points z_1, z_2, z_3 are vertices of an equilateral triangle if and only if $z_1z_2 + z_1z_3 + z_2z_3 = z_1^2 + z_2^2 + z_3^2$. (The triangles (z_1,z_2,z_3) and (z_2,z_3,z_1) are similar. Use Problem 7.)

9. As stated in the text, $a = b \bmod 2\pi$ means that $a - b = 2\pi k$ for some integer k. If $a = b \bmod 2\pi$ and $c = d \bmod 2\pi$, show that $a + c = b + d \bmod 2\pi$ and $2a = 2b \bmod 2\pi$. Show, however, that $(1/2)a = (1/2)b \bmod 2\pi$ need not hold.

10. A *fixed point* of a transformation $z' = \alpha z + \beta$ is a point z_0 such that $z_0' = z_0$. (a) Show that there is a fixed point unless $\alpha = 1$, in which case the transformation is a translation (Section 2). (b) If $|\alpha| = 1$ and z_0 is a fixed point, subtract $z_0' = \alpha z_0 + \beta$ and $z' = \alpha z + \beta$ to get $(z' - z_0) = \alpha(z - z_0)$. Hence, show that the transformation is a rotation about the point z_0 through an angle arg α.

11. The *signed area* of the triangle with vertices $0, \alpha, \beta$, in that order, is $\frac{1}{2}\operatorname{Im} \alpha\bar{\beta}$ (see Section 2, Problem 15). The signed area of a polygon with vertices $z_1, z_2, z_3, \ldots, z_{n-1}, z_n = z_1$, in that order, is

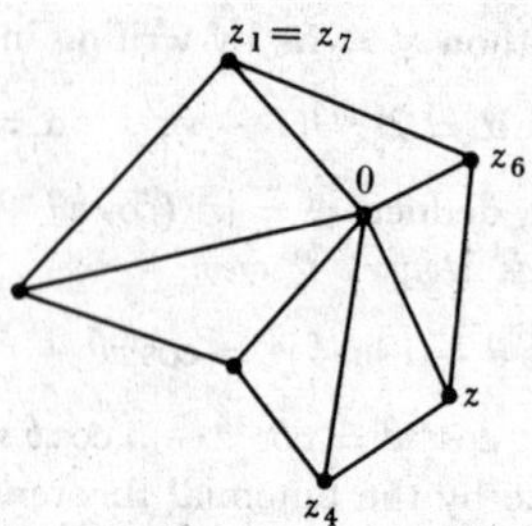

Figure 3-6

$$A = \tfrac{1}{2}\mathrm{Im}(z_1\bar{z}_2 + z_2\bar{z}_3 + \cdots + z_{n-1}\bar{z}_n).$$

(a) Interpret this formula as a sum of triangular areas (Figure 3-6) and test it by application to squares and rectangles. (b) Show that A changes sign under the reflection $z' = \bar{z}$, but is unchanged by the rotation $z' = \alpha z$, $|\alpha| = 1$. (c) Show that A is unchanged by the translation $z' = z + \beta$. (d) If the base point is z instead of 0, the formula is defined as above, with $z_k - z$ replacing z in each instance. Show that this result is independent of z. (e) Try to interpret the formula as applied to self-intersecting polygons, for example, a figure 8.

4. Regions and functions. Frequently it is necessary to restrict z to some region in the plane. Examples of common regions shown in Figure 4-1 are: (1) the disk $|z - z_0| < \rho$, which is the interior of the circle with

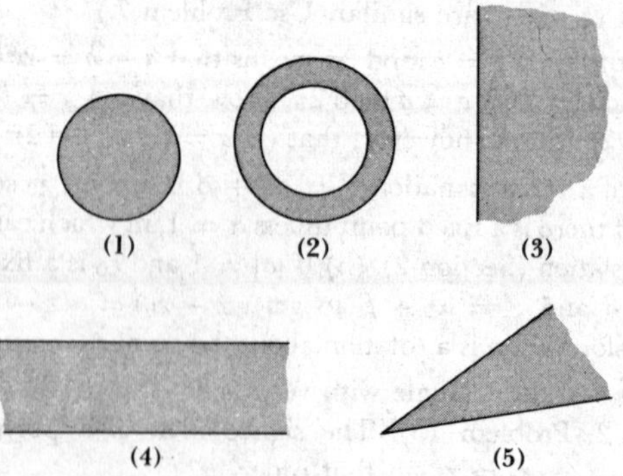

Figure 4-1

center at z_0 and radius ρ; (2) the annulus $\rho_1 < |z - z_0| < \rho_2$ with center at z_0, inner radius ρ_1 and outer radius ρ_2; (3) the right half-plane $\operatorname{Re} z > 0$; (4) the strip $0 < \operatorname{Im} z < 1$; and (5) the sector $\theta_0 < \arg z < \theta_1$. In (4) it is understood that $x = \operatorname{Re} z$ assumes all values $-\infty < x < \infty$, and in (5) that $|z|$ assumes all values $0 < |z| < \infty$. Notice that in these examples the region does not include its boundary; that is, each region consists of interior points.

We now discuss the problem of giving analytic definitions to the terms "region," "boundary," and so on, that were introduced informally above.

As stated in Example 2.3, a collection of points in the z plane is called a point set or a set of points. Each of the regions (1)–(5) described above is an example of a set of points. An ϵ *neighborhood* of a point z_0 is defined as the disk $|z - z_0| < \epsilon$, with center at z_0 and positive radius ϵ. A point z_0 of a set S is said to be an *interior* point of the set if there is an ϵ neighborhood of z_0 which is contained in the set S. If all the points of a set S are interior points, the set S is said to be an *open* set. The regions (1)–(5) described above are open sets.

Let $A_0, A_1, \ldots, A_n$ be $n + 1$ points in the plane. Then the n line segments $A_0A_1, A_1A_2, \ldots, A_{n-1}A_n$ taken in sequence form a *broken line* which is designated by $A_0A_1 \cdots A_n$ and is said to join A_0 to A_n.

An open set S is said to be *connected* if any[1] two of its points z_1 and z_2 can be joined by a broken line which is contained in the set S, as in Figure 4-2. An open connected set of points is called a *domain*. The regions (1)–(5) are domains.

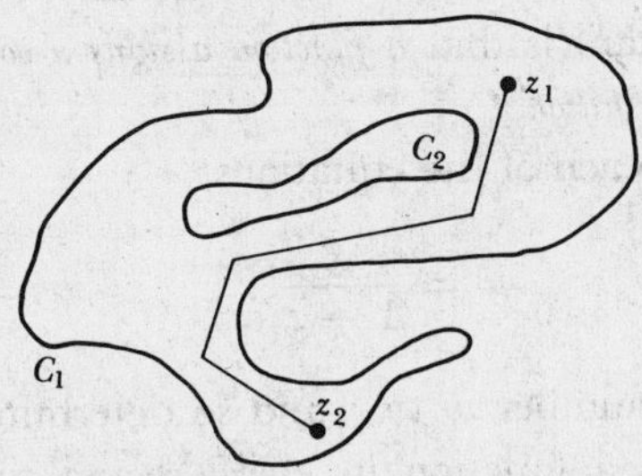

Figure 4-2

[1] In the sense of "every."

A point z_0 is called a *boundary* point of a set S if every ϵ neighborhood of the point z_0 contains at least one point in the set S and at least one point not in the set S. Because a domain is an open set, it follows that no boundary point of a domain can be in the domain. Figure 4-2 shows a domain D in which the two points z_1 and z_2 are joined by a broken line. Any two points could be joined in this way and so the domain is connected. However, the boundary consists of two separate pieces C_1 and C_2.

By a *region* is meant a domain together with none, some, or all of its boundary points. Thus a domain is a region, but the converse is not necessarily true. A region which contains all of its boundary points is said to be *closed.* Thus the square ($|x| \leq 1, |y| \leq 1$) is a closed region while ($|x| < 1, |y| < 1$) is a domain. The quadrant $x \geq 0, y \geq 0$ is also a closed region.

If a region is designated as R, then the closed region consisting of R and all of its boundary points is designated by $\bar{R}$. When R is closed, then $\bar{R} = R$.

A region is said to be *bounded* or *finite* if there is a constant M such that all points z of the region satisfy $|z| \leq M$, that is, lie in the disk of radius M and center at the origin. The regions (1) and (2) are bounded and the others are not. A region which is both closed and bounded is called *compact.*

If for each z in some region R of the plane there is a rule which assigns a complex number w to z, then w is said to be a *function* of the complex variable z in R and this relationship is designated by $w = f(z)$. The function f is said to be defined on R. The precise interpretation of the word "rule" is not important, and the choice of letters f, w, z, R is not important. What is important is that *a function assigns a complex number to each point of its region of definition.*

As an illustration, each of the equations

$$w = z^2, \qquad w = \frac{z}{1-z}, \qquad w = \frac{3}{1-|z|}$$

assigns the complex number w to z and so determines a function. The first equation defines a function in every region, the second in every region that does not contain the point $z = 1$, and the third in every region that does not contain any point of the circle $|z| = 1$.

In analysis the symbol $f(z)$ is used in two different senses. On the one hand, $f(z)$ may denote the value of the function f at z, and on the other hand, $f(z)$ may stand for the function itself. For example, the expression z^2 assigns a complex number to each point z, and so has the basic property of functions as expressed above. This function is called "the function z^2." Other examples will occur to the reader.

As in real analysis, algebraic operations with functions are defined by corresponding operations on the functional values. For example, the product fg of two functions f and g is the function F whose value at z is $F(z) = f(z)g(z)$. When functions are written in the form $f(z)$ and $g(z)$ with z variable the sum, product and composite function are, respectively,

$$f(z) + g(z), \qquad f(z)g(z), \qquad f[g(z)].$$

By separating into real and imaginary parts, any function $f(z)$ can be expressed in terms of two real functions. For example, if $w = u + iv$ and $z = x + iy$, the equation $w = z^2$ gives

$$u + iv = (x + iy)^2 = x^2 - y^2 + 2ixy,$$

which is equivalent to

$$u = x^2 - y^2, \qquad v = 2xy. \tag{4.1}$$

In the general case, $f(z)$ has a real part u and imaginary part v which are functions of z and therefore also functions of (x,y). Thus $w = f(z)$ is equivalent to

$$w = u(x,y) + iv(x,y).$$

A function $f(z)$ can be regarded as providing a transformation of points of the z plane, on which f is defined, onto points of the w plane. Such transformations are also called mappings. For example, with $w = z^2$, the equations (4.1) provide the correspondence between the point (x,y) of the z plane and the point (u,v) of the w plane. It is said that (u,v) is the *image* of (x,y) under the mapping or, equivalently, that w is the *image* of z. Similarly one can speak of the image of a set of points, such as a region.

While $w = z^2$ assigns a unique value of w to each value of z, it is not necessarily the case that the map of a region of the z plane covers its image in the w plane exactly once. Thus one and the same value w may be taken

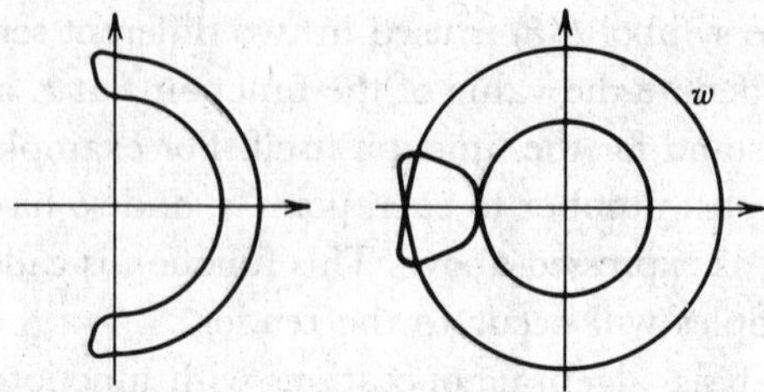

Figure 4-3

on at each of two distant points in the z plane since $(z)^2$ and $(-z)^2$ are equal. The shaded portion in the w plane of Figure 4-3 is in fact the image of two distinct parts of the region shown in the z plane. This matter is discussed more fully in Example 4.3.

So far, all points have been required to be complex numbers in the sense of Section 1. Sometimes it is convenient to adjoin an "ideal" point, which is called the *point at infinity* and is denoted by the symbol ∞. The possibility of doing this uses the fact that a point is determined as soon as all of its ϵ neighborhoods are known (Figure 4-4). An ϵ neighborhood of ∞ cannot be defined by an inequality $|z - z_0| < \epsilon$, but is defined by $|z| > 1/\epsilon$. Geometrically, an ϵ neighborhood of ∞ is the exterior of the disk with radius $1/\epsilon$ and center at the origin. A region is said to contain the point at ∞ if it has a point in common with every neighborhood of ∞. This happens if and only if the region is not bounded, and the region is then said to be *unbounded* or *infinite.*

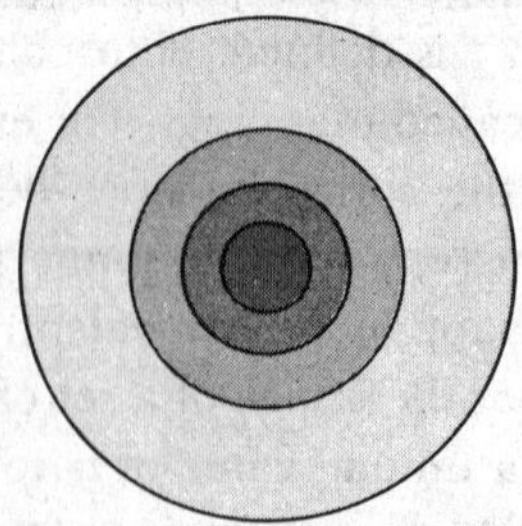

Figure 4-4

A function is defined in an ϵ neighborhood of ∞ if it is defined at all points z outside of some disk. The question whether such a function can be sensibly defined at ∞ hinges on the concept of limit introduced in the next section.

The complex plane to which the point at infinity has been adjoined is called the *extended* complex plane. Sometimes the complex plane without the point $z = \infty$ is called the finite complex plane and is designated by $|z| < \infty$. The extended complex plane can be mapped by stereographic projection onto the surface of a sphere (Figure 4-5). Let the unit sphere have its center at the origin $z = 0$ of the complex plane. The sphere is cut by the z plane along its equator, which coincides with the circle $|z| = 1$. A point P in the z plane is mapped onto the point P' on the sphere by joining the north pole N of the sphere to P by a line. The intersection of the line NP with the sphere is P'. Points for which $|z| < 1$ map onto the lower hemisphere, while those for which $|z| > 1$ map onto the upper hemisphere. The point $z = \infty$ is put in correspondence with N, the north pole of the sphere, and the origin $z = 0$ is mapped onto the south pole of the sphere. In this way a one-to-one correspondence is established between the sphere and the extended complex plane, that is, to each P there corresponds a unique P', and conversely. If P_1 is near P, then P_1' is near P'. This applies even to the point at infinity; that is, an ϵ neighborhood of ∞ in the z plane corresponds to a small region on the sphere if ϵ is small.

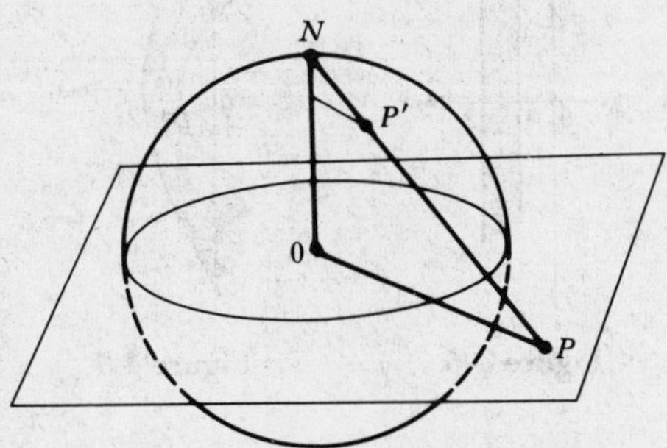

Figure 4-5

Example 4.1. Find the image of the strip $1 \le x \le 2$ under the mapping $w = z^2$. We first get the image of the line $x = x_0 \neq 0$ (Figure 4-6). The image (u,v) of a point (x_0,y) on this line is given by (4.1) as

$$u = x_0^2 - y^2, \qquad v = 2x_0 y.$$

The image of the line is given by the same equations except that y is variable, $-\infty < y < \infty$. Since $y = v/(2x_0)$, we can eliminate y to get part or all of the parabola

$$u = x_0 - \frac{v^2}{(2x_0)^2}.$$

The condition $-\infty < y < \infty$ indicates that v is unrestricted and the image of the line is therefore the whole parabola (Figure 4-7). If x_0 is now allowed to vary from 1 to 2, the lines sweep out the strip and the parabolas sweep out the shaded region shown in the figure. Its boundary consists of the two parabolas

$$u = 1 - \left(\frac{v}{2}\right)^2, \qquad u = 2 - \left(\frac{v}{4}\right)^2.$$

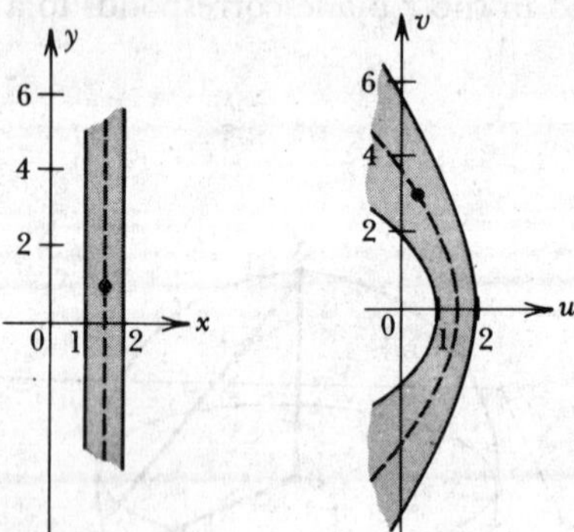

Figure 4-6 Figure 4-7

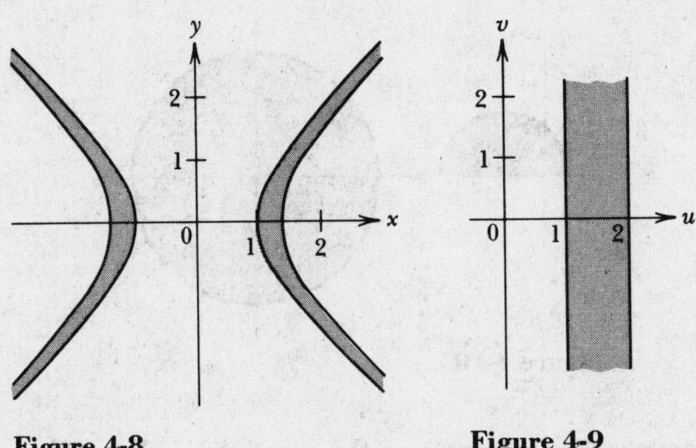

Figure 4-8 Figure 4-9

Example 4.2. Under $w = z^2$, what points of the z plane are mapped onto the strip $1 < u < 2$ of the w plane? Setting $u = u_0 > 0$, gives

$$x^2 - y^2 = u_0, \qquad -\infty < y < \infty \tag{4.2}$$

and a point (x,y) is mapped onto the line $u = u_0$ if and only if it lies on the hyperbola (4.2). Letting u_0 vary, we see that the desired set is bounded by the two hyperbolas

$$x^2 - y^2 = 1, \qquad x^2 - y^2 = 2,$$

as shown in Figure 4-8. The image of each one of the shaded regions is the whole strip of Figure 4-9 so that the strip is covered twice.

Example 4.3. The function $w = z^2$ maps the upper half of the z plane onto a region of the w plane. Interpret by using polar coordinates. Since $|w| = |z|^2$ and $\arg w = 2 \arg z \bmod 2\pi$, the circles $|z| = r$ are mapped onto circles $|w| = r^2$ and a ray making an angle θ with the x axis is mapped onto a ray making twice that angle with the u axis. The upper half-plane is fanned out by the mapping so that its image just covers the w plane, as shown in Figure 4-10.

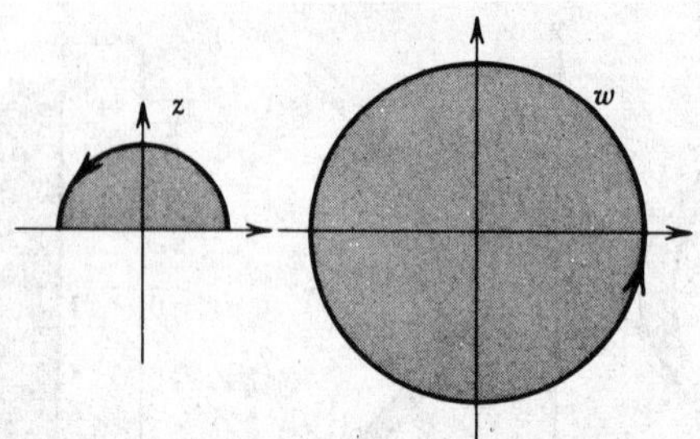

Figure 4-10

If we map a larger region than the upper half-plane, some points will be covered twice, and the image of the whole z plane covers the w plane, except for the origin, exactly twice. This results from the fact that the equation $z^2 = w$ has two solutions for $w \neq 0$; in other words, $\sqrt{w}$ has two values.

Problems

1. Find $f(z+1), f(1/z)$ and $f[f(z)]$ if $f(z) = z + 1,\ z^2,\ 1/z,\ (1+z)(1-z)^{-1}$.
2. Which of the following inequalities describe domains?

   $\operatorname{Re} z > 1,\ |z| \leq 1,\ 0 < |z| < 1,\ \operatorname{Im} z < 2|z|,\ |z-1| < |z+i|,\ 2|z^2-1| < 1.$

3. After the manner of Example 4.3, discuss the mapping of the sector

   $$0 \leq \operatorname{Arg} z < \theta$$

   under the mappings $w = z^3$, $w = z^4$, and $w = z^6$. How large must θ be so that the w plane is just covered once? (A region in the z plane whose image just covers the w plane once is called a *fundamental region* for the function $w = f(z)$.)
4. Suppose $w = f(z)$ maps a region R of the z plane onto a region R^* of the w plane in such a way that the equation $w = f(z)$ can be solved uniquely for z, thus: $z = g(w)$ for w in R^*. The function g is called the *inverse function* for f, and satisfies $f[g(w)] = w$ and $g[f(z)] = z$ for w in R^* and z in R, respectively. Find inverse functions $z = g(w)$ on appropriate regions for:

   $$w = -z, \quad w = \frac{1}{z}, \quad w = \frac{1-z}{1+z}, \quad w = z^2, \quad w = z^3, \quad w = (z-1)^4 + i.$$

   (In the last three cases, use the concept of "fundamental region" introduced in Problem 3.)

5. Find the image of the strip $|\text{Re } z| < 1$ and of the strip $1 < \text{Im } z < 2$ under the following transformations:

$$w = 2z + i, \quad w = (1 + i)z + 1, \quad w = 2z^2, \quad w = z^{-1}, \quad w = (z + 1)(z - 1)^{-1}.$$

6. Find the image of the disk $|z| < 1$ under the transformations above.
7. Let α, β, γ be three points in the plane not on a line. (a) Referring to Example

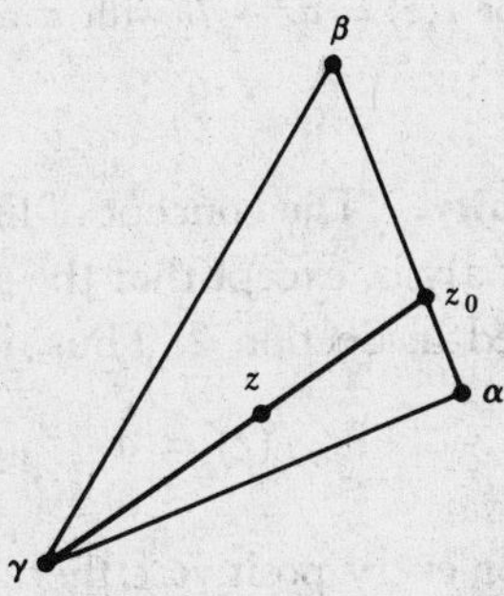

Figure 4-11

3.4 and Figure 4-11, interpret the equations

$$z_0 = r_0\alpha + s_0\beta, \qquad z = r_1 z_0 + s_1\gamma,$$

where r_0, s_0, r_1, s_1 are nonnegative real numbers with $r_0 + s_0 = r_1 + s_1 = 1$. (b) Show that z can be represented as in (a) if and only if

$$z = r\alpha + s\beta + t\gamma, \qquad r + s + t = 1, \qquad r \geq 0, s \geq 0, t \geq 0$$

and that the choice of r, s, t is unique. (c) The set in (b) is called the *triangular region,* or triangle, determined by α, β, γ. Discuss conditions on r, s, t which give the vertices, boundary, and interior of the triangle.

Isometries of the plane

8. A function $f(z)$ is called an *isometry* if it preserves distance, that is, if

$$|f(z_1) - f(z_2)| = |z_1 - z_2|$$

for all complex numbers z_1 and z_2. If $f(z)$ is an isometry and α and β are constant with $|\alpha| = 1$, prove that $g(z) = \alpha f(z) + \beta$ is also an isometry. Deduce that the function

$$g(z) = \frac{f(z) - f(0)}{f(1) - f(0)}$$

is an isometry satisfying $g(0) = 0, g(1) = 1$.

9. Let $w = g(z)$ where $g(z)$ is an isometry satisfying $g(0) = 0, g(1) = 1$. Note that $|w|^2 = |z|^2$, $|1 - w|^2 = |1 - z|^2$, since g preserves distance. Deduce Re $w =$ Re z, and hence $g(i) = i$ or $-i$.
10. If $g(i) = i$ in Problem 9, show that $|w - i|^2 = |z - i|^2$ makes Im $w =$ Im z, and hence $g(z) = z$ for all z. Similarly, if $g(i) = -i$, prove that $g(z) = \bar{z}$ for all z. Referring to Problem 8, deduce that every isometry of the plane has the form $f(z) = \alpha z + \beta$ or $f(z) = \alpha\bar{z} + \beta$ with α and β constant and $|\alpha| = 1$.

***5. Limits and continuity.** The concept of limit is defined in complex analysis just as in real analysis, except that the symbol for absolute value has the meaning assigned in Section 2. Thus, in general,

$$\lim_{z \to z_0} f(z) = \alpha \tag{5.1}$$

means the following: For every positive ϵ there is a positive δ such that

$$0 < |z - z_0| < \delta \qquad \text{implies} \qquad |f(z) - \alpha| < \epsilon. \tag{5.2}$$

The value of δ may depend, of course, on ϵ. Less precisely stated, (5.1) means that if z is near z_0, then $f(z)$ is close to α.

The condition (5.2) can hold only if $f(z)$ is defined throughout some neighborhood of z_0, with the possible exception of z_0 itself. This is of no moment when z_0 is interior to the region of definition R for f, but if z_0 is a boundary point of R, we now agree that the conclusion

$$|f(z) - \alpha| < \epsilon$$

is required only for z in R. Finally, if z_0 is neither an interior nor a boundary point of R, the limit is undefined.

With $w = f(z)$ and $w_0 = \alpha$ it is seen that (5.2) is expressed in terms of δ and ϵ neighborhoods, $|z - z_0| < \delta$ and $|w - w_0| < \epsilon$. This formulation extends to the ideal point, ∞. The statement

$$\lim_{z \to \infty} f(z) = \alpha \tag{5.3}$$

means: For every positive ϵ there is a positive δ such that

$$|z| > \frac{1}{\delta} \qquad \text{implies} \qquad |f(z) - \alpha| < \epsilon.$$

Other statements involving ∞ are interpreted similarly.

The symbol $z \to z_0$ (read "z approaches z_0") is not always written under the lim as in (5.1) but is sometimes written separately. If the fact that $z \to z_0$ is clear from context, the symbol $z \to z_0$ may be omitted altogether. Similar remarks apply to (5.3).

The above definitions are only occasionally needed, but constant use is made of properties that follow from these definitions. If f and g are functions with the same region of definition such that

$$\lim f(z) = \alpha \qquad \text{and} \qquad \lim g(z) = \beta,$$

then it follows that

$$\lim\,[f(z) + g(z)] = \alpha + \beta, \qquad \lim\,[f(z)g(z)] = \alpha\beta, \qquad \lim \frac{f(z)}{g(z)} = \frac{\alpha}{\beta},$$

provided, in the last case, $\beta \neq 0$. Here "lim" is interpreted throughout in either of the senses (5.1) or (5.3). The proof is similar to that in real analysis.

By induction it follows that

$$\lim\,[f(z) + g(z) + \cdots + p(z)] = \lim f(z) + \lim g(z) + \cdots + \lim p(z)$$
$$\lim\,[f(z)g(z) \cdots p(z)] = \lim f(z) \lim g(z) \cdots \lim p(z)$$

where $f, g, \ldots, p$ are functions with the same region of definition such that the limits on the right exist. Since also $\lim cf(z) = c \lim f(z)$ for any constant c, we conclude that

$$\lim F[f(z), g(z), \ldots, p(z)] = F[\lim f(z), \lim g(z), \ldots, \lim p(z)] \tag{5.4}$$

for polynomials $F(\alpha, \beta, \ldots, \sigma)$. The same holds for rational F provided the right-hand side is well-defined.

As in real analysis, a function f is said to be *continuous* at z_0 if

$$\lim_{z \to z_0} f(z) = f(z_0).$$

Clearly this can hold only if z_0 is in the region of definition R for f. If z_0 is an interior point of R, the condition of continuity is equivalent to the following: For any positive ϵ there is a positive δ such that

$$|z - z_0| < \delta \qquad \text{implies} \qquad |f(z) - f(z_0)| < \epsilon.$$

If (5.3) holds for some complex number α, one can define $f(\infty) = \alpha$. It is then said that f is continuous at ∞. Referring to the definition of (5.3) with α replaced by $f(\infty)$, we see that continuity at ∞ is equivalent to the following: For every positive ϵ there is a positive δ such that

$$|z| > \frac{1}{\delta} \quad \text{implies} \quad |f(z) - f(\infty)| < \epsilon.$$

This condition can hold only if f is defined in some neighborhood of ∞, that is, defined for all large $|z|$.

The point $z = \infty$ and its neighborhood are often represented by their images under the transformation $\zeta = 1/z$. The point $z = \infty$ corresponds to $\zeta = 0$ and an ϵ neighborhood of $z = \infty$ corresponds to an ϵ neighborhood of $\zeta = 0$. Hence the continuity of a function $f(z)$ at $z = \infty$ can be described in terms of the continuity of $f(1/\zeta)$ at $\zeta = 0$. Indeed, if $z = 1/\zeta$ and $g(\zeta) = f(1/\zeta)$ and if $g(0) = f(\infty)$, then

$$f(z) - f(\infty) = g(\zeta) - g(0)$$

and $|z| > 1/\delta$ coresponds to $|\zeta| < \delta$. Hence the continuity of $f(z)$ at $z = \infty$ and of $f(1/\zeta)$ at $\zeta = 0$ are equivalent.

To illustrate, let $P(z)$ be a polynomial

$$P(z) = a_0 + a_1 z + \cdots + a_n z^n \tag{5.5}$$

and let

$$f(z) = \frac{P(z)}{z^n} = \frac{a_0}{z^n} + \frac{a_1}{z^{n-1}} + \cdots + a_n, \qquad z \neq 0.$$

Then $\lim f(z) = a_n$ as $z \to \infty$ and so the definition $f(\infty) = a_n$ makes $f(z)$ continuous at ∞. Since

$$g(\zeta) = f\left(\frac{1}{\zeta}\right) = a_0 \zeta^n + a_1 \zeta^{n-1} + \cdots + a_n, \qquad z \neq 0,$$

the condition $f(\infty) = a_n$ corresponds to $g(0) = a_n$.

As the reader will recall from real analysis, theorems on limits lead to corresponding theorems on continuity. For example, the function z is continuous and hence $cz^k = czz \ldots z$ is continuous for any constant c.

Since a polynomial is a sum of such terms, we conclude that polynomials are continuous.

More generally, let $f, g, \ldots, p$ be functions which have the same region of definition and are all continuous at a given point z_0. The conditions

$$\lim f(z) = f(z_0), \qquad \lim g(z) = g(z_0), \ldots, \qquad \lim p(z) = p(z_0)$$

immediately give

$$\lim F[f(z), g(z), \ldots, p(z)] = F[f(z_0), g(z_0), \ldots, p(z_0)]$$

for functions F of the type considered in (5.4). This shows that the function

$$\phi(z) = F[f(z), g(z), \ldots, p(z)]$$

satisfies $\lim \phi(z) = \phi(z_0)$ and so is continuous. Familiar theorems concerning continuity of sums, products, quotients, and positive or negative powers follow by specializing the function F.

Since $|f(z) - f(z_0)| = |\overline{f(z)} - \overline{f(z_0)}|$, it is evident that the conditions

$$\lim f(z) = f(z_0) \qquad \text{and} \qquad \lim \overline{f(z)} = \overline{f(z_0)}$$

are equivalent. Thus, $\overline{f(z)}$ is continuous at z_0 if $f(z)$ is. Considering

$$\frac{f(z) + \overline{f(z)}}{2}, \qquad \frac{f(z) - \overline{f(z)}}{2i}, \qquad [f(z)\overline{f(z)}]^{1/2},$$

we see that the three functions

$$\operatorname{Re} f(z), \qquad \operatorname{Im} f(z), \qquad |f(z)|$$

are continuous at any point z_0 where $f(z)$ is continuous.

A function is said to be continuous in a region if it is continuous at every point of the region. When continuity is considered in a region, the δ needed for a given ϵ usually depends not only on ϵ but also on the point z_0. If a δ can always be chosen independently of z_0, the function is said to be uniformly continuous. More explicitly, $f(z)$ is *uniformly continuous* in a region R if for each positive ϵ there is a positive δ, depending only on ϵ, such that the three conditions

$$z_1 \text{ in } R, \qquad z_2 \text{ in } R, \qquad |z_1 - z_2| < \delta$$

together imply $|f(z_1) - f(z_2)| < \epsilon$.

As stated in the last section, a closed bounded region is called *compact.* Essentially the same proof as in real analysis shows that a function continuous in a compact region is uniformly continuous. It is also a familiar result of real analysis that a continuous function defined on a compact region is *bounded,* that is, it satisfies $|f(z)| \leq M$ for some constant M. Furthermore, such a function actually attains its maximum and minimum at some point of the region. Since $|f(z)|$ is continuous when $f(z)$ is, we conclude that *a continuous complex function on a compact region attains its maximum and minimum modulus.*

The reader will recall that the historical reason for the introduction of complex numbers is to solve polynomial equations. It is of considerable interest that the theory developed in this chapter enables one to prove the fundamental theorem of algebra, which reads as follows:

FUNDAMENTAL THEOREM OF ALGEBRA. *If $P(z)$ is a nonconstant complex polynomial, the equation $P(z) = 0$ has a complex root.*

We do not give the proof here[1] because a shorter proof can be based on more powerful methods introduced later. However, we use the fundamental theorem to establish the following:

FACTORIZATION THEOREM. *If $P(z)$ is a polynomial of degree $n \geq 1$, then $P(z)$ can be expressed in the form*

$$P(z) = c(z - z_1)(z - z_2) \cdots (z - z_n)$$

where c and z_k are constant.

Indeed, the fundamental theorem gives a value z_n such that $P(z_n) = 0$. Hence, by the result of Section 1, Problem 12,

$$P(z) = (z - z_n)Q(z)$$

where the degree of Q is $n - 1$. The factorization theorem now follows by induction.

Example 5.1. Can the following function be defined at z_0 so as to be continuous there?

[1] See Problem 5.3 at the end of this chapter.

$$g(z) = \frac{z^3 - z_0{}^3}{z - z_0}, \qquad z \neq z_0.$$

By a familiar identity (Example 1.5),

$$g(z) = z^2 + zz_0 + z_0{}^2,$$

and so $\lim g(z) = z_0{}^2 + z_0{}^2 + z_0{}^2 = 3z_0{}^2$ as $z \to z_0$. This shows that $g(z_0) = 3z_0{}^2$ makes g continuous at z_0 and no other choice will do.

A function which can be defined or redefined at a single point z_0 in such fashion as to be continuous there is said to have a *removable singularity* at z_0. The function $g(z)$ considered above has a removable singularity at z_0. When z and z_0 are real, $g(z)$ is the familiar difference quotient for getting the derivative of z^3 at z_0 and the result, $3z_0{}^2$, is the derivative. The same definition is used in the complex case, as seen presently.

Example 5.2. Let $f(z) = u(z) + iv(z)$ and $\alpha = a + ib$. Prove that $\lim f(z) = \alpha$ is equivalent to two real equations involving limits,

$$\lim u(z) = a \qquad \text{and} \qquad \lim v(z) = b. \tag{5.6}$$

The two inequalities

$$|u(z) - a| \leq |f(z) - \alpha|, \qquad |v(z) - b| \leq |f(z) - \alpha|$$

show that $\lim f(z) = \alpha$ implies (5.6). The converse follows from

$$|f(z) - \alpha| \leq |u(z) - a| + |v(z) - b|$$

together with the theorem concerning the limit of the sum for real functions. This result shows that the "complex" theory of limits and continuity follows from the "real" theory, which is assumed in this book.

Problems

1. (a) Where are the following functions not continuous?

$$z, \quad \frac{1}{z}, \quad \frac{z^2+1}{z+1}, \quad \frac{2}{z^3+1}, \quad \frac{z+1}{z+1}, \quad \frac{z^4+16}{z^4-16}, \quad \frac{z^4-z^2}{z^8+z^5-z^4-z}.$$

(b) Which of the above functions have removable discontinuities at some point or points of the finite z plane, and how should the function be defined there so as to remove the discontinuity? (c) Same question for $z = \infty$. (Replace z by $1/\zeta$ and consider whether the new function has a limit as $\zeta \to 0$.)

2. What definition at z_0 makes the following functions continuous there?

$$\frac{z - z_0}{z - z_0}, \qquad \frac{z^4 - z_0^4}{z - z_0}, \qquad \frac{1}{z - z_0}\left(\frac{1}{z} - \frac{1}{z_0}\right), \qquad \frac{1}{z - z_0}\left(\frac{1}{z^2} - \frac{1}{z_0^2}\right).$$

3. If $f(z) = (a_0 + a_1 z + a_2 z^2)/(b_0 + b_1 z + b_2 z^2)$ where the a_k and b_k are constant, discuss the behavior of $f(z)$ at 0 and ∞.
4. The set of points z for which a function f is defined is often called the *domain* of f. When f is defined by a formula with no specification as to the domain, it is assumed that the domain is the set of all z for which the formula gives complex values $f(z)$. Sketch the domain of the functions given below and state whether it is a *domain* in the technical sense of Section 4. Also discuss the nature of the discontinuities for $|z| < \infty$ and $z = \infty$:

$$\frac{|z|^2 - 1}{z^2 - 1}, \qquad \frac{1 + z}{1 - |z|}, \qquad \frac{(\operatorname{Re} z)(\operatorname{Im} z)}{(\operatorname{Im} z)(\operatorname{Re} z)}, \qquad \frac{1 + 5z^2 + 4z^3}{(|\operatorname{Im} z|^2 - 1)(|z| - 4)}.$$

5. Discuss the problem of giving a suitable meaning to $\alpha^{\sqrt{2}}$ for $\alpha \neq 0$ by using exponents 1, 1.4, 1.41, 1.414, and so on. How many values do you think $\alpha^{\sqrt{2}}$ should have?

Further results on polynomials

6. Let two polynomials $a_0 + a_1 z + \cdots + a_n z^n$ and $b_0 + b_1 z + \cdots + b_m z^m$ agree in the sense that they give the same value for each real number z in some neighborhood of 0. Use the concept of limit to prove that corresponding coefficients are equal, and hence the polynomials agree for all complex z. (Setting $z = 0$ gives $a_0 = b_0$. Dividing by z and letting $z \to 0$ through real values gives $a_1 = b_1$, and so on. A stronger theorem of the same type is given in Section 1, Problem 17. However, the proof given here also applies to infinite series for which the concept of *degree* used in Section 1 is not available.)
7. Show that a real polynomial can always be expressed as a product of real linear or quadratic factors, and express $z^8 - 1$ in this way. (In the factorization theorem, complex roots occur in conjugate pairs.)
8. Express the sum and product of the roots of a polynomial in terms of the coefficients, and apply to the equation $z^n - 1 = 0$. (Use Problem 6 and the factorization theorem.)
9. By considering $P(\alpha,\beta) = \alpha\beta$ and $Q(\alpha,\beta) = 0$, show that two polynomials in (α,β) can agree at infinitely many values (α,β) without being identical. Show, however, that if $P(\alpha,\beta) = Q(\alpha,\beta)$ for all real α and β in some neighborhood of $\alpha = 0$ and $\beta = 0$, then they are identical, and so agree for all complex α and β. (Collecting terms gives

$$P(\alpha,\beta) = a_0(\alpha) + a_1(\alpha)\beta + \cdots + a_n(\alpha)\beta^n$$

where each $a_k(\alpha)$ is a polynomial. Use Problem 6 twice.)

10. By induction extend Problem 9 to polynomials $P(\alpha,\beta,\ldots,\sigma)$.

***6. Complex numbers as number pairs.** The discussion of Section 1 was more concerned with algebraic techniques than with the logical problem of providing a precise definition. Euler's notation $a + ib$ is not quite satisfactory for a definition because it raises the question: Is the + in the symbol $a + ib$ the same as the + that occurs elsewhere in the algebraic development? A somewhat different approach due to the Irish mathematician William Rowan Hamilton is free of this difficulty and shows that no distinction need be made in the two uses of +. Furthermore, it is found that ib in Euler's symbol actually represents multiplication of i and b, so that $a + ib = a + bi = bi + a = ib + a$. Since Hamilton's approach is also used when complex numbers are inserted into a computer, a brief discussion is given in Problem 1.1. Additional topics pertaining to Chapter 1 are taken up in the remaining problems of this set.

Additional problems on Chapter 1

1.1. According to Hamilton, complex numbers are defined as ordered pairs (a,b) of real numbers, where $(a,b) = (c,d)$ means $a = c$ and $b = d$. Furthermore, by definition,

$$(a,b) + (c,d) = (a + c, b + d), \qquad (a,b)(c,d) = (ac - bd, bc + ad).$$

(a) Show that (0,0) acts as the zero and (1,0) as the unit. (b) Obtain $-(a,b)$ by considering $z + (a,b) = (0,0)$, where $z = (x,y)$. (c) Show that $(a,0) + (c,0) = (a + c,0)$ and $(a,0)(c,0) = (ac,0)$, and hence for algebraic purposes the complex number $(a,0)$ can be identified[1] with the real number a. (d) If -1 is identified with $(-1,0)$, show that $-z = (-1)z$ for $z = (x,y)$. (e) Show that the special complex number $i = (0,1)$ satisfies $i^2 = (-1,0) = -1$. (f) Show that $(a,b) = (a,0) + (0,1)(b,0) = a + ib$, where the first equality results from the laws of addition and multiplication and the second from certain abbreviations. (g) Discuss whether the commutative, associative and

[1] The correspondence $a \leftrightarrow (a, 0)$ is an *isomorphism* between the algebra of reals and the subalgebra of complex numbers having the form $(a, 0)$.

distributive laws must be verified in the notation (a,b), or whether use of $a + ib$ is logically permissible.

1.2. *Complex numbers as matrices.* This problem requires familiarity with 2-by-2 matrices. Let the complex number $a + ib$ of the text or (a,b) of Problem 1.1 be represented by the matrix

$$\begin{pmatrix} a & b \\ -b & a \end{pmatrix}$$

and define equality, sum, and product as for matrices. Following the pattern of Problem 1.1, develop the algebra of complex numbers in this setting. (Apart from the commutative law of multiplication, the algebraic laws follow automatically from the algebra of matrices and need not be verified separately. Hence, even if not used as the definition, these matrices give an efficient way to verify the axioms.)

1.3. Show that the binomial coefficients c_k form a symmetric sequence, that is, $c_k = c_{n-k}$ for $k = 0, 1, 2, \ldots, n$. (Consider $(1 + z)^n = (z + 1)^n$ obtained by suitable choices of α and β in Problem 7.)

1.4. For $n = 1, 2, 3, \ldots$ show there is a unique polynomial P_n of degree n such that

$$z^n + \frac{1}{z^n} = P_n\left(z + \frac{1}{z}\right).$$

(Consider $z^n + 1/z^n - (z + 1/z)^n$ and use induction together with Problem 1.3.)

2.1. *Enestrom's theorem.* Let $n \geq 1$ and $a_0 > a_1 > \cdots a_n > 0$. Prove that all roots of $P(z) = a_0 + a_1 z + \cdots + a_n z^n$ satisfy $|z| > 1$. (Write $(1 - z)P(z)$ as a polynomial and thus show that

$$|(1 - z)P(z)| > a_0 - [(a_0 - a_1)|z| + \cdots + (a_{n-1} - a_n)|z|^n + a_n|z|^{n+1}]$$

unless $z \geq 0$. If $|z| \leq 1$, the expression in brackets is largest when $|z| = 1$, but even in this case the inequality gives $|(1 - z)P(z)| > 0$.)

2.2. A *polynomial with positive coefficients* is obtained from $\alpha, \beta, \ldots, \sigma$ and from the set of positive numbers by repeated addition and multiplication. Such a polynomial can be represented as a sum of terms each of which has the form $c\alpha^m\beta^n \cdots \sigma^k$ where $m, n, \ldots, k$ are integers and $c > 0$. If F is a polynomial with positive coefficients, prove that

$$|F(\alpha,\beta,\ldots,\sigma)| \leq F(|\alpha|,|\beta|,\ldots,|\sigma|).$$

3.1. *Cubic equations.* Show that a cubic equation $z^3 + az^2 + bz + c = 0$ can be put in the form

$$w^3 + b'w + c' = 0$$

by setting $z = w + h$ and choosing h suitably. Find conditions on a, b, c that ensure $b' = 0$ and solve completely in that case.

3.2. If $b' \neq 0$ in Problem 3.1, reduce to the form $s^3 - 3s - \gamma = 0$ by setting $w = ks$ for suitable k. If $\gamma = 2$ or $\gamma = -2$, show that the equation has a double root and factor the left side.

3.3. (a) Show that $z = p + q$ satisfies $z^3 - 3z = \gamma$ if

$$p^3 + q^3 = \gamma \qquad \text{and} \qquad pq = 1. \tag{*}$$

(b) Reduce (*) to a quadratic in p^3 and thus get a value $p = \alpha$ where, for some definite choice of the square root and cube root,

$$\alpha = \left(\frac{\gamma + (\gamma^2 - 4)^{1/2}}{2}\right)^{1/3}$$

(c) If ω is a complex root of 1, show that each pair

$$(p, q) = \left(\alpha, \frac{1}{\alpha}\right), \qquad \left(\omega\alpha, \frac{\omega^2}{\alpha}\right), \qquad \left(\omega^2\alpha, \frac{\omega}{\alpha}\right)$$

satisfies (*) and thus get three roots $z = p + q$ of $z^3 - 3z = \gamma$. (d) If $\alpha \neq 1$, ω, ω^2, -1, $-\omega$, $-\omega^2$, show that these roots $p + q$ are distinct, while $\alpha = 1$, ω, ω^2 lead to the case $\gamma = 2$ of Problem 3.2, and $\alpha = -1, -\omega, -\omega^2$ lead to the case $\gamma = -2$.

4.1. *Stereographic projection.* Let the point $z = x + iy$ correspond to the point (X,Y,Z) of the sphere under stereographic projection so that the three points

$$(0,0,1), \qquad (X,Y,Z), \qquad (x,y,0)$$

are colinear. Thus, $(X,Y,Z-1) = t(x,y,-1)$ where t is a real scalar. Express X,Y,Z in terms of x,y,t; use $X^2 + Y^2 + Z^2 = 1$ to get t, and so get

$$X = \frac{2x}{|z|^2 + 1}, \qquad Y = \frac{2y}{|z|^2 + 1}, \qquad Z = \frac{|z|^2 - 1}{|z|^2 + 1}.$$

4.2. In Problem 4.1, show that a circle on the sphere corresponds to a circle in the z plane unless the former circle contains the pole (0,0,1), in which case it corresponds to a line in the z plane. (The given circle is the intersection of the sphere with some plane $AX + BY + CZ = D$. Express this equation in terms of x and y.)

4.3. If P: (X,Y,Z) and P': (X',Y',Z') are two points of the sphere, their chordal distance is the length of the straight line from P to P'. If P corresponds to z and P' to z', show that the chordal distance is

$$d(z,z') = \frac{2|z - z'|}{\sqrt{1 + |z|^2}\sqrt{1 + |z'|^2}}.$$

4.4. This problem requires knowledge of differential geometry. The differential of arc on the sphere and plane are given respectively by

$$dS^2 = dX^2 + dY^2 + dZ^2, \qquad ds^2 = dx^2 + dy^2.$$

Show that $ds^2 = (1 - Z)^{-2}\, dS^2$, and hence the mapping from the plane to the sphere is conformal.

5.1. In the theory of rational functions, all removable singularities are assumed to be actually removed; for example, z/z is considered to have the value 1 at $z = 0$. (a) Using the fundamental theorem of algebra, show that every bounded rational function is constant. (b) Show conversely that the fundamental theorem of algebra follows from the result (a).

5.2. It is said that $\lim f(z) = \infty$ if $\lim [1/f(z)] = 0$. (a) Express in terms of ϵ neighborhoods of ∞. (b) In what sense and to what extent do the limit theorems of Section 5 remain valid when one or both of the limits are ∞?

5.3. *Fundamental theorem of algebra.* (a) If $P(z)$ is a nonconstant polynomial, show that $|P(z)| > |P(0)|$ holds outside some disk $|z| \le R$. Conclude that if the minimum of $|P(z)|$ for $|z| \le R$ occurs at z_0, then $z = z_0$ gives the minimum of $|P(z)|$ with respect to the whole plane. (b) If $P(z_0) \neq 0$, define a polynomial Q by $Q(z) = P(z + z_0)/P(z_0)$. Show that $Q(z) = 1 + cz^m + \cdots$ where $c \neq 0$, $m \geq 1$, and the terms not written are of higher degree. (c) If α is a root of $c\alpha^m = -1$, show that $Q(\alpha z)$ has the form $1 - z^m + \cdots$ and that its minimum modulus with respect to the plane occurs at $z = 0$. (d) Obtain a contradiction by taking z positive but small.

Chapter 2

The complex derivative

It turns out that a function of a complex variable can have a derivative only if the real and imaginary parts satisfy a system of partial differential equations called the Cauchy-Riemann equations. In this chapter the Cauchy-Riemann equations are used to establish the differentiability of the elementary functions when the latter are suitably extended to allow a complex variable z. In contrast to the real-analysis case it is found that e^z is periodic, and hence $\log z$ has infinitely many values. Representation of $\log z$ in terms of single-valued functions leads to the concept of Riemann surface, developed here insofar as needed for the theory of the logarithm. Further study of the Cauchy-Riemann equations shows a connection between complex analysis and harmonic functions, and a connection between complex analysis and the theory of ideal fluid flow in the plane.

1. Analytic functions. The function f defined on a domain D is said to be *differentiable* at a point z_0 of D if

$$\lim_{z \to z_0} \frac{f(z) - f(z_0)}{z - z_0}$$

exists. The derivative is denoted by $f'(z_0)$. From

$$f(z) = \frac{f(z) - f(z_0)}{z - z_0}(z - z_0) + f(z_0) \qquad (z \neq z_0)$$

it follows that $\lim f(z) = f'(z_0) \cdot 0 + f(z_0) = f(z_0)$ as $z \to z_0$, and hence a function which is differentiable at z_0 is also continuous there. However the converse is not true; differentiability is a far stronger condition than continuity.

Writing $z - z_0 = h$ in the above definition, we get the equivalent form

$$\lim_{h\to 0} \frac{f(z+h)-f(z)}{h} = f'(z) \tag{1.1}$$

after replacing z_0 by z. It is important that h is complex. We denote h by Δz, so that

$$\lim_{\Delta z\to 0} \frac{f(z+\Delta z)-f(z)}{\Delta z} = f'(z).$$

If $w = f(z)$, we sometimes define

$$\Delta w = f(z+\Delta z) - f(z) \tag{1.2}$$

and write the derivative as dw/dz or $(d/dz)w$; thus

$$\frac{dw}{dz} = \lim_{\Delta z\to 0} \frac{\Delta w}{\Delta z}.$$

This formulation is meaningful, however, only in conjunction with (1.2).

If the limit (1.1) exists for each z in a given region R, then the process assigns the complex number $f'(z)$ to each z in R and so determines a function. The function is denoted by f' or by $f'(z)$. Equations involving derivatives such as (1.3) below can thus be interpreted either as equality of functions or as equality of functional values.

As might be expected from these definitions, complex derivatives behave much like real derivatives. For example, if $f(z) = z^n$, with n a positive integer, then

$$\frac{(z+h)^n - z^n}{h} = nz^{n-1} + c_2 z^{n-2}h + \cdots + c_n h^{n-1}$$

where the c_k are the binomial coefficients introduced in Chapter 1, Example 1.6. Letting $h \to 0$ gives

$$\frac{d}{dz} z^n = nz^{n-1}. \tag{1.3}$$

With the same proofs as in real analysis, if f and g are differentiable, then

$$(f+g)' = f' + g', \qquad (fg)' = fg' + f'g, \qquad \left(\frac{f}{g}\right)' = \frac{gf' - fg'}{g^2}$$

under the understanding, in the last case, that $g(z) \neq 0$ in the region considered. By induction

$$(f + g + \cdots + p)' = f' + g' + \cdots + p'$$

if the derivatives on the right exist. When the formula for $(fg)'$ is divided by fg, the result suggests that

$$\frac{(fg \cdots p)'}{fg \cdots p} = \frac{f'}{f} + \frac{g'}{g} + \cdots + \frac{p'}{p}. \qquad (1.4)$$

This is easily proved by induction if the denominators are not zero and the derivatives in the numerators all exist. As an illustration, (1.4) with $f = z, g = z, \ldots, p = z$ gives (1.3) again for $z \neq 0$.

These results show that polynomials and rational functions can be differentiated as in real analysis. It is also true that the chain rule holds; if g is differentiable at z and if f is differentiable at $g(z)$, then $F(z) = f[g(z)]$ is differentiable at z and

$$F'(z) = f'[g(z)]g'(z). \qquad (1.5)$$

A proof is given in Problem 10. With $\zeta = g(z)$ and $w = f(\zeta)$, the chain rule can be written

$$\frac{dw}{dz} = \frac{dw}{d\zeta}\frac{d\zeta}{dz}.$$

So far we have stressed similarities of differentiation in the real and complex cases. Actually, however, the fact that h in (1.1) is complex puts a very severe restriction on the class of functions which have complex derivatives. For example, let $f(z) = \bar{z}$ so that

$$\frac{f(z+h) - f(z)}{h} = \frac{\overline{z+h} - \bar{z}}{h} = \frac{\bar{h}}{h} \qquad (h \neq 0).$$

If h is real, then $\bar{h} = h$ and the limit as $h \to 0$ is 1. But if $h = ik$ is pure imaginary, then $\bar{h} = -h$ and the limit is -1. This function is not differentiable anywhere, although its real and imaginary parts, x and $-y$, are very well behaved.

In the above illustration the nonexistence of $f'(z)$ was proved by letting

$h \to 0$ first through real values and then through imaginary values. These two possibilities lead to the main condition which a complex function must satisfy in order to have a derivative.

Indeed, if $f(z) = u(x,y) + i\,v(x,y)$ and h is real, the derivative $f'(z)$ is the limit of

$$\frac{f(z+h)-f(z)}{h} = \frac{u(x+h,y)-u(x,y)}{h} + i\frac{v(x+h,y)-v(x,y)}{h}.$$

However, if $h = ik$ where k is real, then $f'(z)$ is the limit of

$$\frac{f(z+ik)-f(z)}{ik} = \frac{u(x,y+k)-u(x,y)}{ik} + i\frac{v(x,y+k)-v(x,y)}{ik}.$$

Thus, if $h \to 0$ in the first expression and $k \to 0$ in the second, we get the two formulas

$$f'(z) = u_x + iv_x, \qquad f'(z) = -iu_y + v_y \tag{1.6}$$

where the subscripts denote partial differentiation and the functions on the right are evaluated at (x,y). For example,

$$u_x = u_x(x,y) = \frac{\partial u}{\partial x}.$$

What has also been shown in the proof of (1.6) is that the existence of $f'(z)$ implies the existence of u_x, u_y, v_x and v_y.

The two expressions (1.6) for $f'(z)$ are consistent only if u and v satisfy the *Cauchy-Riemann equations,*

$$u_x = v_y, \qquad u_y = -v_x. \tag{1.7}$$

Here we have a striking difference between the real and complex derivatives. Existence of a real derivative is just a mild smoothness condition, but *existence of a complex derivative leads to a pair of partial differential equations.*

In view of the derivation it is hardly to be expected that the Cauchy-Riemann equations are sufficient for the existence of $f'(z)$ and, as a matter of fact, they are not sufficient. However, if continuity of the partial derivatives is assumed, they are sufficient, as shown next.

At a given point (x,y) we define $\Delta z = \Delta x + i\,\Delta y$,

$$\Delta u = u(x + \Delta x, y + \Delta y) - u(x,y),$$

and Δv similarly with v instead of u. As the reader will recall from real analysis, if u_x and u_y are continuous at the point (x,y), then

$$\Delta u = u_x \, \Delta x + u_y \, \Delta y + \epsilon_1 |\Delta z|, \qquad \lim_{\Delta z \to 0} \epsilon_1 = 0. \quad (1.8)$$

Similarly, if v_x and v_y are continuous at (x,y), then

$$\Delta v = v_x \, \Delta x + v_y \, \Delta y + \epsilon_2 |\Delta z|, \qquad \lim_{\Delta z \to 0} \epsilon_2 = 0. \quad (1.9)$$

When functions u and v satisfy (1.8) and (1.9), respectively, they are said to be differentiable at (x,y). To give a similar criterion for differentiability in the complex sense, at a given point z let η be defined by

$$\eta = \frac{\Delta w}{\Delta z} - f'(z) \qquad (\Delta z \neq 0),$$

where Δw is defined by (1.2). Then $\lim \Delta w/\Delta z$ exists and equals the complex number $f'(z)$ if and only if $\lim \eta = 0$. We now multipy through by Δz and define ϵ by $\epsilon|\Delta z| = \eta \, \Delta z$. The result is that $dw/dz = f'(z)$ holds at a given point z if and only if

$$\Delta w = f'(z) \, \Delta z + \epsilon |\Delta z|, \qquad \lim_{\Delta z \to 0} \epsilon = 0. \quad (1.10)$$

This is used in the proof of the following:

THEOREM 1.1. *The function $f(z) = u(x,y) + iv(x,y)$ is differentiable at a given point $z = x + iy$ if the partial derivatives u_x, u_y, v_x, v_y are continuous and satisfy the Cauchy-Riemann equations there.*

Proof. Let $w = f(z)$. Then $\Delta w = \Delta u + i \, \Delta v$, since $f(z + \Delta z) - f(z)$ equals

$$u(x + \Delta x, y + \Delta y) - u(x,y) + i[v(x + \Delta x, y + \Delta y) - v(x,y)].$$

By (1.8) and (1.9),

$$\Delta u + i \, \Delta v = u_x \, \Delta x + u_y \, \Delta y + i(v_x \, \Delta x + v_y \, \Delta y) + (\epsilon_1 + i\epsilon_2)|\Delta z|.$$

Replacing $\Delta u + i \, \Delta v$ by Δw on the left and using the Cauchy-Riemann equations on the right, we get

$$\Delta w = (u_x + iv_x)(\Delta x + i\,\Delta y) + (\epsilon_1 + i\epsilon_2)|\Delta z|.$$

If $f'(z)$ is defined by $f'(z) = u_x + iv_x$, the criterion (1.10) is fulfilled with $\epsilon = \epsilon_1 + i\epsilon_2$ and hence w is differentiable at z.

The most important result of the foregoing discussion is that a certain system of partial differential equations (the Cauchy-Riemann equations) must hold at every point where $f'(z)$ exists. The reader familiar with applications of analysis will know that partial differential equations are seldom of interest at a single point—it is when they hold throughout a region that interesting consequences follow. This remark may serve to motivate the following definition:

DEFINITION 1.1. *A function $f(z)$ is analytic at a point z_0 if f is differentiable throughout some ϵ neighborhood of z_0. A function is analytic in a region if it is analytic at every point of the region.*

In the case of a domain, differentiability and analyticity are equivalent, but in other cases the differentiability is required in a larger set. For example, f is analytic for $|z| \leq 1$ if f is differentiable for $|z| < 1 + \delta$, where $\delta > 0$. As another illustration, let $f(z) = |z|^2$ so that

$$u(x,y) = x^2 + y^2, \qquad v(x,y) = 0.$$

Here the Cauchy-Riemann equations require $2x = 0$, $2y = 0$ and hence, they require $z = 0$. Since $f'(z)$ does not exist throughout a neighborhood of $z = 0$, this function is not analytic at $z = 0$. Nevertheless $f'(0)$ exists, as shown by the equation

$$\lim \frac{f(0 + \Delta z) - f(0)}{\Delta z} = \lim \frac{|\Delta z|^2}{\Delta z} = 0.$$

Example 1.1. If $f(z) = u(x,y) + iv(x,y)$, show that the curves

$$u(x,y) = \text{const} \qquad \text{and} \qquad v(x,y) = \text{const}$$

are orthogonal at every point where $f'(z)$ exists and is not zero.

If $z = x + iy$ is such a point, then (1.6) and (1.7) give

$$u_x^2 + u_y^2 = v_x^2 + v_y^2 = |f'(z)|^2$$

and so the vectors $\mathbf{i}u_x + \mathbf{j}u_y$ and $\mathbf{i}v_x + \mathbf{j}v_y$ are not zero. By real analysis, these vectors are normal to the curves $u(x,y) = \text{const}$ and $v(x,y) = \text{const}$. When we use (1.7) again, it is seen that the scalar product of these vectors is

$$(\mathbf{i}u_x + \mathbf{j}u_y) \cdot (\mathbf{i}v_x + \mathbf{j}v_y) = u_xv_x + u_yv_y = v_yv_x - v_xv_y = 0.$$

Hence the normals are perpendicular to one another, and the curves intersect at right angles.

Example 1.2. If a function is analytic in a domain D and has a constant real part, a constant imaginary part, or a constant modulus, prove that the function itself is constant.

Let $f = u + iv$. If u is constant, then $u_x = u_y = 0$; hence by the Cauchy-Riemann equations $v_x = v_y = 0$, and so v is constant by a familiar theorem of real analysis. Hence f is constant. If f has a constant modulus, then $|f|^2 = u^2 + v^2$ is constant and differentiation gives

$$uu_x + vv_x = 0, \qquad uu_y + vv_y = 0.$$

Replacing v_x by $-u_y$ and v_y by u_x, we get a system of linear equations in the two unknowns u_x and u_y with determinant $u^2 + v^2$. If the constant value of $|f|$ is not 0, then $u^2 + v^2 > 0$; hence $u_x = u_y = 0$, and so $u = \text{const}$. This case has already been considered. On the other hand, if $|f| = 0$, then f is obviously constant, namely $f = 0$.

Example 1.3. If all the zeros of a polynomial $P(z)$ have negative real parts, prove that all the zeros of $P'(z)$ also have negative real parts.

Polynomials with this property are called Hurwitz polynomials. They are encountered in the theory of stable mechanical systems and stable electrical networks. To prove the theorem let $P(z)$ have zeros z_k, not necessarily distinct, so that the factorization is

$$P(z) = c(z - z_1)(z - z_2) \cdots (z - z_n), \qquad c \text{ const.}$$

By (1.4), with $f = c(z - z_1)$, $g = (z - z_2)$, $\ldots$, $p = (z - z_n)$, or by direct calculation, we get

$$\frac{P'(z)}{P(z)} = \frac{1}{z - z_1} + \frac{1}{z - z_2} + \cdots + \frac{1}{z - z_n}, \qquad (z \neq z_k).$$

Suppose, now, that $\operatorname{Re} z_k < 0$ for each k but that $\operatorname{Re} z \geq 0$. In this case, $\operatorname{Re}(z - z_k) > 0$ and therefore

$$\operatorname{Re}\left(\frac{1}{z - z_k}\right) > 0. \tag{1.11}$$

(The latter follows from the formula $1/\alpha = \bar{\alpha}/|\alpha|^2$, which shows that $\operatorname{Re} \alpha$ and $\operatorname{Re}(1/\alpha)$ have the same sign for all α.) Since $P'(z)/P(z)$ is a sum of terms (1.11), it too has a positive real part, and hence is not 0. In other words, $P'(z) \neq 0$ for $\operatorname{Re} z \geq 0$, and so the zeros of $P'(z)$ must have $\operatorname{Re} z < 0$.

Problems

1. Differentiate, and indicate which of the differentiation formulas are used in each case: $(2z + 3)^5$, $(z - i)/(z + i)$, $(2z + z^2)^4(1 + z^3)^2$.
2. Which of the following satisfy the Cauchy-Riemann equations?

   (a) $f(z) = x^2 - y^2 - 2ixy, \quad x^3 - 3y^2 + 2x + i(3x^2y - y^3 + 2y)$;

   (b) $\frac{1}{2}\log(x^2 - y^2) + i\tan^{-1} y/x, \quad \frac{1}{2}\log(x^2 + y^2) + i\cot^{-1} x/y$.

3. (a) Differentiate by (1.1), and (b) verify agreement of your result with both formulas (1.6): z, $(1 - i)z$, z^2, iz^3, $(1 + z)/z$.
4. If $z \neq 0$, show directly from (1.1) that $(z^{-1})' = -z^{-2}$. Deduce the formula for differentiating $z^{-n} = (z^{-1})^n$ from this and (1.3) if n is a positive integer. (Use the chain rule.)
5. (a) Let $F(z) = c_1 f(z) + c_2 g(z)$ where c_1 and c_2 are constant and $f'(z)$ and $g'(z)$ exist. Divide the identity

   $$F(z + h) - F(z) = c_1[f(z + h) - f(z)] + c_2[g(z + h) - g(z)]$$

   by h and deduce that $F'(z) = c_1 f'(z) + c_2 g'(z)$. By appropriate choice of c_i, get formulas for differentiating $f + g, f - g$ and cf, c being constant. (b) Deduce the formula for differentiating a product by considering

   $$f(z + h)[g(z + h) - g(z)] + g(z)[f(z + h) - f(z)].$$

6. Show that a real-valued analytic function is constant.
7. Let it be required to define an analytic function e^z such that $e^0 = 1$ and such that $e^{x+iy} = e^x e^{iy}$, where e^x is the familiar exponential of real analysis. If

   $$e^{iy} = c(y) + is(y),$$

   show by the Cauchy-Riemann equations that $c(y) = s'(y)$, $s(y) = -c'(y)$.

Deduce that the function

$$\phi(y) = [c(y) - \cos y]^2 + [s(y) - \sin y]^2$$

satisfies $\phi'(y) = 0$, and hence is constant. From $e^0 = 1$, get $\phi(y) = 0$, and hence the only candidate for e^z, satisfying the above conditions, is

$$e^{x+iy} = e^x(\cos y + i \sin y).$$

8. (a) Sketch the curves u = const and v = const for the following functions. (b) Verify agreement with the result of Example 1.2:

$$z, \qquad (1-2i)z, \qquad z^2, \qquad z^{-1}, \qquad z+z^2, \qquad (z+i)(z-i)^{-1}.$$

9. If $P(z) = a_0 + a_1(z-a) + \cdots + a_n(z-a)^n$, the a's being constant, show that

$$a_k = \frac{P^{(k)}(a)}{k!}, \qquad k = 0, 1, \ldots, n.$$

(Differentiate k times an then set $z = a$.) Applying this result to

$$(1+z)^n = c_0 + c_1 z + \cdots + c_n z^n,$$

show that the kth binomial coefficient, c_k, satisfies

$$c_k = \frac{n!}{k!(n-k)!}, \qquad k = 0, 1, \ldots, n.$$

10. *Chain rule.* Let F, f and g satisfy the hypothesis of (1.5), with $\zeta = g(z)$ and $w = f(\zeta)$. From the two equations

$$\Delta w = f'(\zeta)\,\Delta\zeta + \epsilon_1|\Delta\zeta|, \qquad \Delta\zeta = g'(z)\,\Delta z + \epsilon_2|\Delta z|$$

deduce the equation

$$\Delta w = f'(\zeta)g'(z)\,\Delta z + f'(\zeta)\epsilon_2|\Delta z| + \epsilon_1|\Delta\zeta|.$$

Dividing by Δz and letting $\Delta z \to 0$, obtain the chain rule. (Note that $\Delta z \to 0$ implies $\Delta\zeta \to 0$, hence $\epsilon_1 \to 0$ as well as $\epsilon_2 \to 0$.)

2. Exponential and trigonometric functions.

If the real-variable property $e^{x_1+x_2} = e^{x_1}e^{x_2}$ of the exponential function can be extended to complex values of the variables, then

$$e^z = e^{x+iy} = e^x e^{iy},$$

which suggests[1] as the definition of e^z,

[1] See Problem 7 of the preceding section.

$$e^z = e^x(\cos y + i \sin y). \tag{2.1}$$

(The notation e^z implies certain properties which actually must be shown to be a consequence of the definition.) With $u = e^x \cos y$ and $v = e^x \sin y$, the partial derivatives are continuous and the Cauchy-Riemann equations are satisfied. Thus e^z is an analytic function in the z plane. A function analytic in the finite z plane, $|z| < \infty$, is said to be an entire function. Clearly, e^z is an entire function, as are the polynomials in z. From (1.6),

$$\frac{d}{dz} e^z = \frac{\partial}{\partial x}(e^x \cos y) + i \frac{\partial}{\partial x}(e^x \sin y) = e^z, \tag{2.2}$$

so that e^z is its own derivative.

Setting $x = 0$ in (2.1) and using results of Chapter 1, Section 3, we get

$$e^{iy_1}e^{iy_2} = (\cos y_1 + i \sin y_1)(\cos y_2 + i \sin y_2) = e^{i(y_1+y_2)}$$

and hence

$$(e^{x_1}e^{iy_1})(e^{x_2}e^{iy_2}) = (e^{x_1}e^{x_2})(e^{iy_1}e^{iy_2}) = e^{x_1+x_2}e^{i(y_1+y_2)}.$$

Since $e^z = e^x e^{iy}$ by definition, the foregoing equation gives

$$e^{z_1}e^{z_2} = e^{z_1+z_2} \tag{2.3}$$

for arbitrary complex numbers $z_1 = x_1 + iy_1$ and $z_2 = x_2 + iy_2$. The result (2.3) is called the *addition theorem* for the exponential function. In general, an addition theorem for a function f is a formula which expresses $f(z_1 + z_2)$ in terms of functions of z_1 alone and z_2 alone.

The choice $z_1 = z$ and $z_2 = -z$ in (2.3) gives $e^z e^{-z} = 1$, which shows that e^z never vanishes. The same conclusion follows from

$$|e^z| = e^x(\cos^2 y + \sin^2 y)^{1/2} = e^x,$$

which is obtained by equating absolute values in (2.1).

Replacing z by $iz = -y + ix$ in (2.1) gives

$$e^{iz} = e^{-y}(\cos x + i \sin x)$$

and, since $\cos(-x) = \cos x$ and $\sin(-x) = -\sin x$, the substitution $-z$ for z in the latter equation leads to

$$e^{-iz} = e^y(\cos x - i \sin x).$$

Setting $y = 0$ and subtracting or adding, we get, respectively,

$$\sin x = \frac{e^{ix} - e^{-ix}}{2i}, \qquad \cos x = \frac{e^{ix} + e^{-ix}}{2}. \tag{2.4}$$

These real-variable formulas suggest the definitions

$$\sin z = \frac{e^{iz} - e^{-iz}}{2i}, \qquad \cos z = \frac{e^{iz} + e^{-iz}}{2} \tag{2.5}$$

for complex z. Since e^{iz} and e^{-iz} are entire functions, the same is true for $\sin z$ and $\cos z$ and, indeed, from (2.5) it follows that

$$(\sin z)' = \cos z, \qquad (\cos z)' = -\sin z. \tag{2.6}$$

From (2.5) it also follows that

$$\sin^2 z + \cos^2 z = 1, \qquad \sin 2z = 2 \sin z \cos z,$$

and that the other familiar formulas of trigonometry can be extended to the complex plane. The proof involves nothing more than the algebra of complex numbers together with the basic formula (2.3), and is discussed in the problems at the end of this section.

Instead of carrying out these calculations here we mention a more efficient method which follows from a general theorem of complex analysis. This theorem asserts that *if two entire functions agree on a segment of the real axis, no matter how small, then they agree on the whole complex plane.* Suppose, for example, that we want to prove the addition formula

$$\sin(z_1 + z_2) = \sin z_1 \cos z_2 + \cos z_1 \sin z_2.$$

By real analysis, the result holds when both z_1 and z_2 are real. But for any given real z_1 both sides are entire functions of z_2, and so the above-mentioned theorem shows that the equation holds for all complex z_2. Now give any complex value to z_2. Since the equation holds for real z_1, and both sides are entire functions of z_1, it follows that the equation also holds for complex z_1.

When suitably refined and extended, this type of argument leads to a branch of complex analysis known as the theory of *analytic continuation.* The above discussion cannot be considered complete, because the theorem upon which it is based is deferred to Chapter 3, Section 7. Nevertheless, it illustrates the power of complex-variable methods.

The hyperbolic functions are defined as in real analysis by

$$\sinh z = \frac{e^z - e^{-z}}{2}, \qquad \cosh z = \frac{e^z + e^{-z}}{2}.$$

Comparing (2.5), we see that

$$\sinh z = -i \sin iz, \qquad \cosh z = \cos iz \tag{2.7}$$

and hence any trigonometric identity leads to a corresponding identity for hyperbolic functions and vice versa. The analogy between the trigonometric and hyperbolic functions is extended by the definitions

$$\tanh z = \frac{\sinh z}{\cosh z}, \ \coth z = \frac{\cosh z}{\sinh z}, \ \operatorname{sech} z = \frac{1}{\cosh z}, \ \operatorname{csch} z = \frac{1}{\sinh z},$$

which correspond to the definitions

$$\tan z = \frac{\sin z}{\cos z}, \qquad \cot z = \frac{\cos z}{\sin z}, \qquad \sec z = \frac{1}{\cos z}, \qquad \operatorname{cosec} z = \frac{1}{\sin z}.$$

None of these eight functions are entire since they are discontinuous at the points where the denominators are zero.

As in real analysis, a function which satisfies $f(-z) = f(z)$ for all z in its region of definition is said to be even, while a function satisfying $f(-z) = -f(z)$ is odd. Clearly cos and cosh are even and sin and sinh are odd.

Although the above formulas have the same algebraic structure as in the real case, the details of analytic behavior can be quite different. For example, $|\sin x| \leq 1$ for all real x, but (2.5) gives

$$|\sin iy| = |\sinh y|, \qquad -\infty < y < \infty.$$

Hence $|\sin iy|$ is almost as large as $e^{|y|}/2$. Similarly, e^x is large when x is real and large, but (2.1) gives

$$|e^{iy}| = 1, \qquad -\infty < y < \infty.$$

One of the most important differences between the real and complex cases is that the function e^z is periodic. In general, a function $f(z)$ is *periodic* if $f(z + \omega) = f(z)$ holds for a nonzero constant ω and for all z in the region of definition of f. Any constant ω with this property is called a *period* of f. Since

$$e^{2\pi i} = \cos 2\pi + i \sin 2\pi = 1,$$

(2.3) gives

$$e^{z+2\pi i} = e^z$$

and so e^z has the complex period $2\pi i$. It is also readily verified that any other period of e^z must be a multiple of $2\pi i$. Indeed, $e^{z+\omega} = e^z$ implies $e^\omega = 1$ and conversely. Letting $\omega = a + ib$ and using (2.1), gives

$$e^a(\cos b + i \sin b) = 1.$$

Taking the magnitude of each side of the equation, we get $e^a = 1$ and hence $a = 0$, since a is real. From $\cos b = 1$ and $\sin b = 0$ it follows that $b = 2\pi k$ where k is an integer. Hence $\omega = 2\pi k$.

The fact that e^z is periodic has an interesting bearing on the mapping properties of the transformation $w = e^z$. If $z = x + iy$, the equation $w = e^z$ becomes

$$w = e^x(\cos y + i \sin y)$$

by (2.1). Comparing with the polar form of w as given in Chapter 1, Section 3, we see that

$$|w| = e^x, \qquad \arg w = y + 2\pi k.$$

It is convenient to consider the choice $\arg w = y$ corresponding to $k = 0$, and to add the further restriction that $-\pi < y \leq \pi$. Thus, $\operatorname{Arg} w = y$. The restriction $-\pi < y \leq \pi$ means that z lies in the horizontal strip of

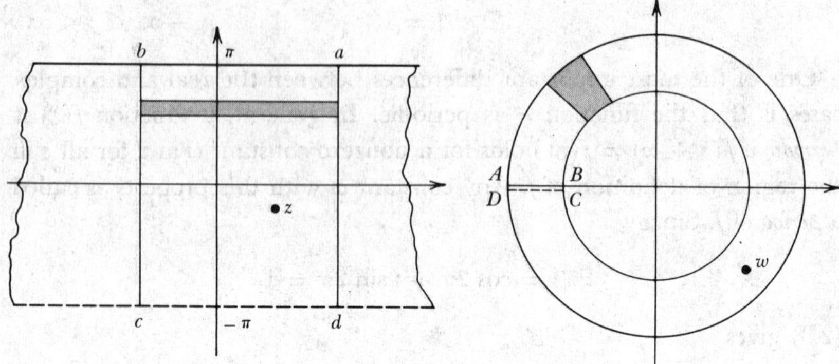

Figure 2-1

Figure 2-1, which includes its upper boundary but not the lower boundary. The restriction $-\pi < \operatorname{Arg} w \leq \pi$ can be indicated in the w plane by cutting the w plane along the negative real axis, that is, on $v = 0$, $u < 0$. The upper edge of this cut corresponding to $\operatorname{Arg} w = \pi$ is regarded as being in the cut plane, but the lower edge of the cut, corresponding to $\operatorname{Arg} w = -\pi$, is not regarded as belonging to the cut plane. The points $w = 0$ and $w = \infty$ are also excluded.

Under these conditions the horizontal line $y = y_0$, $-\infty < x < \infty$, in the z plane maps into the radial line $\arg w = y_0$, $0 < |w| < \infty$, in the w plane. The vertical line segment $x = x_0$, $-\pi < y \leq \pi$, in the z plane maps into the circle $|w| = e^{x_0}$, $-\pi < \arg w \leq \pi$, in the w plane. As $x_0 \to -\infty$ the radius of the circle in the w plane tends to zero, and as x_0 increases to $+\infty$ the radius of the circle also increases to $+\infty$. The rectangle in the z plane shown in Figure 2-1 is mapped into the annulus in the w plane cut open on the negative real axis. The points a, b, c and d go into A, B, C and D. Polar coordinates ρ, ϕ in the w plane are given by $\rho = |w|$, $\phi = \arg w$. The strip $-\infty < x < \infty$, $-\pi < y \leq \pi$, maps onto the w plane cut along the negative real axis, $0 < \rho < \infty$, $-\pi < \phi \leq \pi$. The mapping $|w| = e^x$, $\operatorname{Arg} w = y$, $-\infty < x < \infty$, $-\pi < y \leq \pi$, is one-to-one, that is, each point z in the strip has a unique image in the cut w plane and conversely. A region of the z plane which, under the mapping $w = f(z)$, is mapped one-to-one onto the whole w plane (which

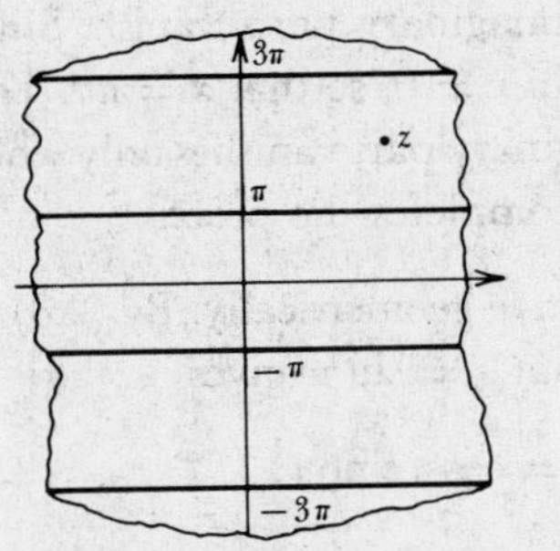

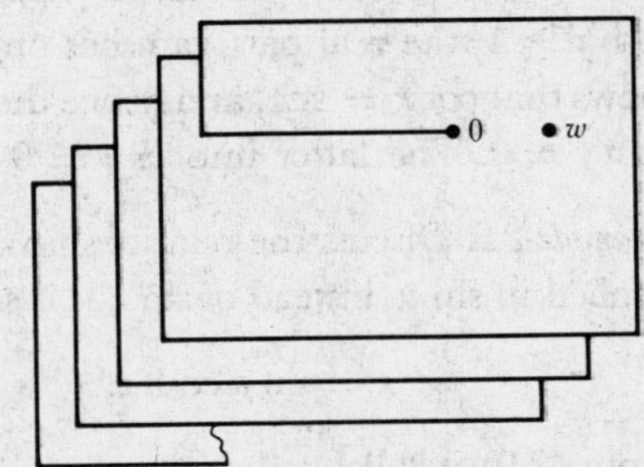

Figure 2-2

may, however, have cuts), is called a *fundamental region* for f. Hence the strip $-\infty < x < \infty$, $-\pi < y \leq \pi$, is a fundamental region for e^z.

Because of the periodicity of e^z, the mapping properties on each of the other horizontal strips in the figure are the same as for the central strip. Thus the strips

$$-\infty < x < \infty, \qquad (2n-1)\pi < y \leq (2n+1)\pi \quad (n = 0, \pm 1, \pm 2, \ldots)$$

are also fundamental regions for e^z, so that $w = e^z$ maps the nth strip, one-to-one, onto the cut plane $0 < |w| < \infty$, $(2n-1)\pi < \arg w \leq (2n+1)\pi$.

If the cut planes are stacked one above the other, the mapping $w = e^z$ is one-to-one from the whole z plane to the stack of cut w planes. As will be seen in Section 4, the cut planes can be connected by joining one side of the cut on one plane to the other side of the cut on the next plane. The resulting surface has a helical structure something like a spiral ramp and is called the Riemann surface for e^z. See Figure 2-2.

Example 2.1. Show that the only zeros of sin z are at $n\pi$, where n is an integer. Setting $z = y$ in (2.7) gives

$$\sin iy = i \sinh y, \qquad \cos iy = \cosh y$$

and hence the formula for $\sin(z_1 + z_2)$ gives

$$\sin(x + iy) = \sin x \cosh y + i \cos x \sinh y. \tag{2.8}$$

The result is 0 only if both the real and imaginary parts vanish. Since $\cosh y \geq 1$, the real part vanishes only if $\sin x = 0$, so that $x = n\pi$. This shows that $\cos x = \pm 1$, and hence the imaginary part vanishes only when $\sinh y = 0$. The latter implies $y = 0$ and completes the proof.

Example 2.2. Discuss the relationship $z = \sin w$ geometrically. By (2.8) as applied to $\sin w$ instead of $\sin z$ it is seen that $z = \sin w$ gives

$$x = \sin u \cosh v, \qquad y = \cos u \sinh v. \tag{2.9}$$

From (2.9) it follows that

$$\frac{x^2}{\cosh^2 v} + \frac{y^2}{\sinh^2 v} = 1 \tag{2.10}$$

$$\frac{x^2}{\sin^2 u} - \frac{y^2}{\cos^2 u} = 1. \tag{2.11}$$

These equations show that the lines $v = v_0$ of the w plane map onto ellipses (2.10) in the z plane and that the lines $u = u_0$ of the w plane map onto hyperbolas (2.11) in the z plane. More precisely it follows from (2.9) that the line segment $v = v_0 > 0$, $-\frac{1}{2}\pi < u \leq \frac{1}{2}\pi$, of the w plane maps onto the upper half of an ellipse in the z plane with semimajor axis $\cosh v_0$ and semiminor axis $\sinh v_0$ and with foci at $z = \pm 1$. The line segment $v = -v_0$, $-\frac{1}{2}\pi \leq u < \frac{1}{2}\pi$ maps onto the lower half of the same ellipse. As v_0 is increased from $+0$ to $+\infty$ the semiellipses grow outward and fill out the upper half z plane. From (2.9) it follows that the vertical line $u = u_0$, $-\infty < v < \infty$, for fixed u_0, $-\frac{1}{2}\pi < u < \frac{1}{2}\pi$, maps onto the right half of a hyperbola if $u_0 > 0$ and the left half if $u_0 < 0$. The foci as shown by (2.11) are again at $z = \pm 1$.

It is readily ascertained that the system of semiellipses and semihyperbolas form an *orthogonal curvilinear coordinate system* (confocal ellipses and hyperbolas) in the z plane cut open in the real axis for $-\infty < x < -1$ and $1 < x < \infty$. Each point in the cut z plane is the image of a unique point in the strip $|u| < \frac{1}{2}\pi$, $-\infty < v < \infty$, of the w plane. Hence this strip of width π is a fundamental region of the w plane for the function $\sin w$ (Figure 2-3).

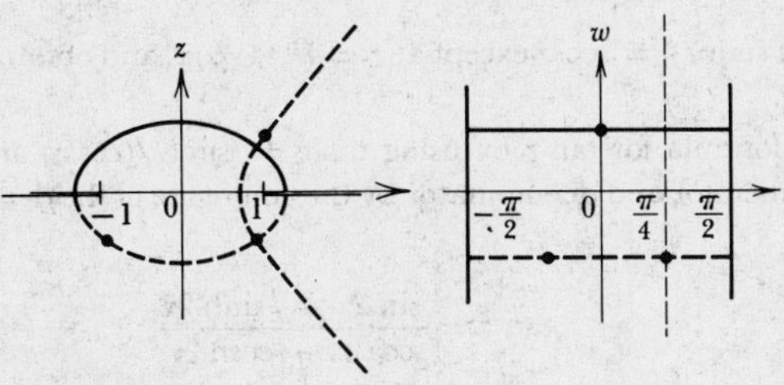

Figure 2-3

Problems

1. Find real and imaginary parts of $\cosh \pi i$, e^i, $\sin i\pi$, $\sinh(1 + i)$.
2. Using $f'(0) = \lim [f(z) - f(0)]/z$, show that as $z \to 0$

$$\lim \frac{e^z - 1}{z} = 1, \qquad \lim \frac{\sin z}{z} = 1, \qquad \lim \frac{\cos z - 1}{z} = 0.$$

3. (a) Show that $z = e^{i\theta}$, $0 \le \theta < 2\pi$, represents the unit circle described once. (b) If z_0 and $R > 0$ are constant, what does $z = z_0 + Re^{i\theta}$ represent, $0 \le \theta < 2\pi$?
4. Using exponentials, prove $\cos 2z = \cos^2 z - \sin^2 z$.
5. Show that $f(z) = e^z$ satisfies $f(\bar{z}) = \overline{f(z)}$, and from this get similar results for $\sinh z$, $\cosh z$, $\tanh z$, $\sin z$, $\cos z$, $\tan z$.
6. From (2.5) show that $e^{iz} = \cos z + i \sin z$ when z is complex.
7. Using Problem 6 and the addition theorem for e^z, get

$$\cos(z_1 + z_2) + i \sin(z_1 + z_2) = (\cos z_1 + i \sin z_1)(\cos z_2 + i \sin z_2).$$

Changing the sign of z_1 and z_2, get

$$\cos(z_1 + z_2) - i \sin(z_1 + z_2) = (\cos z_1 - i \sin z_1)(\cos z_2 - i \sin z_2).$$

Multiply out the right-hand sides and add or subtract these two equations to get the formula of the text for $\sin(z_1 + z_2)$ and also

$$\cos(z_1 + z_2) = \cos z_1 \cos z_2 - \sin z_1 \sin z_2.$$

8. Prove that $\cos(x + iy) = \cos x \cosh y - i \sin x \sinh y$. (See Problem 7.)
9. Prove $|\cos z|^2 = \cosh^2 y - \sin^2 x$, $\qquad |\sin z|^2 = \cosh^2 y - \cos^2 x$.
10. Find (a) all the zeros, and (b) all the periods:

$$\cos z, \qquad \sinh z, \qquad \cosh z, \qquad \tan z, \qquad \tanh z.$$

11. Show that $(\tan z)' = \sec^2 z$ except at $z = (k + \frac{1}{2})\pi$, and obtain a similar result for $\tanh z$.

12. Obtain a formula for $\tan z$ by using $\tan z = (\sin z)/(\cos z)$ and (2.5). Multiplying numerator and denominator by the conjugate of the denominator, show that

$$\tan z = \frac{\sin 2x + i \sinh 2y}{\cos 2x + \cosh 2y}.$$

13. Show that $2\sqrt{2}\exp(\pi i/12) = (\sqrt{3} + 1) + i(\sqrt{3} - 1)$. (Use $\frac{1}{12} = \frac{1}{3} - \frac{1}{4}$.)

14. Find all z for which: $e^z = i$, $e^z = 1 + i\sqrt{3}$, $\sin z = 2$, $\cos^2 z = -1$.

15. Show that the image of the line $z = (1 + i)t$, $-\infty < t < \infty$, under the transformation $w = e^z$ is a logarithmic spiral, and sketch. (Write $w = re^{i\theta}$ and note that the polar coordinate θ can be identified with t.)

16. Graph the real-variable functions e^x, $\sinh x$, $\cosh x$. Then sketch the image of the following point sets of the z plane (a) under the mapping $w = e^z$ and (b) under $w = \sinh z$:

$$0 < x < 2, \quad y = \frac{\pi}{2}; \qquad x = 1, \quad |y| < \frac{\pi}{2}; \qquad x < 0, \quad \frac{\pi}{3} < y < \pi.$$

3. The logarithm and power functions. With $r = |z| > 0$ and any value $\theta = \arg z$, the polar representation introduced in Chapter 1 is

$$z = r(\cos\theta + i \sin\theta).$$

By results of the last section this becomes

$$z = re^{i\theta} \qquad (z \neq 0). \tag{3.1}$$

If the familiar properties $\log ab = \log a + \log b$ and $\log e^a = a$ for the real logarithm of positive numbers can be extended to complex values of the variables, (3.1) suggests that at least one value of $\log z$ should be

$$\log z = \log r + i\theta. \tag{3.2}$$

There may be other values, however, because θ is determined by z only within a multiple of 2π.

To investigate (3.2) let $w = \log z$ be defined to mean that $e^w = z$. When $w = u + iv$ and z is expressed in polar form, the equation $e^w = z$ becomes

$$e^u e^{iv} = re^{i\theta}.$$

Equating absolute values, we get $e^u = r$, so that $u = \text{Log}\, r$, the capital letter being used to denote the logarithm in the sense of real-variable theory. When e^u and r are canceled from both sides it is seen that $e^{iv} = e^{i\theta}$ and so, by the periodicity established in the preceding section, $v = \theta \bmod 2\pi$. Thus all values of $\log z$ are given by

$$\log z = \text{Log}\,|z| + i \arg z \tag{3.3}$$

in essential agreement with (3.2). Both $\text{Log}\,|z|$ and $\arg z$ are meaningless for $z = 0$, and the number 0 has no logarithm. This agrees with the fact that e^w never vanishes.

Since $\arg z$ has infinitely many values, the same is true of $\log z$. The values of $\arg z$ differ by $2\pi k$, where k is an integer, and so the values of $\log z$ differ by $2\pi ik$. In the notation of Chapter 1, Section 3

$$\log z = \text{Log}\,|z| + i\,\text{Arg}\, z + 2\pi ik,$$

or equivalently, $\log z = \text{Log}\,|z| + i\,\text{Arg}\, z \bmod 2\pi i$. Familiar properties of the real-variable logarithm continue to hold but are complicated by the fact that $\log z$ has infinitely many values. Thus, when $z \neq 0$

$$e^{\log z} = z \qquad \text{but} \qquad \log e^z = z + 2\pi ik.$$

Applying the first equality with $z = \alpha, \beta$, and $\alpha\beta$, where $\alpha\beta \neq 0$, we get

$$e^{\log(\alpha\beta)} = \alpha\beta = e^{\log\alpha} e^{\log\beta} = e^{\log\alpha + \log\beta}$$

and hence

$$\log \alpha\beta = \log\alpha + \log\beta + 2\pi ik.$$

That a multiple of $2\pi i$ may be required to validate the above equation is seen by the example $1 = (-1)(-1)$, where one can choose $\log 1 = 0$ and[1] $\log(-1) + \log(-1) = \pi i + \pi i \neq 0$.

If α is complex and $z \neq 0$, then by definition

$$z^\alpha = e^{\alpha \log z} = e^{\alpha(\text{Log}\,|z| + i \arg z)} = e^{\alpha\,\text{Log}\,|z|} e^{i\alpha \arg z}. \tag{3.4}$$

For instance, since $\log 2 = \text{Log}\, 2 + 2\pi ik$, we have

[1] Since $e^{i\pi} = \cos\pi + i \sin\pi = -1$, a logarithm of -1 is $i\pi$.

$$2^i = e^{i \log 2} = e^{i \operatorname{Log} 2} e^{-2\pi k}, \qquad k = 0, \pm 1, \pm 2, \ldots.$$

There are infinitely many values of 2^i, including values of arbitrarily large magnitude and values of arbitrarily small magnitude.

When α is real, $\alpha \operatorname{Log} |z| = \operatorname{Log} |z|^\alpha$ by real analysis, and (3.4) reduces to

$$z^\alpha = |z|^\alpha e^{i\alpha \arg z} = |z|^\alpha[\cos(\alpha \arg z) + i \sin(\alpha \arg z)].$$

If $\alpha = p/q$ is rational as well as real, the result agrees with the results of Chapter 1, Section 3, as it should. All of the above equations are interpreted to mean that the set of numbers given by one side of the equation is the same as the set given by the other side. Only when α is an integer do these equations between sets of numbers reduce to equations between single numbers.

The expression $\log z$ has infinitely many values and is an example of a *many-valued function.* The whole italicized phrase is relevant here; the word "function" without any qualification is always used in the sense of Chapter 1. Functions in the sense of Chapter 1 are sometimes called single-valued functions, but this is only for emphasis.

Since equations between many-valued functions can be interpreted with ease, as equality of sets, it might be thought that analytic operations with many-valued functions could be interpreted with equal ease. However, this is not the case. For example, if we agree that $\sqrt{z}$ is a two-valued function for $z \neq 0$, an equation as simple as

$$\sqrt{1} + \sqrt{1} = 2\sqrt{1}$$

fails to hold. The left side has *three* values -2, 0 and 2 while the right side has *two* values -2 and 2. The analytic study of many-valued functions usually requires that the many-valued function be expressed in terms of single-valued functions.

One way of doing this is to consider the many-valued function in a restricted region of the plane and choose a value at each point in such a way that the resulting single-valued function is continuous. A continuous function obtained from a many-valued function in this way is called a *branch* of the many-valued function.

For example, the logarithm is expressed in terms of single-valued func-

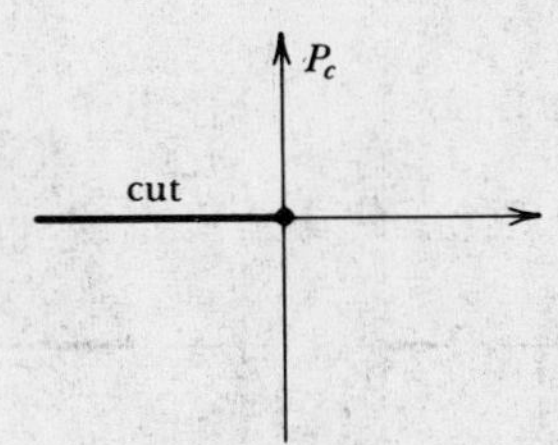

Figure 3-1

tions by using Arg z instead of arg z. The restriction of $\theta = \text{Arg } z$ to $-\pi < \theta \leq \pi$ can be indicated by cutting the z plane along the negative real axis, that is, on $y = 0$, $x < 0$. In the next section the upper edge of the cut, corresponding to $\theta = \pi$, is regarded as being in the cut plane, while the lower edge, corresponding to $\theta = -\pi$, is not regarded as belonging to the cut plane. However, for present purposes an open region of the z plane is preferable, and so the whole negative real axis is deleted. The points 0 and ∞ are also deleted. The cut plane is then an open region, which we denote by P_c (see Figure 3-1).

The function Log z defined by

$$\text{Log } z = \text{Log } |z| + i \text{ Arg } z \tag{3.5}$$

is analytic in P_c, as shown next. Hence it is certainly continuous there, and so represents a branch of log z in the sense of the above definition.

To carry out this program we must study Arg z as a function of the complex variable z. As is suggested by a graph of tan θ, and as the reader will recall from real analysis, the equation $\tan \theta = t$, with t real, has a unique solution θ on the interval $(-\pi/2, \pi/2)$. The solution is denoted by $\text{Tan}^{-1} t$ and satisfies

$$\frac{d}{dt} \text{Tan}^{-1} t = \frac{1}{1 + t^2} \qquad -\infty < t < \infty. \tag{3.6}$$

In the half-planes $y < 0$, $x > 0$ and $y > 0$ of P_c we define $\theta = \text{Arg } z$ by the formulas

$$\theta = -\text{Tan}^{-1} \frac{x}{y} - \frac{\pi}{2}, \quad \theta = \text{Tan}^{-1} \frac{y}{x}, \quad \theta = -\text{Tan}^{-1} \frac{x}{y} + \frac{\pi}{2}, \tag{3.7}$$

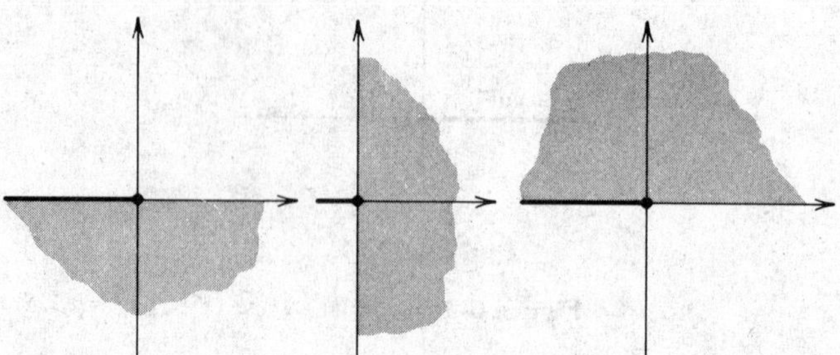

Figure 3-2

respectively. The regions $x > 0$ and $y > 0$ overlap in the first quadrant, and $x > 0$ and $y < 0$ overlap in the fourth quadrant (Figure 3-2). However, the above definitions are consistent in the overlapping parts, as shown in Section 4, Problem 6, or as can be seen by a sketch.

We now use (3.7) to establish the analyticity of Log z. If Log $z = u + iv$, then (3.5) gives $u = \frac{1}{2}\,\mathrm{Log}(x^2 + y^2)$ and hence

$$u_x = \frac{x}{x^2 + y^2}, \qquad u_y = \frac{y}{x^2 + y^2}.$$

By (3.5) it is seen that $v = \mathrm{Arg}\, z = \theta$, and each expression (3.7) gives

$$v_x = -\frac{y}{x^2 + y^2}, \qquad v_y = \frac{x}{x^2 + y^2}$$

as follows from (3.6). Hence the Cauchy-Riemann equations hold and Theorem 1.1 shows that Log z is analytic in P_c. Since

$$u_x + iv_x = \frac{x - iy}{x^2 + y^2} = \frac{1}{x + iy},$$

(1.6) gives the formula

$$\frac{d}{dz}\mathrm{Log}\, z = \frac{1}{z}, \qquad z \text{ in } P_c. \tag{3.8}$$

The function Log z considered above is called the *principal branch* of the logarithm. Any other branch in the cut plane is given by

$$\mathrm{Log}_n\, z = \mathrm{Log}\, z + 2\pi i n$$

where n is an integer. Instead of using the notation $\text{Log}_n\, z$, one can convey the same information by $\log z$ together with a statement that a definite branch has been chosen.

To be sure, one could get more complicated functions by using the above formula with one value of n in one part of the cut plane and with another value of n in another part. However, such a function would not be a branch, because it violates the condition of continuity. Since $2\pi in$ is constant, every branch $\text{Log}_n\, z$ satisfies the fundamental relation (3.8). It is natural to inquire whether the different branches can be connected to give a single function which has a right to be called $\log z$ and which also satisfies (3.8). In the following section it is seen that such a construction is possible and leads to the concept of Riemann surface.

Example 3.1. The principal branch of z^α is obtained by using $\text{Log}\, z$ in the equation $z^\alpha = \exp(\alpha \log z)$, and other branches are obtained by using other branches of $\log z$. Show that each branch satisfies

$$\frac{d}{dz} z^\alpha = \alpha z^\alpha \frac{1}{z} \tag{3.9}$$

in the cut plane. We can write $\alpha z^{\alpha-1}$ if it is understood that the same branch of z^α is used on both sides. By the chain rule

$$(z^\alpha)' = e^{\alpha \log z}(\alpha \log z)'$$

and the result follows from (3.8), since $\log z = \text{Log}\, z + \text{const.}$

When z is real and negative the result fails because then z lies on the cut used in the definition of $\text{Log}\, z$. However, the result could be restored for such z by moving the cut. Instead of defining $\text{Log}\, z$ in the cut plane obtained by deleting the line $x < 0, y = 0$, one could just as easily define $\text{Log}\, z$ in the cut plane obtained by deleting the line $x > 0, y = 0$. In that case the formula (3.9) would hold on the negative real axis and would fail on the positive real axis.

Similar remarks apply in general. If z is given beforehand, one can usually introduce branch cuts which avoid z, and thus get a single-valued branch in a neighborhood of z. It is sometimes said that a branch is determined *locally,* to emphasize that the results obtained may be valid only near the given z, and not at distant points.

In the following section $\log z$ and z^α are defined *globally,* so that (3.9) holds under the sole condition that $z \neq 0$, $z \neq \infty$.

Example 3.2. The equation $w = \sin^{-1} z$ means that $z = \sin w$. Show that all values of $\sin^{-1} z$ are given by

$$\sin^{-1} z = -i \log[iz + (1 - z^2)^{1/2}]. \tag{3.10}$$

From $z = \sin w$ it follows that $2iz = e^{iw} - e^{-iw}$ or, multiplying by e^{iw},

$$e^{2iw} - 2ize^{iw} - 1 = 0.$$

Solving this quadratic for e^{iw} and then taking the logarithm gives the desired result, the square root being two-valued, as usual. Since either choice of the square root gives

$$[iz + (1 - z^2)^{1/2}][iz - (1 - z^2)^{1/2}] = -1 \tag{3.11}$$

neither bracket is 0, and so the expression (3.10) is well-defined for all z. The fact that each value w satisfies $\sin w = z$ is readily verified by use of[2] $\exp(\log \alpha) = \alpha$, $\alpha \neq 0$.

Equation (3.11) shows that for any given choice of $(1 - z^2)^{1/2}$ the logarithms can be selected so that

$$-i \log[iz + (1 - z^2)^{1/2}] - i \log[iz - (1 - z^2)^{1/2}] = \pi.$$

Thus if one value of $\sin^{-1} z$ is w, another is given by $\pi - w$. This corresponds to the familiar relation $\sin(\pi - w) = \sin w$. Since the logarithm is determined only to within a multiple of $2\pi i$, it follows further from (3.10) that $w + 2n\pi$ is a value of $\sin^{-1} z$ if w is, n being any integer. This expresses the periodicity of $\sin w$. Since (3.10) gives all solutions of $\sin w = z$, these two cases exhaust the values of $\sin^{-1} z$.

Example 3.3. If $z \neq 1$ and $z \neq -1$, show that any branch of $\sin^{-1} z$ satisfies

$$\frac{d}{dz} \sin^{-1} z = \frac{1}{(1 - z^2)^{1/2}} \tag{3.12}$$

[2] As in real analysis, the notation $\exp z \equiv e^z$ is sometimes used for convenience in printing.

locally, the square root on the right being the same choice as that used in $\sin^{-1}z$. Since $z^2 \neq 1$, a branch of $\sin^{-1}z$ can be determined locally by choosing first one of the square roots and then choosing a branch of the logarithm. If $\zeta = 1 - z^2 \neq 0$, then

$$\sin^{-1}z = -i\log(iz + \zeta^{1/2}).$$

We can differentiate $\zeta^{1/2}$ by Example 3.1 and the chain rule. By (3.11) the term $iz + \zeta^{1/2}$ does not vanish, and the chain rule gives

$$\frac{d}{dz}\sin^{-1}z = \frac{-i}{iz + (1 - z^2)^{1/2}}\left[i + \frac{1}{2}\frac{-2z}{(1 - z^2)^{1/2}}\right].$$

This reduces to (3.12).

Problems

1. If $\alpha \neq 0$, there are infinitely many functions $\alpha^z = e^{z\log\alpha}$, one for each determination of $\log\alpha$. Show that each of these functions satisfies
$$\frac{d}{dz}\alpha^z = \alpha^z\log\alpha$$
provided one and the same value of $\log\alpha$ is used throughout.
2. From $\operatorname{Arg} i = \pi/2$ obtain $\operatorname{Log} i = \operatorname{Log} 1 + i\pi/2$. Thus get
$$i^i = e^{i\log i} = e^{i(\operatorname{Log} i + 2\pi ik)} = e^{-(4k+1)\pi/2}, \qquad k = 0, \pm 1, \ldots.$$
Plot the values of i^i in the complex plane.
3. (a) Compute all values and (b) plot in the complex plane:
$\log(-i)$, $\log(1 + i)$, $3\log(1 + i\sqrt{3})$, $\log(1 + i\sqrt{3})^3$, $\log(1 + i)^{\pi i}$,
$(-i)^{-i}$, i^2, 3^π, $2^{\pi i}$, $(1 + i)^{1+i}$, $(1 + i)^i(1 + i)^{-i}$.
4. Find all solutions of the following equations:
$\operatorname{Log} z = \frac{1}{4}\pi i$, $e^z = i$, $\sin z = i$, $\cos z = \sin z$, $\tan^2 z = -1$.
5. (a) Explain why $\log z^\alpha = \alpha\log z + 2\pi ik$. (b) If $\alpha\beta \neq 0$, solve $z^\alpha = \beta$ and $\alpha^z = \beta$.
6. If $w = \sinh^{-1}z$ means $z = \sinh w$, obtain the formula
$$\sinh^{-1}z = \log[z + (z^2 + 1)^{1/2}]$$
by the procedure and also by the result of Example 3.2. (Use $\sin iw = i\sinh w$.)
7. Following the pattern of Example 3.2, obtain the formulas

$$\cos^{-1}z = -i\log[z + (z^2 - 1)^{1/2}], \qquad \tan^{-1}z = \frac{i}{2}\log\frac{i+z}{i-z}$$

$$\cosh^{-1}z = \log[z + (z^2 - 1)^{1/2}], \qquad \tanh^{-1}z = \frac{1}{2}\log\frac{1+z}{1-z}.$$

8. Obtain the following local formulas, valid for any branch defined throughout a neighborhood of the point z:

$$\frac{d}{dz}\tan^{-1}z = \frac{1}{1+z^2}, \qquad \frac{d}{dz}\sinh^{-1}z = \frac{1}{(1+z^2)^{1/2}}.$$

9. If $z = x + iy = re^{i\theta}$, compute all values of the following:

$$\text{Re } z^i, \quad \text{Im } z^i, \quad |z^i|, \quad \text{Re } i^z, \quad \text{Im } i^z, \quad |i^z|, \quad \text{Re } z^z, \quad \text{Im } z^z, \quad |z^z|.$$

10. In what sense is it true that $\alpha^{\beta+\gamma} = \alpha^\beta\alpha^\gamma$? That $(\alpha^\beta)^\gamma = \alpha^{\beta\gamma}$?

***4. The Riemann surface for log z.** Each branch of a many-valued function is defined, in general, in some region of the z plane. If the regions for different branches do not overlap, there is no conflict in the assignment of values, but if they do overlap, the points of the regions must be distinguished from each other in the common parts. One way of doing this is to consider that the regions lie in different planes, one above the other. Thus the variable is thought to be not just z but (z,n) where the integer n specifies the plane in which z lies. Each plane is called a *sheet* of the surface.

It is a triviality that the many-valued functions of z commonly encountered in analysis can be made into single-valued functions of (z,n) in this way. However, it is less trivial that the different regions lying in the different planes can be joined, in general, to give a single region of definition which is actually connected in the sense of Chapter 1. This matter is discussed now.

To illustrate the idea, let us discuss the multivalued function arg z in a region which surrounds but does not contain the origin, as shown in Figure 4-1. (For subsequent purposes this figure has been drawn so as to suggest a region which overlaps itself.) It is impossible to pick out a single-

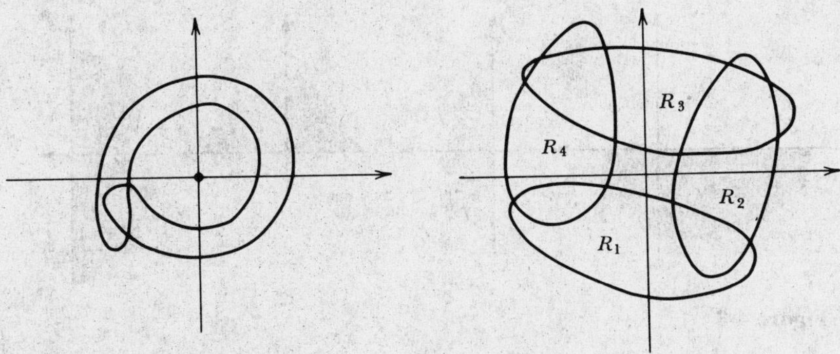

Figure 4-1 **Figure 4-2**

valued branch, because any given choice in the overlapping part inevitably changes to another choice when we return to the overlapping part via a path which encircles the origin. However, it is possible to cover the region R by regions R_1, R_2, R_3, R_4, each of which overlaps some of the others but none of which surrounds the origin (Figure 4-2). A branch of arg z is now selected in R_1. By continuity this branch is uniquely determined in R_2 if arg z is to be single-valued in the overlapping parts of R_1 and R_2. Similarly the branch of arg z is uniquely determined in R_3 from its values in the common part of R_2 and R_3. Proceeding from R_3 to R_4, we see that the values of arg z are again uniquely determined in R_4. However, the values in the portions of R_1 and R_4 that overlap differ by 2π.

To account for this we agree to distinguish the points of R_4 from those of R_1, even in the overlapping part. It is as if R_4 lies in a copy of the z plane above the original z plane, so that R_1 and R_4 do not intersect. Clearly, the region formed by the R_i is connected, even though, by this agreement, R_1 and R_4 are considered not to overlap. The fact that R_1 is connected to R_4 follows, not directly, but by way of R_2 and R_3.

The same method is now applied to discussion of the logarithm. The process of adding successive half-planes need not stop with the three half-

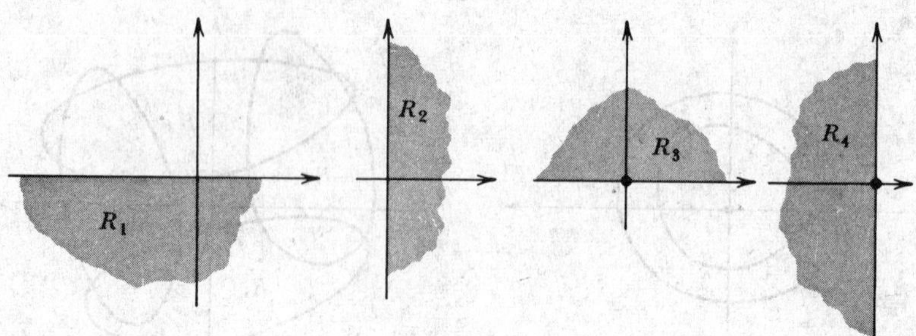

Figure 4-3

planes of Figure 3-2, but is continued as suggested in Figure 4-3. The half-plane $x < 0$, added next, is provided with the function

$$\arg z = \theta = \operatorname{Tan}^{-1}\frac{y}{x} + \pi. \tag{4.1}$$

This agrees with the value which was assigned in the plane $y > 0$; that is, at points common to both planes, the two formulas give the same value for θ. However, the new formula does not agree with the value formerly assigned in the half-plane $y < 0$.

To account for this we number the successive half-planes as R_1, R_2, R_3, and R_4, and we agree to distinguish the points of R_4 from those of R_1. It is as if the part of R_4 with $y < 0$ lies in a copy of the z plane above the original z plane, so that R_1 and R_4 do not intersect. The half-planes R_i in this discussion play the same role as the regions R_i in Figure 4-2.

Evidently the process can be continued both upward and downward. The half-planes overlapping as described form a spiral surface which is called the *Riemann surface* for $\log z$ (see Figure 4-4). If a precise notation is needed, one can specify the points of the Riemann surface by (z,n) where n indicates that z lies in the nth sheet. Then z itself is the projection of the point (z,n) onto the ordinary z plane. We convey the same idea by stating that "z is a point of the Riemann surface," or that a given function is considered "on its Riemann surface."

To give an analytic description of $\log z$ as a single-valued function on its Riemann surface we define θ on R_1, R_2 and R_3 by (3.6) and (3.7). The

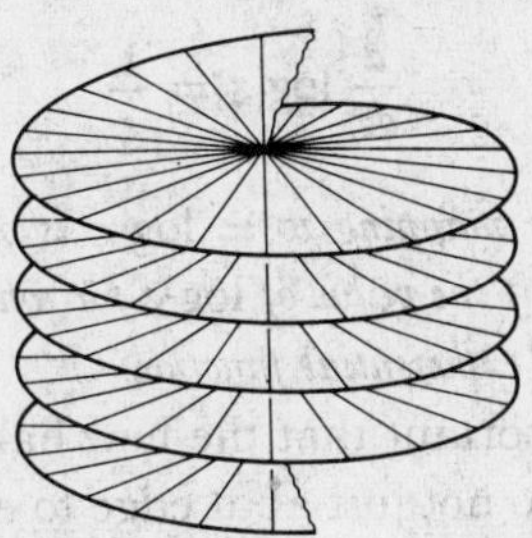

Figure 4-4

definition on the nth half-plane, R_n, is made by the rule that if n increases or decreases by 2, the assigned value of θ increases or decreases, respectively, by π. In the common part of R_n and R_{n+1} two formulas are available for θ. However, both formulas give the same value, as follows from elementary properties of the function $\tan \theta$.

It is not hard to establish the analyticity of $\log z$ on its Riemann surface. Any point $z = (z,n)$ of the surface is interior to one of the half-planes used in the construction, and so $\arg z$ has one of the values used in Section 3, apart from an additive constant. Hence v_x and v_y have the same values as they did there, and the derivative of $\log z$ not only exists, but equals $1/z$.

We now discuss the sense in which $\log z$ and e^z are inverse functions. Since $\exp(\log z) = z$ for every determination of $\log z$ we need only inquire whether, with the present definition, $\log(e^z) = z$. If $\phi(z) = \log(e^z)$, the chain rule gives $\phi'(z) = 1$, and so $\phi(z) - z$ is constant on each half-plane of the Riemann surface. Hence it is constant on the surface. If the determination of $\log z$ at 1 is such that $\log 1 = 0$, then it follows that the constant is 0, and so $\log e^z = z$. (Other determinations would give $z + 2\pi ik$, where k has one and the same value throughout the Riemann surface.)

Since $\log z$ has an inverse function, as just seen, the mapping given by $w = \log z$ from the Riemann surface to the w plane must be one-to-one. The discussion is summarized as follows:

THEOREM 4.1. *The function* $\log z$ *is analytic at every point of its Riemann surface and satisfies*

$$\frac{d}{dz}\log z = \frac{1}{z}$$

throughout the surface. The mapping $w = \log z$ *is one-to-one from the Riemann surface to the* w *plane, and if the value of* $\log z$ *for which* $\log 1 = 0$ *is chosen, the inverse map is given by the exponential function.*

In Figure 4-3 it is important that the new half-planes actually overlap with the old ones, and do not just abut edge to edge. The objection to an edge is that analyticity at an edge is hard to verify, and there are also problems about uniqueness of the definition as one crosses the edge. Although a general development of Riemann surfaces is not given here, it can be said that the general development, like the special case, depends on the concept of a chain of overlapping regions, in each of which the function considered is single-valued.

The geometric visualization of a general Riemann surface can be difficult, and it may even be impossible to imbed the surface satisfactorily in three-dimensional space (see Example 4.1). However, the geometric visualization is of much lesser importance than the analytic description by means of a chain of regions, together with a rule stating which regions overlap and which do not.

It should be mentioned that one and the same Riemann surface can be described in several ways. For example, consider a stack of cut planes P_n, each of which is a copy of the cut plane used in Section 3, except that here the upper edge of the cut, corresponding to $\theta = \pi$, belongs to the cut plane and the lower edge does not. If the upper edge of P_n is joined to the lower edge of P_{n+1}, and the planes are otherwise disjoint, the result is equivalent to the Riemann surface of Figure 4-4. Here, however, n actually specifies the nth sheet, $n = 0, \pm 1, \pm 2, \ldots$.

Introduction of the Riemann surface is not the only way in which to treat a many-valued function such as arg z. By introducing suitable cuts the many-valued function can always be described by single-valued functions, just as we described the log by means of Log. Another way is to use a chain of regions as in Figure 4-2 but never to consider two regions together when such a consideration would lead to inconsistency. Thus, when we use R_4 in Figure 4-2, we can consider that R_1 is thereupon obliterated. A third way is to follow a definite curve in the complex plane and insist that the chosen value be a continuous function of the arc s on that curve. The result need not be a single-valued function of the point z in the plane, but it is a single-valued function of s.

This third method of dealing with many-valued functions leads to the interesting concept of branch point. If values of $\log z$ are chosen so as to give a continuous function of s on the circle

$$z = \epsilon e^{is}, \qquad 0 \le s \le 2\pi, \quad \epsilon > 0$$

in the z plane, the values of $\log z$ at $s = 0$ and at $s = 2\pi$ differ by 2π, even though the $s = 0$ and $s = 2\pi$ both correspond to the same point, $z = \epsilon$. The origin is called a *branch point* of $\log z$ because by following a small circle enclosing the origin we can pass from one branch of $\log z$ to another. On a sufficiently small circle enclosing any other point z, every branch of $\log z$ returns to its original value. Thus $z = 0$ is the only finite branch point of $\log z$. (In the extended plane $z = \infty$ is also a branch point.)

In the cut plane it is not possible to follow a circle all the way around the origin, and there is also no closed path around the origin on the Riemann surface. The cut plane prevents closed paths by the cut which joins the branch point at 0 to that at ∞, while the Riemann surface prevents closed paths by its geometrical structure. On the Riemann surface one can pass continuously from one branch of the logarithm to another, while on the cut plane one cannot. Each branch lies in its own sheet of the Riemann surface, and these sheets are connected.

Example 4.1. Describe the Riemann surface for $w = z^{1/3}$. Since $z^\alpha = \exp(\alpha \log z)$, the function $w = z^\alpha$ is single-valued on the Riemann surface for $\log z$. However, the mapping of the surface to the w plane is not one-to-one, if α is rational, because the values assumed on the Riemann surface repeat. In the present case, as $\arg z$ increases by 6π, the value of $\frac{1}{3} \arg z$ increases by 2π, and so $\exp(\frac{1}{3} \log z)$ returns to its original value.

To account for this the Riemann surface is described by successive cut planes similar to those used in the Riemann surface for $\log z$. However, when we get to the third sheet, P_3, the upper edge of P_3 is joined to the lower edge of P_1 and no new cut plane P_4 is added. The resulting Riemann surface has three sheets, as it should, since $z^{1/3}$ has three values. The fact that the map from the Riemann surface to the w plane is one-to-one follows from the periodicity properties of e^z.

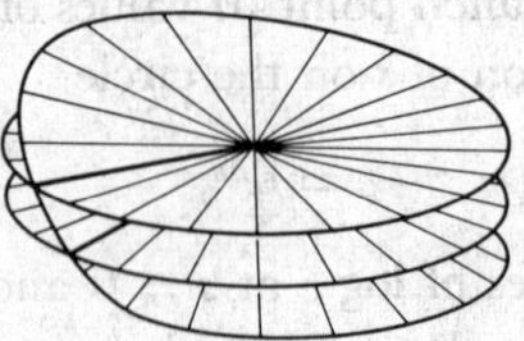

Figure 4-5

Although the appearance of this surface is suggested in Figure 4-5, the surface cannot be faithfully imbedded in three-dimensional space, because the cut plane P_3 intersects the other sheets of the surface as it returns to P_1 in the first sheet. We agree in such cases to ignore the intersection, precedence being given to the explicit statement about overlapping rather than to the geometric appearance of the figure.

Example 4.2. Discuss Riemann surfaces for $w = z^{5/3}$. The equation is equivalent, by definition, to $w^3 = z^5$. If $\arg z$ ranges from 0 to 6π, then $\arg w$ ranges from 0 to 10π and hence z belongs to the Riemann surface illustrated in Figure 4-5, while w belongs to a similar surface of five sheets rather than three. The mapping from one surface to the other is one-to-one.

As suggested by this example, the study of multivalued functions $w = f(z)$ often requires two Riemann surfaces, one in the z plane and one in the w plane. Only if the function or its inverse is single-valued does the corresponding Riemann surface reduce to a surface with a single sheet.

Problems

1. Describe fundamental regions and Riemann surfaces for:

$$w = z, \quad w = z^2, \quad w = z^{1/4}, \quad w = z^{3/2}, \quad w = z^e, \quad w = \log(1 - z), \quad w^2 = \log z.$$

2. Sketch the image of the circle $w = e^{2it} - 1, 0 \leq t < 2\pi$, as given by $z = w^2$ in the Riemann surface. Imagine that you are looking down on the surface and

use dotted lines to show the image in the lower sheet, solid lines in the upper sheet.

3. If α and β are constant, investigate the validity of the following equations on the Riemann surface for log z:

$$z^{\alpha+\beta} = z^\alpha z^\beta, \qquad \log(z^\alpha) = \alpha \log z, \qquad z(z^\alpha)' = \alpha z^\alpha, \qquad z^{\alpha\beta} = (z^\alpha)^\beta.$$

Which, if any, require $\log 1 = 0$ and fail for $\log 1 = 2\pi i$?

4. If $a \neq 0$ is constant, show by the chain rule that the function

$$\phi(z) = \log(az) - \log z$$

satisfies $\phi'(z) = 0$ on the Riemann surface for log z, hence is constant. Assuming $\log 1 = 0$, determine the constant by setting $z = 1$. Thus get

$$\log az = \log a + \log z, \qquad az \neq 0.$$

Analytic definition of arg *z

5. Let the function $\text{Tan}^{-1}t$ and the constant π be defined by

$$\text{Tan}^{-1}t = \int_0^t \frac{d\tau}{1+\tau^2}, \qquad \pi = \int_{-\infty}^{\infty} \frac{d\tau}{1+\tau^2}.$$

(a) Evaluate $(d/dt)\text{Tan}^{-1}t$ by inspection. (b) Deduce

$$\text{Tan}^{-1}t + \text{Tan}^{-1}\frac{1}{t} = (\text{sign } t)\frac{\pi}{2}, \qquad t \neq 0, \quad (*)$$

where sign $t = +$ for $t > 0$ and sign $t = -$ for $t < 0$. (By the chain rule the left side of (*) has 0 derivative, hence is constant on any interval in which $t \neq 0$. Determine the constant on $(-\infty,0)$ and on $(0,\infty)$ by letting $t \to -\infty$ or $t \to \infty$, respectively.)

6. Show that the definition of $\theta = \text{Arg } z$ given in (3.7) is consistent at points where two different formulas apply. (The value assigned in the half-plane $x > 0$ minus that assigned in the half-plane $y < 0$ is

$$\text{Tan}^{-1}\frac{y}{x} + \text{Tan}^{-1}\frac{x}{y} + \frac{\pi}{2}.$$

Use Problem 5 with $t = y/x < 0$. Discussion of the half-planes $x > 0, y > 0$ is similar; in fact the method applies to any two overlapping half-planes of the Riemann surface.)

5. Harmonic functions.

At any point $z = x + iy$ where $f'(z)$ exists, (1.6) and the Cauchy-Riemann equations give

$$f'(z) = u_x - iu_y, \qquad f'(z) = v_y + iv_x,$$

the functions on the right being evaluated at (x,y). If not only $f'(z)$ but also $f''(z)$ exists, we can apply the Cauchy-Riemann equations again to each of these expressions. The result is

$$u_{xy} = u_{yx}, \qquad u_{xx} + u_{yy} = 0, \qquad v_{xy} = v_{yx}, \qquad v_{xx} + v_{yy} = 0. \quad (5.1)$$

It is known from real analysis that $u_{xy} = u_{yx}$ at any point where the second partial derivatives are continuous, and so this condition is automatically satisfied by functions of the type commonly encountered in calculus. However, (5.1) also asserts that u and v satisfy an equation of the form

$$\phi_{xx} + \phi_{yy} = 0, \qquad (5.2)$$

which is an extremely restrictive condition. Equation (5.2) is known as *Laplace's equation* and plays a fundamental rôle in pure and applied mathematics. A continuous function $\phi(x,y)$ that satisfies Laplace's equation in a domain D is said to be *harmonic* in D. Thus the real and imaginary parts of $f(z)$ are harmonic in any domain throughout which $f''(z)$ exists.

In the next chapter we prove that $f'(z)$ is analytic at any point where $f(z)$ is analytic, and hence both existence and continuity of $f''(z)$ are assured. We also show, in Chapter 6, that a function which is harmonic in a domain necessarily has derivatives of all orders, and hence, the second derivatives are continuous. These results are taken for granted here and in Chapter 5.

A harmonic function v related to u by the Cauchy-Riemann equations is said to be *conjugate*[1] to u, or to be the *harmonic conjugate* of u. When it exists, a harmonic conjugate is clearly unique apart from an additive constant. If v is a conjugate for u, then $-u$ is a conjugate for v, since if is analytic whenever f is analytic.

It is a matter of considerable interest that every harmonic function u admits a conjugate, in a suitably restricted domain, so that u is the real part of an analytic function $u + iv$. This subject is discussed now, under the assumption that the domain D has the properties suggested by

[1] This use of the term is different from its use in reference to complex numbers, as in $\bar{z} = x - iy$.

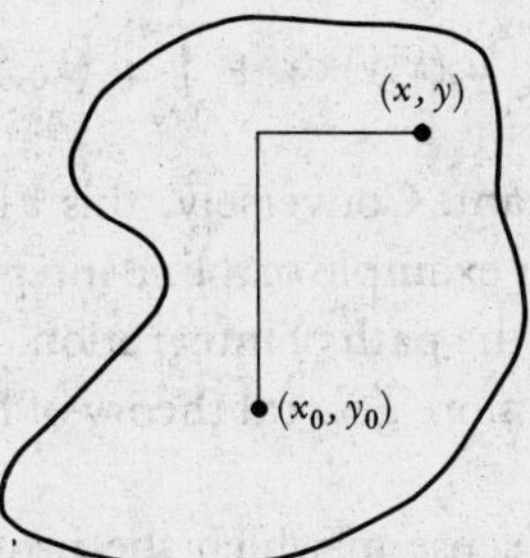

Figure 5-1

Figure 5-1. The intent of the figure is as follows: There is a fixed point (x_0,y_0) such that the line segment joining (x_0,y_0) to (x_0,y) and the line segment joining (x_0,y) to (x,y) both lie in D whenever (x,y) is a point of D.

Let u be harmonic in a domain D, as described above, and let it be required to construct a conjugate function v, such that

$$v_x = -u_y, \qquad v_y = u_x. \tag{5.3}$$

On integrating the first of these with respect to x, we get

$$v(x,y) = -\int_{x_0}^{x} u_y(x,y)\,dx + c(y).$$

For every choice of $c(y)$ it is clear that $\partial v/\partial x = -u_y$, by the fundamental theorem of calculus. To make $\partial v/\partial y = u_x$ we require

$$-\frac{\partial}{\partial y}\int_{x_0}^{x} u_y(x,y)\,dx + c'(y) = u_x(x,y). \tag{5.4}$$

By a familiar theorem of real analysis the order of differentiation and integration can be interchanged here, since u_{yy} is continuous, and so the first term can be expressed as an integral of $-u_{yy}$. Since u is harmonic, the result is

$$\int_{x_0}^{x} (-u_{yy})\,dx = \int_{x_0}^{x} u_{xx}\,dx = u_x(x,y) - u_x(x_0,y).$$

Substitution into (5.4) gives $c'(y) = u_x(x_0,y)$; integration gives $c(y)$, and

$$v(x,y) = -\int_{x_0}^{x} u_y(x,y)\,dx + \int_{y_0}^{y} u_x(x_0,y)\,dy + c \tag{5.5}$$

where $c = v(x_0,y_0)$ is constant. Conversely, this v satisfies (5.3).

The expression (5.5) is an example of a line integral, and the rectilinear path shown in Figure 5-1 is the path of integration. This brief introduction may serve as a guide to the more general theory of line integrals presented in the next chapter.

We now discuss another sense in which the problem of constructing a conjugate function is connected with the problem of integration. If u is a given harmonic function in a domain D, the function

$$g(z) = u_x(x,y) - i\,u_y(x,y) \tag{5.6}$$

satisfies the Cauchy-Riemann equations, since the latter reduce to

$$(u_x)_x = (-u_y)_y, \qquad (u_x)_y = (u_y)_x$$

just as in (5.1). The first equation holds because u is harmonic and the second holds by a theorem of real analysis. Hence, g is analytic in D. If u is the real part of an analytic function f, then $g = f'$, since $f'(z)$ agrees with the right side of (5.6) by (1.6) and by the Cauchy-Riemann equations for f. Conversely, if we can find an analytic function f such that $f' = g$ then, as is easily seen, u agrees with the real part of f except possibly for an additive constant. Thus, given a harmonic function u in D, the question as to whether there is an analytic function f in D which has u as its real part is essentially the question: Given g analytic in D, is there an f such that $f' = g$? This question also leads to the theory of integration, as presented in the following chapter.

Example 5.1. Choose the constant a so that the function

$$u = x^3 + axy^2$$

is harmonic and find a conjugate function v in that case.

Here $u_{xx} + u_{yy} = 6x + 2ax = 0$ for $a = -3$, and hence this choice makes u harmonic. When $a = -3$

$$u_x = 3x^2 - 3y^2, \qquad u_y = -6xy$$

and taking $x_0 = y_0 = 0$ in (5.5), gives

$$v(x,y) = -\int_0^x (-6xy)\,dx + \int_0^y (-3y^2)\,dy + c = 3x^2y - y^3 + c.$$

Alternatively, by (5.6)

$$g(z) = 3x^2 - 3y^2 + 6ixy = 3(x+iy)^2 = 3z^2.$$

Hence $f' = g$ gives $f(z) = z^3 + C$, where C is constant. Setting $\operatorname{Im} C = c$ we get, as before,

$$v(x,y) = \operatorname{Im}(z^3 + C) = \operatorname{Im}[(x+iy)^3 + C] = 3x^2y - y^3 + c.$$

Example 5.2. Let $f(z) = u(x,y) + i\,v(x,y)$ be a rational function and let v be modified by an additive constant, if necessary, so that $f(0)$ is real. Show that

$$f(z) = 2u\left(\frac{z}{2}, \frac{z}{2i}\right) - u(0,0). \tag{5.7}$$

By means of the theory of power series introduced later, it can be proved that the result is valid, in a sufficiently small neighborhood, under the sole assumption that u is harmonic; the rationality of f is not needed.

To deduce (5.7) when f is rational, define another rational function f_1 by $f_1(z) = \overline{f(\bar{z})}$. Then

$$2u(x,y) = f(z) + \overline{f(z)} = f(z) + f_1(\bar{z})$$

and so

$$2u(x,y) = f(x+iy) + f_1(x-iy).$$

By Chapter 1, Section 5, Problem 10, this holds for complex as well as real x and y. Replacing x by $z/2$ and y by $z/(2i)$, we obtain

$$2u\left(\frac{z}{2}, \frac{z}{2i}\right) = f(z) + f_1(0).$$

Since $f(0)$ is real, $f_1(0) = f(0) = u(0,0)$ and (5.7) follows.

As an illustration, the choice $u(x,y) = x^3 - 3xy^2$ gives

$$f(z) = 2\left[\left(\frac{z}{2}\right)^3 - 3\left(\frac{z}{2}\right)\left(\frac{z}{2i}\right)^2\right] = z^3$$

in agreement with Example 5.1.

Problems

1. Which of the following functions $u(x,y)$ are harmonic?

   $x^2 - y^2 + y, \quad x^3 - y^3, \quad 3x^2y - y^3 + xy, \quad x^4 - 6x^2y^2 + y^4 + x^3y - xy^3.$

2. (a) Let $f(z) = u(x,y) + i\,v(x,y)$ be rational. By Chapter 1, Section 5, Problem 10, the identity $f(x) = u(x,0) + i\,v(x,0)$ continues to hold when x is replaced by a complex number. Replacing x by z, get

   $$f(z) = u(z,0) + i\,v(z,0).$$

   (b) Using (5.5) with $x_0 = y_0 = 0$, get a conjugate v for the harmonic functions of Problem 1. Then get $f(z)$ by the formula of part (a).

3. (a) If $u(x,y)$ is harmonic, results of the text indicate that $u = \text{Re}\, f$ where

   $$f'(z) = u_x(x,y) - i\,u_y(x,y).$$

   Assuming $f'(z)$ rational, set $y = 0$ and replace x by z, as in Problem 2(a), to get

   $$f'(z) = u_x(z,0) - i\,u_y(z,0).$$

   (b) By part (a) get $f'(z)$ for the harmonic functions of Problem 1, and then find $f(z)$, by inspection, from $f'(z)$.

4. Compute $f(z)$ for the following harmonic functions $u(x,y)$ by the method of Problem 2 or 3. Since the formulas were derived for rational functions only, verify that $f(z)$ is analytic and that $u = \text{Re}\, f$:

   $e^x \sin y, \quad \cosh y \sin x, \quad xe^x \cos y - ye^x \sin y, \quad e^{x^2-y^2} \sin 2xy.$

5. Without assuming $a = -3$ in Example 5.1, compute v by (5.5). Then show that $u + iv$ satisfies the Cauchy-Riemann equations only if $a = -3$.

6. (a) Apply the formula of Example 5.2 to appropriate functions in Problem 1, and also to the harmonic function $(x^3 + xy^2 + 2xy)/(x^2 + y^2 + 2y + 1)$. (b) Apply the formula of Example 5.2 to Problem 4, and verify.

7. By the method of Example 5.2 obtain the formula

   $$f(z) = 2u\left(\frac{z + z_0}{2}, \frac{z - z_0}{2i}\right) + \text{const}$$

   and apply it to the function $u(x,y) = y(x^2 + y^2)^{-1}$, taking $z_0 = 1$.

8. It is said that a complex function $f = u + iv$ satisfies the Cauchy-Riemann

equations if u and v satisfy them. By direct calculation prove that if f and g both satisfy the Cauchy-Riemann equations at a given point z, the same is true of $f + g$ and of fg. Also verify that $a + ib$ and $x + iy$ satisfy them at every point, a and b being constant. Thus deduce that polynomials $P(z)$ satisfy the Cauchy-Riemann equations. (This shows, without reference to the theory of complex differentiation, that Re $P(z)$ and Im $P(z)$ are harmonic.)

9. Show that the function ϕ defined by $\phi(0,0) = 0$ and by $\phi(x,y) = \text{Im } z^{-2}$ for $z \neq 0$ has an infinite discontinuity at the origin and yet satisfies Laplace's equation for all z. The reason for requiring continuity in the definition of harmonic functions is to rule out examples such as this.

***6. Application to fluid flow.** The theory of analytic functions is a veritable mine of effective tools for the solution of important problems in electrostatics, heat conduction, diffusion, gravitation, elasticity, and the flow of electric currents. The main reason for the usefuless of complex analysis in such fields is the fact that the real and imaginary parts of an analytic function satisfy Laplace's equation, as explained in the last section.

Derivation of Laplace's equation in connection with fields mentioned above is not difficult, and can be found in many books. These derivations are not given here. Instead we discuss the two-dimensional flow of an ideal fluid. It is found that the fluid-flow model not only leads to Laplace's equation, but provides a physical interpretation of both Cauchy-Riemann equations.

Suppose a fluid flows over the z plane in such a way that the velocity at any point depends only on the position of that point, not on the time. Since the flow pattern is two-dimensional, the sheet of fluid could be considered infinitely thin. More realistically it is assumed to have constant depth, 1, at every point and the nature of the flow is assumed to be independent of the height above the bottom. Thus each point of the fluid which is over a given point z of the plane has the same velocity, $\mathbf{v}(z)$, independent both of the height and of the time. The velocity is parallel to the z plane and so can be represented by a two-dimensional vector,

$$\mathbf{v} = p\mathbf{i} + q\mathbf{j}.$$

We require also that p and q have continuous partial derivatives with

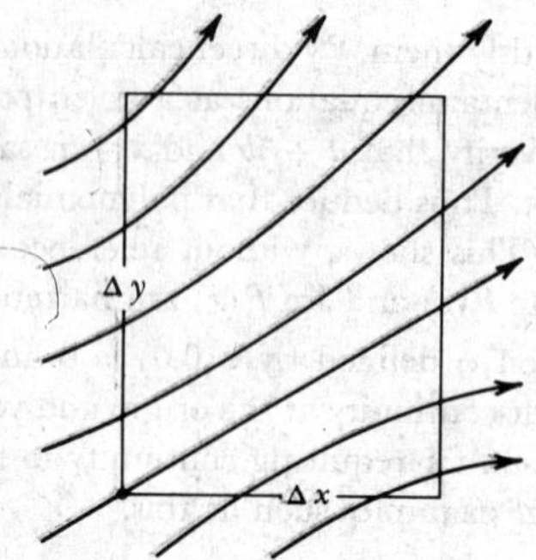

Figure 6-1

respect to x and y. The fact that $\mathbf{v}$ has this form and the fact that the integrals (6.1) and (6.3) given below vanish are viewed here as three basic axioms governing ideal fluid flow in the plane.

By definition, the net rate of flow out of a given region is the line integral[1] of the exterior normal component of $\mathbf{v}$ over the boundary. The calculation is easy when the region is a rectangle as shown in Figure 6-1 and the velocity is given in cartesian coordinates, $\mathbf{v} = p\mathbf{i} + q\mathbf{j}$. Indeed, the net rate of flow out of the rectangle of Figure 6-1 is clearly

$$\int_y^{y+\Delta y} [p(x + \Delta x,\eta) - p(x,\eta)]\, d\eta + \int_x^{x+\Delta x} [q(\xi,y + \Delta y) - q(\xi,y)]\, d\xi \tag{6.1}$$

provided Δx and Δy are so small that the rectangle is entirely interior to the region of ideal fluid flow. In this case the principle of conservation of volume asserts that the net flow is 0.

Since p and q are continuously differentiable, the mean-value theorem gives

$$p(x + \Delta x,\eta) - p(x,\eta) = p_x(\bar{x},\eta)\, \Delta x, \qquad x < \bar{x} < x + \Delta x$$

where p_x denotes the partial derivative. When this equation and a similar equation for q are substituted into (6.1), the result is

[1] Since the result is applied only to rectangles, familiarity with the general concept of line integral is not required in the sequel.

$$\frac{1}{\Delta y}\int_y^{y+\Delta y} p_x(\bar{x},\eta)\,d\eta + \frac{1}{\Delta x}\int_x^{x+\Delta x} q_y(\xi,\bar{y})\,d\xi = 0$$

after division by $\Delta x\,\Delta y$ and equating to 0. Upon letting $\Delta x \to 0$ and $\Delta y \to 0$, we get the *equation of continuity*

$$p_x + q_y = 0 \tag{6.2}$$

In Problem 9 it is shown conversely that (6.2) implies the vanishing of (6.1).

The principle of conservation of volume is not the only condition satisfied by ideal fluid flow. Another condition pertains to the *circulation*. By definition, the circulation along a given curve is the line integral of the tangential component of **v** along the curve. For the closed curve bounding the rectangle of Figure 6-1 the circulation is

$$\int_y^{y+\Delta y} [q(x+\Delta x,\eta) - q(x,\eta)]\,d\eta - \int_x^{x+\Delta x} [p(\xi,y+\Delta y) - p(\xi,y)]\,d\xi. \tag{6.3}$$

If the rectangle (including its interior) is contained in the region of ideal fluid flow it is known from physics that the circulation is 0, and a development parallel to that used for (6.2) gives

$$q_x - p_y = 0. \tag{6.4}$$

Conversely, (6.4) implies the vanishing of (6.3).

By definition, an ideal fluid flow in a plane region is any flow whose velocity $\mathbf{v} = p\mathbf{i} + q\mathbf{j}$ is continuously differentiable there and satisfies (6.2) and (6.4). If we introduce the *complex velocity*

$$V(z) = p(x,y) + i\,q(x,y), \tag{6.5}$$

these equations indicate that the function

$$\overline{V(z)} = p(x,y) - i\,q(x,y)$$

satisfies the Cauchy-Riemann equations and hence is analytic. Conversely, any analytic function is continuously differentiable and satisfies the Cauchy-Riemann equations, hence its conjugate defines the complex velocity of an ideal fluid flow.

A technique for obtaining the paths of the fluid particles is explained next. Since p and q are continuously differentiable it is known from real analysis that (6.2) and (6.4) imply local existence of functions ψ and ϕ such that

$$d\psi = p\,dy - q\,dx, \qquad d\phi = p\,dx + q\,dy,$$

respectively. Since $d\psi = \psi_x\,dx + \psi_y\,dy$ and similarly for ϕ, it follows that

$$p = \phi_x = \psi_y, \qquad q = \phi_y = -\psi_x. \tag{6.6}$$

These equations show that the function

$$f(z) = \phi(x,y) + i\psi(x,y)$$

satisfies the Cauchy-Riemann equations and hence is analytic. Since

$$f'(z) = \phi_x(x,y) + i\psi_x(x,y) = p(x,y) - iq(x,y)$$

by (1.6) and (6.6), the complex velocity (6.5) is

$$V(z) = \overline{f'(z)}. \tag{6.7}$$

According to (6.6), $v = \text{grad}\,\phi$ so that ϕ is the velocity potential. The curves $\phi = \text{const}$ are called *equipotentials.* Since differentiation of f gives the velocity, f is sometimes called the complex potential.

The function ψ is the *stream function* and the curves

$$\psi(x,y) = \text{const}$$

are called streamlines. To interpret these curves, observe that the vector $\mathbf{i}\psi_x + \mathbf{j}\psi_y$ is normal to the curve $\psi = \text{const}$, and it is also normal to $\mathbf{v}$, since (6.6) gives

$$(\mathbf{i}\psi_x + \mathbf{j}\psi_y) \cdot (\mathbf{i}p + \mathbf{j}q) = 0.$$

Hence $\mathbf{v}$ is tangent to the curve $\psi = \text{const}$ at each point of the curve. Physically this means that *the streamlines describe the paths of the fluid particles.*

If $f = \phi + i\psi$ is analytic, it has been seen that the curves $\psi(x,y) =$

const give the streamlines of an ideal fluid flow. Since the function

$$if = -\psi + i\phi$$

is also analytic, the curves $\phi(x,y) = \text{const}$ also give the streamlines of an ideal fluid flow. Two fluid-flow patterns related in this way are said to be *conjugate* because the new stream function ϕ is a conjugate of ψ in the sense of Section 5. For conjugate flows the equipotentials of one flow are streamlines of the other and vice versa.

Since $(\overline{if})' = -\overline{if'}$, (6.7) shows that if the complex velocity for a given flow is V, the complex velocity for the conjugate flow is $-iV$. The equation $|V| = |-iV|$ shows that a flow and its conjugate have the same speed at each point. Multiplication by $-i$ rotates the vector through $-90°$ and hence the direction of the conjugate flow is at right angles to that of the original flow at each point. This indicates that the streamlines for the conjugate flow are the orthogonal trajectories of those for the original flow. Since the streamlines are, respectively,

$$\psi(x,y) = \text{const}, \qquad \phi(x,y) = \text{const},$$

where ϕ and ψ are real and imaginary parts of an analytic function, the orthogonality agrees with the result of Example 1.1.

In conclusion we mention a physical interpretation of the stream function, independent of the foregoing discussion, which makes the basic results intuitively obvious. Since the conditions are, in part, physical, this informal discussion does not replace the mathematical analysis given above. Its purpose is to give added insight.

Let z_0 be a given point which remains fixed throughout the discussion. The stream function $\psi(x,y)$ is defined to be the flow across a curve which joins z_0 to $z = x + iy$. Thus, ψ measures volume per unit time. The assumption that ψ is independent of the particular curve used is equivalent to the assumption that the flow is incompressible and that there are no sources in the region considered. Thus, in Figure 6-2 the flow across the two curves must be the same, since the amount of fluid in the region between the curves remains constant in the steady state.

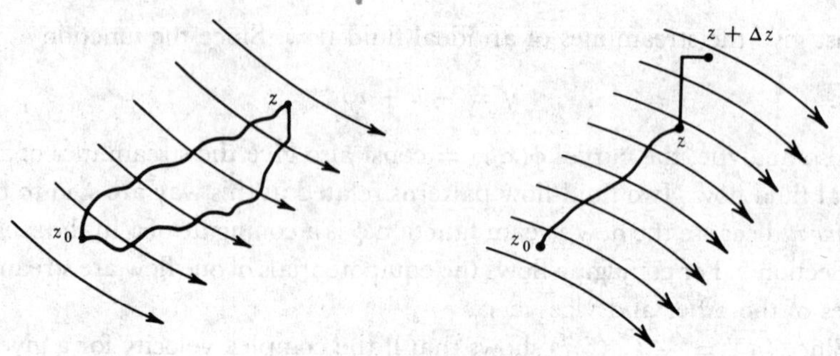

Figure 6-2 **Figure 6-3**

In Figure 6-3 the flow across a curve from z_0 to $z + \Delta z$ can be computed by addition of the flows across the curve from z to z_0 plus that across the two line segments. With a suitable choice of sign for ψ the result is

$$\psi(x + \Delta x, y + \Delta y) = \psi(x,y) + \tilde{p}\,\Delta y - \tilde{q}\,\Delta x$$

where $\tilde{p}$ is the x coordinate of $\mathbf{v}$ averaged over the vertical segment and $\tilde{q}$ is the y coordinate averaged over the horizontal segment. Upon setting $\Delta x = 0$ or $\Delta y = 0$ and rearranging, we get, respectively,

$$\frac{\psi(x,y + \Delta y) - \psi(x,y)}{\Delta y} = \tilde{p}, \qquad \frac{\psi(x + \Delta x,y) - \psi(x,y)}{\Delta x} = -\tilde{q}.$$

If $\mathbf{v}$ is continuous, the mean values $\tilde{p}$ and $\tilde{q}$ tend to the values at (x,y) and hence, as before,

$$\psi_y = p, \qquad \psi_x = -q.$$

It should be noted that this approach requires only that $\mathbf{v}$ be continuous. The analysis shows, then, that ψ has partial derivatives which equal the continuous functions p and $-q$. Hence, ψ is continuously differentiable.

If the conjugate flow also admits a stream function ϕ, we get

$$\phi_y = q, \qquad \phi_x = p$$

by applying the foregoing to the vector $(-i)(p + iq) = q - ip$, with ϕ instead of ψ. This shows that

$$\phi_x = \psi_y, \qquad \phi_y = -\psi_x$$

and so the function $f = \phi + i\psi$ is analytic, as before.

Example 6.1. Discuss the flow represented by $f(z) = \log z$. Setting $z = re^{i\theta}$ gives

$$f(z) = \operatorname{Log} r + i\theta + 2\pi ik$$

and hence the streamlines are the curves $\theta =$ const. Since the streamlines are the paths of the fluid particles, the flow is radial as indicated in Figure 6-4(a). This conclusion is confirmed by computation of the complex velocity,

$$V = \overline{f'(z)} = \overline{\left(\frac{1}{z}\right)} = \frac{1}{r}e^{i\theta}. \tag{6.8}$$

Equation (6.8) not only shows that the flow is radial but gives the speed,

$$|V| = |f'(z)| = \frac{1}{r}.$$

The total flow across a circle $|z| = r$ is obtained by integrating the normal component of V over the boundary. Since (6.8) shows that V is directed away from the origin along a radius, the normal component is $|V| = 1/r$. Hence the total flow across the circle is

$$\int_0^{2\pi} \frac{1}{r} r\, d\theta = 2\pi \tag{6.9}$$

and the circulation is 0. It is said, briefly, that $f(z) = \log z$ represents a *source* of strength 2π at the origin.

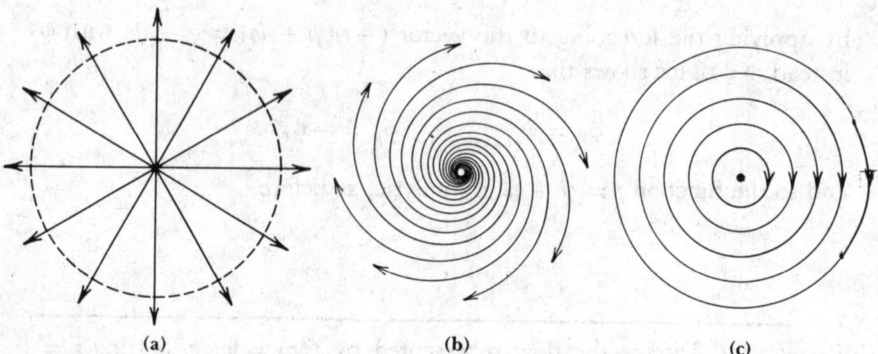

Figure 6-4

The conjugate flow is given by $i \log z$, so that the roles of the streamlines and equipotentials are interchanged (Figure 6-4(c)). The streamlines for the conjugate flow are concentric circles and the integral (6.9) now pertains to the tangential rather than the normal component of the velocity. The circulation around any circle centered at the origin is -2π and the total flow across the circle is 0. It is said that $i \log z$ represents a *vortex* of strength -2π centered at the origin.

If $f(z) = \alpha \log z$ where $\alpha = a + ib$ is constant, the equation

$$f(z) = a \log z + b(i \log z)$$

shows that $\alpha \log z$ represents the superposition of a source[3] of strength $2\pi a$ and a vortex of strength $-2\pi b$, both at the origin. The combination of radial and circular motion leads to spiral streamlines (Figure 6-4(b)).

Example 6.2. Flow at a wall. Discuss the flow pattern for a source of strength $2\pi c$ at $z = a > 0$ in the presence of a barrier along the y axis (Figure 6-5). This problem is solved by the method of images. Let a second source of strength $2\pi c$ be placed at the point $-a$ as shown in Figure 6-6. The corresponding complex potential is

$$f(z) = c \log(z - a) + c \log(z + a) = c \log(z^2 - a^2) + 2k\pi i c.$$

[3] If a is negative, a source of strength $2\pi a$ is also called a sink of strength $2\pi|a|$.

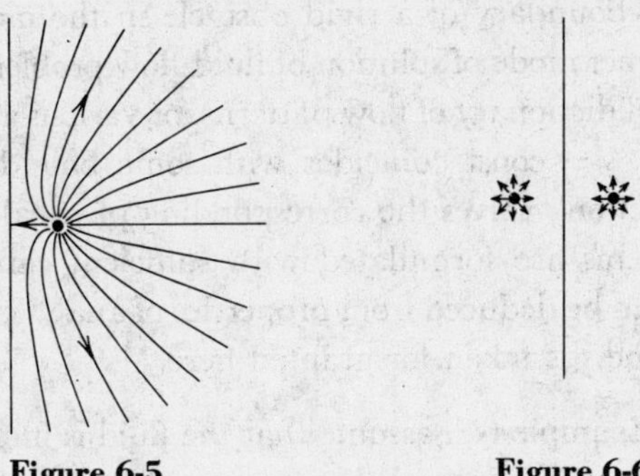

Figure 6-5 Figure 6-6

From symmetry it is clear that the y axis must be a streamline, and this surmise is confirmed by the calculation

$$f(iy) = c \log(-y^2 - a^2) + 2k\pi ic = c \operatorname{Log}(y^2 + a^2) + (2k + 1)\pi ic.$$

The imaginary part is constant and so the y axis is a streamline. Physically, this means that a barrier can be introduced along the y axis without changing the flow pattern. Hence, if the new problem with two sources is solved in the right half-plane, it solves the original problem pertaining to a source and a wall.

Since the foregoing results give

$$f'(z) = \frac{2cz}{z^2 - a^2}, \tag{6.10}$$

the complex velocity for $z = iy$ is

$$V = \overline{f'(iy)} = \frac{2icy}{a^2 + y^2}. \tag{6.11}$$

The velocity is directed along the y axis, so that the normal component of the velocity at the wall is 0. This confirms the fact, already known from the streamlines, that the flow satisfies appropriate boundary conditions at the wall. The flow pattern is shown in Figure 6-5.

In this example the boundary conditions at the wall were not introduced directly, but were satisfied by construction of a fluid flow in which the line $x = 0$ is a streamline. Quite generally, any streamline can be regarded as the boundary of a rigid obstacle in the moving fluid. This suggests an indirect mode of solution of fluid-flow problems, in which one compiles a sort of dictionary of flow patterns for various analytic functions $f(z)$. If a curve $\psi =$ const coincides with some boundary of technical interest, the function f solves the corresponding physical problem. When fluid-flow problems are formulated with sufficient care, uniqueness of their solutions can be deduced from properties of analytic functions. However the uniqueness is taken for granted here.

Example 6.3. In Example 6.2. assume that the fluid is initially at rest over the whole z plane, then the wall is constructed, and finally the source is turned on. After the system has come to equilibrium show that the total force on the wall is directed toward the right and has magnitude $\pi\rho c^2/a$ where ρ is the density of the fluid.

According to Bernoulli's law the pressure p, density ρ and speed $|V|$ are related by the equation

$$\frac{p}{\rho} + \frac{|V|^2}{2} = \text{const}$$

when the density is constant and p is measured at points of constant height above the level bottom. Bernoulli's law can be deduced from Newton's laws of motion but is taken for granted here, p being measured at depth ½, that is, halfway from the surface to the bottom. When $V \to 0$ as $|z| \to \infty$ the equation yields

$$p_0 - p = \rho\frac{|V|^2}{2}$$

where p_0 is the pressure at ∞, or the pressure in the fluid at rest. In an ideal fluid the force due to pressure is considered to be normal to the boundary at every point.

Applying these principles to Example 6.2 shows that the force has 0 y coordinate and that its x coordinate is

$$\frac{\rho}{2}\int_{-\infty}^{\infty} |V|^2\, dy = 2\rho c^2 \int_{-\infty}^{\infty} \frac{y^2}{(a^2 + y^2)^2}\, dy = \pi\rho\frac{c^2}{a}.$$

The result would be unchanged if the source were replaced by a sink of the same strength.

Problems

1. The following functions give the complex potential $f(z)$ for an ideal fluid flow. Find the velocity potential ϕ, the stream function ψ, the complex velocity V, the speed $|V|$, and plot the streamlines:
$$z, \quad iz, \quad (1+i)z, \quad z^{-1}, \quad iz^{-1}, \quad z^2, \quad iz^2.$$
2. *Flow at a corner.* Let $f(z) = z^c$ where c is constant, $c > \frac{1}{2}$. Writing $z = re^{i\theta}$, show that the rays $\theta = 0$ and $\theta = \pi/c$ are streamlines, and hence can be replaced by barriers. Plot the streamlines in the sector $0 \leq \theta \leq \pi/c$ in the three cases $c = 4$, $c = 1$, $c = \frac{2}{3}$. Show that the speed of flow is cr^{c-1} where $r = |z|$ is the distance to the corner.
3. (a) Show that the maximum speed at the wall of Example 6.2 equals the maximum speed that would be observed on the y axis for the single source at $z = a$ without the wall. (b) Let a source of strength $2\pi c$ be placed at $z = a > 0$, with no other singularities or boundaries. If a barrier is now introduced along the y axis, as in Example 6.2, find the locus of all points in the right half-plane at which the speed of flow is increased by introduction of the barrier.
4. Let $f(z) = \alpha \log z$ where $\alpha \neq 0$ is a complex constant. Show that the velocity at $z = re^{i\theta}$ makes a constant angle with the radius from 0 to z, and hence the streamlines are logarithmic spirals (Figure 6-4(b)). Find the polar equation of the streamlines, r vs θ.
5. Describe flow patterns for $f(z) = e^z$ and $f(z) = \sinh z$, $0 \leq y \leq \pi$.
6. Interpret the potential $f(z) = \log[(z-1)/(z+1)]$ by sources and sinks, and show that $|V||z^2 - 1| = 2$. Also show that the streamlines are circles, and plot. (At all points on a given streamline, the segment $-1 \leq x \leq 1$ subtends a constant angle.)
7. Discuss the flow conjugate to that of Problem 6.

7. The foundations of complex analysis. The Cauchy-Riemann equations provide one of three avenues of approach to complex analysis, the other two being the theory of complex integration and the theory of power series. These different ways of getting started are associated with the names of Riemann, Cauchy and Weierstrass, respectively. An introduction to Cauchy's method is given in the following two chapters, which are based on a theorem of complex calculus known as the Cauchy integral theorem. The rudiments of Weierstrass' method are given in Chap. 6.

Additional problems on Chapter 2

1.1. *Partial fraction expansion.* Let $a_1, a_2, \ldots, a_n$ be unequal complex numbers, let

$$Q(z) = (z - a_1)(z - a_2) \cdots (z - a_n),$$

and let $P(z)$ be any polynomial of degree $\leq n - 1$. Prove that

$$\frac{P(z)}{Q(z)} = \frac{A_1}{z - a_1} + \frac{A_2}{z - a_2} + \cdots + \frac{A_n}{z - a_n} \qquad (*)$$

holds for all $z \neq a_i$ if the constants A_i are defined by

$$A_i = \lim_{z \to a_i} (z - a_i) \frac{P(z)}{Q(z)} = \lim_{z \to a_i} P(z) \frac{z - a_i}{Q(z) - Q(a_i)} = \frac{P(a_i)}{Q'(a_i)}.$$

(To show that the expansion is possible, multiply (*) by $Q(z)$ and determine A_i so that the resulting equation between polynomials of degree $\leq n - 1$ holds at the n points a_i. For instance, when $n = 3$ one considers

$$P(z) = A_1(z - a_2)(z - a_3) + A_2(z - a_1)(z - a_3) + A_3(z - a_1)(z - a_2).$$

Clearly, A_i can be chosen so that this equality between polynomials of degree ≤ 2 holds at the three values a_i, hence is an identity. Once (*) is seen to be possible, the formula for A_i follows by inspection.)

1.2. Referring to Problem 1.1, expand in partial fractions:

$$\frac{z^2 + 1}{z(z^2 - 1)}, \qquad \frac{z^3 + 3z^2}{z(z^2 + 1)(z^2 + 3z + 2)}, \qquad \frac{z^2 - 3z + 1}{(z^4 - 1)(z^2 + 2z + 2)}.$$

1.3. If $\alpha_1, \alpha_2, \ldots, \alpha_n$ are the roots of a polynomial $P(z)$ of degree $n \geq 2$ and $\beta_1, \beta_2, \ldots, \beta_{n-1}$ are the roots of $P'(z)$, show that

$$\frac{\alpha_1 + \alpha_2 + \cdots + \alpha_n}{n} = \frac{\beta_1 + \beta_2 + \cdots + \beta_{n-1}}{n - 1}.$$

1.4. *Theorem of Gauss-Lucas.* By definition, a half-plane is the locus of all points z such that $\operatorname{Re}(\alpha z + \beta) < 0$, where $\alpha \neq 0$ and β are constant. (a) Interpret geometrically. (b) If $Q(z)$ is a given polynomial, define another polynomial $P(z)$ by $P(\tilde{z}) = Q(z)$, where $\tilde{z} = \alpha z + \beta$. Applying Example 1.3 to P, show that if all the zeros of Q lie in the half-plane $\operatorname{Re} \tilde{z} < 0$, the same is true of the zeros of Q'. (c) The intersection of all half-planes containing a given point set S is called the *convex hull* of S. Conclude from (b) that the set of all zeros of

Q' lies in the convex hull of the set of zeros of Q. (The intersection of any number of point sets is the set of points common to all of them.)

1.5. Let $f(z)$ be defined by $(x^6 + y^2)f(z) = x^2yz$ for $z \neq 0$ and by $f(0) = 0$. Show that f is not continuous at $z = 0$ although the limit defining $f'(0)$ exists as $z \to 0$ along any straight line passing through the origin.

1.6. Prove: A necessary and sufficient condition that $f = u + iv$ be differentiable at a given point $z = x + iy$ is that u and v be differentiable at (x,y) and satisfy the Cauchy-Riemann equations there.

1.7. At each point of a domain D, suppose that u and v satisfy the Cauchy-Riemann equations and that at least one partial derivative u_x, u_y, v_x or v_y is continuous. Prove that $f = u + iv$ is analytic in D. (By real analysis, any function $u(x,y)$ is differentiable at a point where both partial derivatives exist and one of them is continuous.)

2.1. *Trigonometric solution of the cubic.* Prove the identity

$$4 \sin^3 z - 3 \sin z + \sin 3z = 0.$$

According to Chapter 1, Problem 3.1, a general cubic can be reduced to the form $w^3 + b'w + c' = 0$. Setting $w = ks$ for suitable k, reduce this to

$$4s^3 - 3s + \gamma = 0.$$

Hence if z is determined by $\sin 3z = \gamma$, a solution is $s = \sin z$.

2.2. Let $\omega_k = \exp(2\pi ik/n)$ for $k = 1, 2, \ldots, n$ where n is a positive integer. For integral m show that

$$\omega_1{}^m + \omega_2{}^m + \cdots + \omega_n{}^m = 0,$$

unless m is a multiple of n, in which case the sum is n. (Note that $\omega_k = \omega_1{}^k$ and use Chapter 1, Example 1.5.)

2.3. If $P(z)$ is a polynomial of degree $\leq 2n - 1$, show, in the notation of Problem 2.2, that

$$\frac{P(\omega_1) + P(\omega_2) + \cdots + P(\omega_n)}{n} = P(0) + \frac{P^{(n)}(0)}{n!}.$$

2.4. Let $f(z) = \exp(-z^{-4})$ for $z \neq 0$ and $f(0) = 0$. Show that the Cauchy-Riemann equations hold for every value of z, although f is not continuous at $z = 0$. (The equations hold for $z \neq 0$ since $f(z)$ is analytic, and hence it suffices to verify them at $z = 0$ only.)

3.1. Let $N(\alpha)$ denote the number of solutions of the equation $z^\alpha = 1$. Discuss the appearance of a graph of $N(\alpha)$ vs α, $0 < \alpha < \infty$.

5.1. *Subharmonic functions.* A real-valued function w which is continuous in a region R and satisfies $w_{xx} + w_{yy} \geq 0$ there is said to be subharmonic. Show that

$w = |f(z)|^2$ is subharmonic in any region throughout which f is analytic. (Set $f = u + iv$, $w = u^2 + v^2$ and compute.)

5.2. *Weak maximum principle.* Let w be subharmonic in a bounded region R, continuous in $\bar{R}$, and such that $w \leq m$ on the boundary, where m is constant. Show that $w \leq m$ in R. (In the contrary case

$$w(x_0, y_0) = m + \delta, \qquad \delta > 0,$$

holds at some interior point (x_0, y_0). Define

$$W(x,y) = w(x,y) + \epsilon x^2$$

where the positive constant ϵ is so small that $\epsilon x^2 < \delta$ in $\bar{R}$. It is easily checked that $W < m + \delta$ on the boundary, but $W \geq m + \delta$ at (x_0, y_0), and hence W assumes its maximum over $\bar{R}$ at an interior point. By elementary calculus, $W_{xx} \leq 0$ and $W_{yy} \leq 0$ at the maximum, and so $w_{xx} + w_{yy} \leq -2\epsilon$ there.)

5.3. *Fundamental theorem of algebra.* Show that every nonconstant polynomial has a root. (If $P(z)$ does not vanish, the function $f(z) = 1/P(z)$ is entire. If P is not constant, then $|P(z)| \to \infty$ as $|z| \to \infty$ and hence $|f(z)| \to 0$ as $|z| \to \infty$. Thus we can find a disk $|z| \leq r$ such that $|f(z)| < |f(0)|$ holds for $|z| = r$. This contradicts Problem 5.2.)

5.4. *Uniqueness.* Let $\tilde{u}$ and u be harmonic in a bounded region R and continuous in $\bar{R}$. If $u = \tilde{u}$ on the boundary, prove that the same equality holds throughout R. (Apply Problem 5.2 to $u - \tilde{u}$ and to $\tilde{u} - u$.)

6.1. *Flow around a circle.* (a) If $f(z) = z + z^{-1}$, show that the streamline $\psi = 0$ consists of the x axis together with the circle $|z| = 1$. Considering the behavior of $f(z)$ as $z \to \infty$, interpret the result as the complex potential for uniform flow around a circular obstacle. Sketch some of the streamlines. (b) Find the speed at points $z = e^{i\theta}$ of the circle and show that the maximum speed is twice the speed of the original uniform flow. (c) Find the pressure at the boundary of the circle and show, by symmetry or otherwise, that the total resultant force on the circle is 0. This curious fact is sometimes called the *hydrodynamic paradox.* (d) If $-ic \log z$ is added to $f(z)$, where $c > 0$ is constant, show that the circle is still a streamline, but that there is now a force on the circle, at right angles to the direction of uniform flow, which is proportional to the circulation $2\pi c$.

6.2. *Rankine's method.* (a) Let a family of curves $f(x,y) = \text{const}$ and another family $g(x,y) = \text{const}$ be such that each f curve is intersected by just one g curve and vice versa. Explain why the family

$$f(x,y) + g(x,y) = \text{const}$$

can be obtained by joining intersection points as suggested in Figure 7-1. (b) Apply this method to sketch the streamlines associated with $f(z) = z + \log z$ and interpret the result as giving the flow around the stern of a ship.

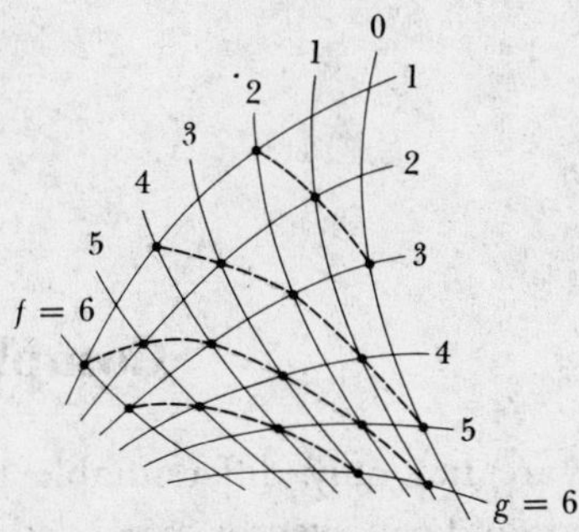

Figure 7-1

6.3. A *stagnation point* is a point where $V = 0$. Find the stagnation point for $f(z) = \alpha(z - z_0)^2$ where $\alpha \neq 0$ and z_0 are constant, and show that two streamlines meet there at right angles. By means of the theory of power series, introduced later, it is found that similar behavior occurs at any stagnation point where $f'(z_0) = 0$ and $f''(z_0) \neq 0$.

6.4. Let p and q be continuously differentiable functions satisfying $p_y = q_x$ in a suitable region (see Figure 5-1). Show that ϕ defined by

$$\phi(x,y) = \int_{x_0}^{x} p(x,y)\,dx + \int_{y_0}^{y} q(x_0,y)\,dy$$

satisfies $\phi_x = p$ and $\phi_y = q$.

6.5. If $p = \phi_x$ and $q = \phi_y$ as in Problem 6.4, evaluate the four integrals (6.1) explicitly in terms of ϕ and thus show that (6.1) vanishes.

6.6. This problem requires knowledge of field theory. (a) Given a fluid flow for which the velocity $\mathbf{v}$ is continuously differentiable, formulate the concepts of total outward flow and total circulation in terms of surface and line integrals, respectively. (b) By the theorems of Gauss and Stokes, respectively, obtain the equations

$$\operatorname{div} \mathbf{v} = 0, \qquad \operatorname{curl} \mathbf{v} = \mathbf{0}$$

at points interior to a region of ideal flow. (c) When $\mathbf{v} = p\mathbf{i} + q\mathbf{j}$ show that the cartesian form of these equations reduces to the Cauchy-Riemann equations for $p - iq$. (d) Using the theory of conservative fields, discuss the existence of the potentials ϕ and ψ.

Chapter 3

Complex integration

Analytic functions are not only differentiable but have continuous derivatives of all orders, and can even be expanded in convergent Taylor series. As a consequence, an analytic function is determined in its whole domain of analyticity by the values on an arbitrarily short curve in that domain. It is also true that an analytic function in a domain cannot attain its maximum modulus unless the function is constant. These facts are troublesome to prove directly, but follow with ease from the theory of complex integration. An integral formula due to Cauchy gives the values of an analytic function in a disk in terms of the values on the boundary, and the above properties of analytic functions follow from the structure of this formula.

***1. Integration on a contour.** The integration of a function of a complex variable is carried out over a curve and leads to results of great importance in pure and applied mathematics. In this section we discuss first complex-valued functions, then curves, and finally integration over curves.

A function ϕ of the real variable t is said to be complex-valued on an interval $a \leq t \leq b$ if

$$\phi(t) = \phi_1(t) + i\ \phi_2(t)$$

holds on this interval, where ϕ_1 and ϕ_2 are real-valued functions. Here ϕ_1 is called the real part of ϕ, and ϕ_2 is the imaginary part. For example, e^{it} is a complex-valued function defined on every interval. The real part of this function is $\cos t$ and the imaginary part is $\sin t$.

The complex-valued function $\phi(t)$ is said to be continuous at a point t_0 if

$$\lim_{t \to t_0} |\phi(t) - \phi(t_0)| = 0$$

holds in the sense of real-variable theory. This can be expressed in (ϵ,δ) terms also. When ϕ is continuous at every point of an interval, ϕ is said to be continuous on the interval, and a similar convention for passing from points to intervals applies to other properties introduced later.

If ϕ is continuous, then so are ϕ_1 and ϕ_2. Indeed, since $|\mathrm{Re}\, z| \leq |z|$, we have

$$|\phi_1(t) - \phi_1(t_0)| \leq |\phi(t) - \phi(t_0)|,$$

and if the right side tends to zero as $t \to t_0$, the left side must. Thus ϕ_1 is continuous at t_0 if ϕ is. A similar proof holds for ϕ_2. Conversely, if ϕ_1 and ϕ_2 are continuous, then ϕ must be. This follows from the inequality

$$|z| \leq |\mathrm{Re}\, z| + |\mathrm{Im}\, z|$$

applied to $\phi(t) - \phi(t_0)$.

The complex-valued function ϕ is said to be differentiable if ϕ_1 and ϕ_2 are, and in that case, by definition,

$$\phi'(t) = \phi_1'(t) + i\ \phi_2'(t). \tag{1.1}$$

It is easily verified that the sum, difference, product and quotient of complex-valued functions can be differentiated as in real analysis. Also the chain rule holds[1] in the following form: If F is analytic at the point $z = \zeta(t)$, and if $\zeta'(t)$ exists, then the complex-valued function $\phi(t) = F[\zeta(t)]$ is differentiable at t, and

$$\phi'(t) = F'[\zeta(t)]\zeta'(t). \tag{1.2}$$

A complex-valued function $\phi = \phi_1 + i\phi_2$ is said to be integrable over an interval $[a,b]$ if ϕ_1 and ϕ_2 are, and then, by definition,

$$\int_a^b \phi(t)\, dt = \int_a^b \phi_1(t)\, dt + i \int_a^b \phi_2(t)\, dt. \tag{1.3}$$

Many familiar rules of integration for real functions can be carried over to the complex case. In particular, the two forms of the fundamental

[1] See Problem 9.

theorem of calculus for real functions give corresponding results for complex functions, namely

$$\frac{d}{dt}\int_a^t \phi(\tau)\, d\tau = \phi(t), \qquad \int_a^b \phi'(t)\, dt = \phi(t)\Big|_a^b \tag{1.4}$$

when ϕ is continuous or ϕ' is continuous, respectively. Here the right-hand member denotes $\phi(b) - \phi(a)$, as in real analysis.

One of the main uses of complex-valued functions is to represent curves in the complex plane. As the reader will recall from real analysis, a curve is given by parametric equations

$$x = \xi(t), \qquad y = \eta(t)$$

where ξ and η are real-valued functions defined on some interval

$$a \leq t \leq b.$$

The curve is said to be continuous if ξ and η are continuous; differentiable if ξ and η are differentiable, and so forth. If we define

$$\zeta(t) = \xi(t) + i\eta(t), \qquad z = x + iy,$$

then both equations $x = \xi(t)$, $y = \eta(t)$ are equivalent to the single equation $z = \zeta(t)$. The function $\zeta(t)$ in this representation is, clearly, complex-valued. It should perhaps be mentioned that no logical distinction is made between the curve and the function ζ so that, from a purely logical standpoint, the curve "is" the function.

As a matter of terminology, the set of points occupied by a curve is called the *trace* of the curve, and the curve is said to lie in a given region if the trace does. A curve is *simple* if it does not intersect itself, that is, if $\zeta(t_1) \neq \zeta(t_2)$ for $t_1 \neq t_2$, with the possible exception that $\zeta(a) = \zeta(b)$ is allowed. In the latter case the curve is said to be closed. Except when $\zeta(a) = \zeta(b)$, there is a one-to-one correspondence between the points of the simple curve and those of the interval $a \leq t \leq b$. When t increases from a to b, the curve is traversed from the initial point $\zeta(a)$ to the terminal point $\zeta(b)$. A continuously differentiable curve is called an *arc*. Thus, an arc is given by $z = \zeta(t)$ on some interval $a \leq t \leq b$ where $\zeta'(t)$ is continuous. The case $\zeta(a) = \zeta(b)$ is not excluded. This holds, for example, for the arc

$$z = \sin t + i \sin 2t, \qquad 0 \leq t \leq \pi.$$

This arc with the direction in which it is traversed for increasing t is shown in Figure 1-1. Further illustration of terminology is given in Figure 1-2.

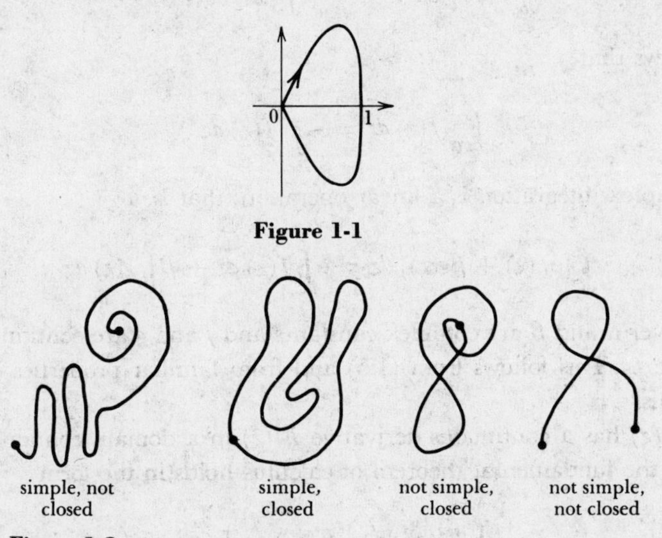

Figure 1-1

Figure 1-2

A complex-valued function $f(z)$ is said to be continuous on the arc C if the function

$$\phi(t) = f[\zeta(t)]$$

is continuous for $a \leq t \leq b$. By familiar theorems pertaining to continuity of composite functions, it is seen that f is continuous on C if f is continuous in a region of the complex plane containing C. When f is continuous on C, the integral of f on C is defined to be

$$\int_C f(z)\, dz = \int_a^b f[\zeta(t)]\zeta'(t)\, dt. \tag{1.5}$$

It should be observed that the integrand on the right would be obtained by formal substitution of

$$z = \zeta(t), \qquad dz = \zeta'(t)\,dt$$

into the left-hand member.

The arc C described in the opposite direction, that is, from $t = b$ to $t = a$, will be designated by $-C$. Since

$$\int_b^a f[\zeta(t)]\zeta'(t)\,dt = -\int_a^b f[\zeta(t)]\zeta'(t)\,dt,$$

it follows that

$$\int_{-C} f(z)\,dz = -\int_C f(z)\,dz. \tag{1.6}$$

Complex integration is a linear operation, that is,

$$\int_C [\alpha f(z) + \beta g(z)]\,dz = \alpha\int_C f(z)\,dz + \beta\int_C g(z)\,dz \tag{1.7}$$

whenever α and β are complex constants and f and g are continuous on the arc C. This follows from (1.5) and from familiar properties of real integrals.

If $F(z)$ has a continuous derivative $F'(z)$ in a domain containing an arc C, the fundamental theorem of calculus holds in the form

$$\int_C F'(z)\,dz = F(z_2) - F(z_1) \tag{1.8}$$

where z_1 is the initial point of C and z_2 is the terminal point. Indeed, the chain rule and the second relation of (1.4) give

$$\int_a^b F'[\zeta(t)]\zeta'(t)\,dt = \int_a^b \frac{d}{dt}F[\zeta(t)]\,dt = F[\zeta(t)]\Big|_a^b.$$

Since $z_1 = \zeta(a)$ and $z_2 = \zeta(b)$, this is (1.8).

These results can be extended to more general curves by addition. For $j = 1, 2, \ldots, n$ let C_j be an arc

$$z = \zeta_j(t), \qquad a_j \le t \le b_j.$$

The arcs are said to form a *contour* C if the terminal point of C_j coincides with the initial point of C_{j+1} for $j = 1, 2, \ldots, n-1$. The terminology introduced above in connection with arcs extends in a natural way to contours. Also, if f is continuous on C, that is, continuous on each C_j, the integral is defined by

$$\int_C f(z)\,dz = \int_{C_1} f(z)\,dz + \int_{C_2} f(z)\,dz + \cdots + \int_{C_n} f(z)\,dz.$$

Applying (1.7) to each C_j and adding, we find that (1.7) holds when C is a contour. Similarly, if $-C$ is defined by using $-C_j$ instead of C_j, it is found that (1.6) holds for contours.

Integration of the familiar functions is facilitated by the following result:

THEOREM 1.1. *Let $F(z)$ be an analytic function with a continuous derivative $f(z) = F'(z)$ in a domain D. Let C be a contour lying in D with initial point z_1 and terminal point z_2. Then*

$$\int_C f(z)\,dz = F(z_2) - F(z_1).$$

The theorem follows by addition of the formulas (1.8) for the arcs C_j. The expression $F(z_2) - F(z_1)$ is called the variation of F along C and is sometimes written as

$$F(z)\Big|_C = F(z_2) - F(z_1)$$

where z_1 is the initial point and z_2 the terminal point of C.

Theorem 1.1 indicates that the result of integration depends only on the end points of C, and hence the notation

$$\int_{z_1}^{z_2} f(z)\,dz = F(z_2) - F(z_1) \tag{1.9}$$

can be used. It is understood that the path of integration is a contour C which lies in D and has initial point z_1 and terminal point z_2, but otherwise C is unrestricted. We say, briefly, that the integral is independent of the path. For closed curves C independence of the path gives

$$\int_C f(z)\,dz = 0, \qquad C \text{ closed}, \tag{1.10}$$

since C can be replaced by another contour, of zero length, joining z_1 and z_2. Conversely, if (1.10) holds for all closed contours lying in D, it is easily seen that the integral (1.9) is independent of the path.

Theorem 1.1 implies (1.10) provided the basic condition $F'(z) = f(z)$ is fulfilled. A major objective of this chapter is to extend this result to analytic functions $f(z)$ for which no F such that $F' = f$ is at hand. An example of such a function is $\sin z^2$. It will be shown later that if

$$F(z) = \int_0^z \sin \zeta^2 \, d\zeta$$

where the integration is along a straight line from 0 to z, then $F(z)$ is analytic for all z and $F'(z) = \sin z^2$. From this it follows that on any closed contour C

$$\int_C \sin z^2 \, dz = 0.$$

Example 1.1. Let C be the curve $z = z_0 + \rho e^{it}$, $a \leq t \leq b$, where z_0 and $\rho \neq 0$ are constant. Show that

$$\frac{1}{2\pi i} \int_C \frac{dz}{z - z_0} = \frac{b - a}{2\pi}$$

and deduce that this expression is an integer N if and only if C is closed.

The first result follows by substitution of

$$z - z_0 = \rho e^{it}, \qquad dz = i\rho e^{it}$$

into the integrand and integrating from a to b. The second follows from the periodicity properties of e^z, which show that

$$e^{ia} = e^{ib}$$

holds if and only if $b - a = 2\pi N$ for some integer N.

When C is closed, the integer N is called the winding number of C with respect to the point z_0. Geometrically, it represents the number of times the point z goes around the point z_0 as z traverses C.

Example 1.2. If n is an integer and $n \neq -1$, show that

$$\int_{z_1}^{z_2} z^n \, dz = \frac{z_2^{n+1} - z_1^{n+1}}{n+1}$$

provided, when $n < 0$, the path of integration avoids the origin. This follows by taking $F(z) = z^{n+1}/(n + 1)$ in Theorem 1.1. Similarly,

$$\int_{z_1}^{z_2} e^z \, dz = e^{z_2} - e^{z_1}, \qquad \int_{z_1}^{z_2} \sin z \, dz = -\cos z_2 + \cos z_1,$$

and so on.

Example 1.3. Let f and g be analytic in a domain D and have continuous derivatives f' and g' there. Show that the formula

$$\int_C f(z)g'(z) \, dz = f(z)g(z)\Big|_C - \int_C f'(z)g(z) \, dz$$

holds whenever C is a contour lying in D; in other words, the procedure of integrating by parts is valid in the complex plane. This follows by integrating

$$\frac{d}{dz}(fg) = f'g + fg'$$

and applying Theorem 1.1 to evaluate the left side.

Problems

1. Let C be a closed contour and let n be an integer. Show that

$$\int_C z^n \, dz = 0 \qquad (n \neq -1)$$

provided, when $n < 0$, the origin is not on C. What is the value of the integral if $n = -1$ and C is the circle $z = e^{it}$, $0 \leq t \leq 2\pi$?

2. (a) Each of these equations describes an arc for $0 \leq t \leq 1$. Sketch the arc and indicate its direction of traversal:

$z = 1 + it, \qquad z = e^{-\pi it}, \qquad z = 3e^{2\pi it}, \qquad z = e^{4\pi it}, \qquad z = 1 + it + t^2.$

(b) Using Theorem 1.1 when appropriate, integrate each of these functions over each of the arcs of part (a): $4z^3$, $\bar{z}$, $1/z$.

3. If $f'(z) = 0$ throughout a domain D, use Theorem 1.1 to prove that f is constant in D.

4. If C_0 is an arc $z = \zeta(t)$ and α is a complex constant, the *translated arc*, C_α, is given by $z = \zeta(t) + \alpha$. Assuming continuity of f, show that

$$\int_{C_0} f(z + \alpha)\, dz = \int_{C_\alpha} f(z)\, dz.$$

5. Let $\alpha = a + ib$ where a and b are constant. By integrating $e^{\alpha t}$ and equating real parts, derive the formula

$$(a^2 + b^2)\int_0^t e^{at} \cos bt\, dt = e^{at}(a \cos bt + b \sin bt) - a.$$

6. Let $F = F_1 + iF_2$ and $G = G_1 + iG_2$ be differentiable complex-valued functions of the real variable t. Obtain formulas

$$(F + G)' = F' + G', \qquad (FG)' = FG' + GF'$$

from known formulas of this type for real functions.

7. Let $z = \zeta(t)$ be an arc, $a \le t \le b$. Using $\zeta' = \xi' + i\eta'$ or

$$\frac{d\zeta}{dt} = \frac{dx}{dt} + i\frac{dy}{dt},$$

interpret $\zeta'(t)$ as being the complex representation of a vector tangent to the arc at any point where $\zeta'(t)$ is not zero. Also interpret $|\zeta'(t)|$ as ds/dt where s is arc.

8. The derivative of a complex-valued function $z = \zeta(t)$ is often defined by

$$\zeta'(t) = \lim_{\Delta t \to 0} \frac{\Delta z}{\Delta t} \qquad \text{where} \qquad \Delta z = \zeta(t + \Delta t) - \zeta(t).$$

(a) Show that this definition is equivalent to that of the text. (b) Interpret Δz, $\Delta z/\Delta t$ and $|\Delta z|$ by a sketch and compare Problem 7.

9. Prove the chain rule. Outline of solution: In (1.2) let

$$F(z) = u(x,y) + i\,v(x,y), \qquad \zeta(t) = \xi(t) + i\,\eta(t)$$

so that $\phi(t) = u(\xi,\eta) + i\,v(\xi,\eta)$ where we have written ξ for $\xi(t)$ and η for $\eta(t)$. The chain rule for real functions followed by the Cauchy-Riemann equations for F gives

$$\phi'(t) = [u_x(\xi,\eta) + i\,v_x(\xi,\eta)](\xi' + i\eta').$$

10. Let $\zeta(t) = t^2$ for $-1 \le t \le 0$ and let $\zeta(t) = it^2$ for $0 \le t \le 1$. Show that the curve $z = \zeta(t)$, $-1 \le t \le 1$, is an arc though it has a corner.

11. *The continuous logarithm.* Let $z = \zeta(t)$, $a \le t \le b$, denote an arc that does not pass through the point α and define

$$L(t) = \int_a^t \frac{\zeta'(t)}{\zeta(t) - \alpha}\, dt, \qquad a \le t \le b.$$

Show that the function $J(t) = e^{-L(t)}[\zeta(t) - \alpha]$ satisfies $J'(t) = 0$, and hence is constant. Determine the constant by setting $t = a$ and deduce

$$e^{L(t)} = \frac{\zeta(t) - \alpha}{\zeta(a) - \alpha}, \qquad a \le t \le b.$$

12. Show that two simple arcs can have infinitely many points of intersection without coinciding on any interval and that a contour can have infinitely many isolated points of self-intersection. (Consider $z = \zeta(t)$ where $\zeta(t) = -t$, $-1 \le t \le 0$, and $\zeta(t) = t + it^3 \sin(1/t)$, $0 < t \le 1$.)

*2. Further properties of integrals. Invariance.

If $F(t)$ is a continuous complex-valued function for $a \le t \le b$, it will be shown that

$$\left| \int_a^b F(t)\, dt \right| \le \int_a^b |F(t)|\, dt, \tag{2.1}$$

that is, the magnitude of an integral does not exceed the integral of the magnitude of the integrand. Equation (2.1) follows from the familiar inequality

$$\int_a^b g(t)\, dt \le \int_a^b G(t)\, dt \tag{2.2}$$

which holds whenever g and G are continuous and satisfy $g(t) \le G(t)$ on $a \le t \le b$.

Indeed, if the integral on the left of (2.1) is zero, the result is obvious. If it is not zero, let it be written in polar form, so that

$$\int_a^b F(t)\, dt = Je^{i\theta}, \qquad J > 0.$$

Since θ is constant, multiplication by $e^{-i\theta}$ gives

$$J = \int_a^b e^{-i\theta} F(t)\, dt = \int_a^b \operatorname{Re} e^{-i\theta} F(t)\, dt + i \int_a^b \operatorname{Im} e^{-i\theta} F(t)\, dt \tag{2.3}$$

where the second equality follows from the definition of integration for complex-valued functions. In (2.3) the last integral is 0 because J is real.

If we now set

$$g(t) = \operatorname{Re} e^{-i\theta}F(t), \qquad G(t) = |F(t)|,$$

then $g(t) \leq G(t)$ and hence (2.3) and (2.2) give

$$J = \int_a^b g(t)\,dt \leq \int_a^b |F(t)|\,dt.$$

This is (2.1). A similar inequality holds for integration along arcs and contours, as seen next.

Let $f(z)$ be continuous on an arc C given by $z = \zeta(t)$, $a \leq t \leq b$. Since by definition

$$\int_C f(z)\,dz = \int_a^b f[\zeta(t)]\zeta'(t)\,dt, \tag{2.4}$$

the choice $F(t) = f[\zeta(t)]\zeta'(t)$ in (2.1) gives

$$\left|\int_C f(z)\,dz\right| \leq \int_a^b |f[\zeta(t)]||\zeta'(t)|\,dt. \tag{2.5}$$

If $|f(z)| \leq M$ on C, where M is constant, then (2.5) yields

$$\left|\int_C f(z)\,dz\right| \leq M\int_a^b |\zeta'(t)|\,dt. \tag{2.6}$$

Since $\zeta = x + iy$ on C, differentiation gives

$$|\zeta'(t)| = \left|\frac{dx}{dt} + i\frac{dy}{dt}\right| = \sqrt{\left(\frac{dx}{dt}\right)^2 + \left(\frac{dy}{dt}\right)^2}.$$

This is recognized as ds/dt, where s is arc measured from the point $\zeta(a)$ of C. Hence the integral on the right of (2.6) is the length L of the arc C.

If s is arc length on C measured from the point $\zeta(a)$, the first and second integrals below are defined by the third integral:

$$\int_C f(z)\,ds = \int_C f(z)|dz| = \int_a^b f[\zeta(t)]|\zeta'(t)|\,dt.$$

With this notation (2.5) is equivalent to

$$\left|\int_C f(z)\,dz\right| \leq \int_C |f(z)||dz|, \tag{2.7}$$

and (2.6) takes the form

$$\left|\int_C f(z)\,dz\right| \leq M\int_C |dz| = ML.$$

These inequalities can be extended from arcs to contours by addition. The extended form of (2.6) reads as follows:

THEOREM 2.1. *Let $f(z)$ be continuous on a contour C. Then*

$$\left|\int_C f(z)\,dz\right| \leq ML$$

where L is the length of C and M is an upper bound for $|f|$ on C.

As already observed, a curve is associated with a definite parametric representation $z = \zeta(t)$, $a \leq t \leq b$. Nevertheless the quantity L in Theorem 2.1 has an obvious geometric meaning which is, to a large extent, independent of the parametric representation. A similar remark applies to the general concept of integration. Although the definition (2.4) involves the particular representation $z = \zeta(t)$ of the curve C, the result of the integration is, to a large extent, independent of this representation.

In illustration of this fact let C_1 be an arc $z = \zeta(t)$, $0 \leq t \leq 2$, and let C_2 be the arc $z = \zeta(2\tau - 4)$, $2 \leq \tau \leq 3$. It is easily verified that C_1 and C_2 have the same initial point, the same end point, and the same trace. Furthermore, the equation

$$\int_{C_1} f(z)\,dz = \int_{C_2} f(z)\,dz \tag{2.8}$$

holds for every function f which is continuous in a region containing C_1 and C_2. To see this, observe, by the chain rule, that

$$\frac{d}{d\tau}\zeta(2\tau - 4) = 2\zeta'(2\tau - 4).$$

Hence

$$\int_{C_2} f(z)\,dz = \int_2^3 f[\zeta(2\tau - 4)]2\zeta'(2\tau - 4)\,d\tau. \tag{2.9}$$

Upon taking $t = 2\tau - 4$ as a new variable of integration and applying familiar rules for change of variable to the real and imaginary parts, we see that (2.9) reduces to

$$\int_0^2 f[\zeta(t)]\zeta'(t)\, dt.$$

This is the integral on the left of (2.8).

The curves C_1 and C_2 in the above example are related by a linear change of parameter, $t = 2\tau - 4$. More generally one could set $t = \phi(\tau)$. For broad classes of functions ϕ the equivalence (2.8) continues to hold, as seen by an argument similar to the above.[1] It is said, briefly, that the integral is invariant under admissible changes of parameter.

In the theory of integration two curves C_1 and C_2 are considered to be equivalent if they have the same trace and if the condition (2.8) holds for all functions f which are continuous in a region containing this trace. In the same spirit one can write

$$C = C_1 + C_2 + \cdots + C_n$$

if it is true that

$$\int_C f(z)\, dz = \int_{C_1} f(z)\, dz + \int_{C_2} f(z)\, dz + \cdots + \int_{C_n} f(z)\, dz$$

for all continuous functions f. This holds, for instance, when C is a contour composed of arcs, as explained in the foregoing section.

It can be shown that two simple arcs which are not closed are equivalent if both have the same trace, the same initial point, and the same terminal point. With the rather obvious qualification that the arcs must be traversed in the same direction, a similar remark applies to simple closed arcs. In the latter case it does not matter what point is considered to be the common initial and terminal point.

The practical importance of these considerations is that they allow integration over simple curves with any convenient, or perhaps unspecified, choice of parameter. Thus one can speak of integration over the circle $|z| = 1$, over the boundary of the unit square, and so on. Some conventions needed to get an unambiguous result are explained in Example 2.1.

[1] A more general case is considered in Problem 8.

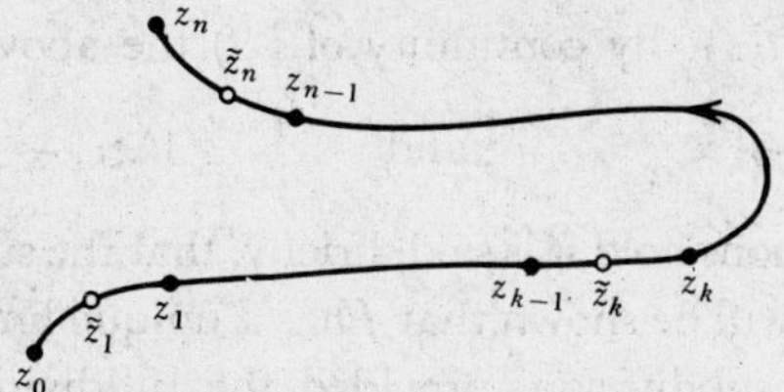

Figure 2-1

We conclude this discussion of invariance by mentioning another definition of the complex integral, which makes the invariance properties rather obvious. While the definition (2.4) is entirely satisfactory for rigorous development of the theory, the following sketchy observations give useful insight.

Let the arc C be divided into subarcs (z_k, z_{k+1}) by points $z_0, z_1, \ldots, z_n$ and let $\tilde{z}_k$ lie on C between z_{k-1} and z_k as shown in Figure 2-1. If C is given by $z = \zeta(t)$, $a \leq t \leq b$, and if

$$z_k = \zeta(t_k), \qquad \tilde{z}_k = \zeta(\tilde{t}_k),$$

the statement that $\tilde{z}_k$ is between z_{k-1} and z_k means that

$$t_{k-1} \leq \tilde{t}_k \leq t_k.$$

We assume also that z_0 is the initial and z_n the terminal point of C, so that

$$z_0 = \zeta(a), \qquad z_n = \zeta(b).$$

Given a continuous function f on C we define

$$J = \sum_{k=1}^{n} f(\tilde{z}_k)\,\Delta z_k \tag{2.10}$$

where $\Delta z_k = z_k - z_{k-1}$. It will be desirable to consider a sequence of subdivisions which are made finer and finer by choosing more and more points z_k. The precise requirement is

$$n \to \infty \qquad \text{and} \qquad \max_k \Delta t_k \to 0$$

where $\Delta t_k = t_k - t_{k-1}$. By continuity of $\zeta(t)$ the above also implies

$$n \to \infty \qquad \text{and} \qquad |\Delta z_k| \to 0.$$

When these conditions hold it is said, briefly, that the subdivision becomes arbitrarily fine. It will be shown that J has a unique limit J_0, independent of the manner of subdivision, provided the subdivision becomes arbitrarily fine. Furthermore, the limit J_0 agrees with the definition of the complex integral given in (2.4).

Let a positive number ϵ be given. If the subdivision is sufficiently fine, the uniform continuity of f on C shows that

$$|f(\tilde{z}_k) - f(z_k)| \le \epsilon, \qquad k = 1, 2, \ldots, n.$$

Hence if $f(\tilde{z}_k)$ in (2.10) is replaced by $f(z_k)$, the error committed does not exceed

$$\epsilon(|\Delta z_1| + |\Delta z_2| + \cdots + |\Delta z_n|).$$

Geometrically $|\Delta z_k|$ represents the length of the straight line joining z_{k-1} and z_k, and this does not exceed the length of the arc of C joining these points. Hence we conclude that

$$\left| \sum f(\tilde{z}_k)\, \Delta z_k - \sum f(z_k)\, \Delta z_k \right| \le \epsilon L$$

where L is the length of C.

For the second step of the proof observe that

$$\Delta z_k = \zeta(t_k) - \zeta(t_{k-1}) = \zeta'(t_k)\, \Delta t_k + \epsilon_k\, \Delta t_k, \tag{2.11}$$

where $|\epsilon_k| \le \epsilon$ provided the subdivision is sufficiently fine. Indeed, if $\zeta = \xi + i\eta$, the mean-value theorem gives relations of form

$$\xi(t_k) - \xi(t_{k-1}) = \xi'(t_k')\, \Delta t_k, \qquad \eta(t_k) - \eta(t_{k-1}) = \eta'(t_k'')\, \Delta t_k.$$

Since ξ' and η' are continuous it is easy to assess the error committed when the intermediate values t_k' and t_k'' in this equation are both replaced by t_k. The result is (2.11). Equation (2.11) shows that if the term Δz_k in J is replaced by $\zeta'(t_k)\, \Delta t_k$, the error committed does not exceed

$$\epsilon M(\Delta t_1 + \Delta t_2 + \cdots + \Delta t_n)$$

where M is a bound for $|f(z)|$ on C. This reduces to

$$\epsilon M(b - a).$$

As the third and final step of the proof we observe that the sum

$$\sum f[\zeta(t_k)]\zeta'(t_k)\,\Delta t_k$$

is as close as desired to the integral

$$J_0 = \int_a^b f[\zeta(t)]\zeta'(t)\,dt \tag{2.12}$$

provided the subdivision is sufficiently fine. This follows from the fact that the real and imaginary parts of (2.12) are real Riemann integrals, the existence of which is assured by results of real analysis.

If the preceding estimates are combined, we conclude that

$$|J - J_0| \leq \epsilon L + \epsilon M(b - a) + \epsilon \tag{2.13}$$

provided the subdivision is sufficiently fine. The term ϵL arises from the replacement of $\tilde{z}_k$ by z_k, the term $\epsilon M(b - a)$ arises from the replacement of Δz_k by $\zeta'(t_k)\,\Delta t_k$, and the final ϵ arises from approximation of the integral J_0 by a finite sum. Since ϵ can be chosen as small as desired, (2.13) shows that $J \rightarrow J_0$ as the subdivision becomes arbitrarily fine and this completes the proof.

Example 2.1. Describe unsuitable and also suitable parametric representations for evaluation of

$$\int_C f(z)\,dz \tag{2.14}$$

where C is the unit circle $|z| = 1$.

When a path of integration is described by just giving its trace, the intended curve is assumed to be simple. Thus the parametrization

$$z = e^{it}, \qquad 0 \leq t \leq 4\pi,$$

is not suitable in (2.14), because the circle is described twice with this representation. As another convention, the direction of traversal, when not otherwise specified, is given by the following rule: A counterclockwise rotation of the tangent through 90° makes the tangent point into the region bounded by the curve. More informally, the direction is such that a person walking along the curve in that direction will always have the region on his left. In (2.14) the representation

$$z = e^{-it}, \qquad 0 \leq t \leq 2\pi$$

is not suitable, because the circle is described in the wrong direction. On the other hand, either of the representations

$$z = e^{it}, \quad 0 \leq t \leq 2\pi \qquad \text{or} \qquad z = e^{\pi it}, \quad -1 \leq t \leq 1$$

is suitable, as are infinitely many others.

The fact that a continuous simple closed curve divides the plane into just two regions, one of which is bounded, is known as the Jordan curve theorem. It is not proved here. However, for curves commonly encountered in applications the verification is trivial, and the above conventions are easy to apply. They are followed throughout this text.

Example 2.2. If C is a contour that avoids the point α, evaluate

$$\int_C \frac{dz}{z - \alpha}$$

and interpret geometrically when C is closed.

We first assume that C is an arc $z = \zeta(t)$, $a \leq t \leq b$. The integral is then

$$\int_a^b \frac{\zeta'(t)\,dt}{\zeta(t) - \alpha} = \int_a^b \frac{d}{dt} \log[\zeta(t) - \alpha]\,dt = \log(z - \alpha)\Big|_C$$

where the symbol on the right denotes the variation of $\log(z - \alpha)$ along C. In this calculation one chooses a branch of $\log(z - \alpha)$ at the initial point $z_1 = \zeta(a)$ of C and one lets $\log(z - \alpha)$ vary continuously as z traverses C of the final point z_2. In other words, the values are chosen so as to make the function

$$\phi(t) = \log[\zeta(t) - \alpha]$$

continuous for $a \leq t \leq b$. With this stipulation

$$\int_C \frac{dz}{z-\alpha} = \log(z-\alpha)\Big|_C = \log[\zeta(t)-\alpha]\Big|_a^b.$$

The result is extended from arcs to contours by addition.

In general the variation of a many-valued function along a given curve is computed, as here, by explicit imposition of a continuity condition. Usually the result depends on the choice of branch at the initial point $\zeta(a)$. In the present case, however, the result is independent of the initial branch.

To get a geometric interpretation recall that

$$\log(z-\alpha) = \operatorname{Log}|z-\alpha| + i\theta$$

where $\theta = \arg(z-\alpha)$. Since $\operatorname{Log}|z-\alpha|$ is single-valued, its variation around C is 0 when C is closed, so that

$$\log(z-\alpha)\Big|_C = i\theta\,\Big|_C \qquad (C \text{ closed}).$$

Now θ is the angle which the line joining α to the variable point $z = \zeta(t)$ makes with the horizontal, and hence the total variation is 2π times the number of times z winds around α as z traverses C (Figure 2-2). Thus

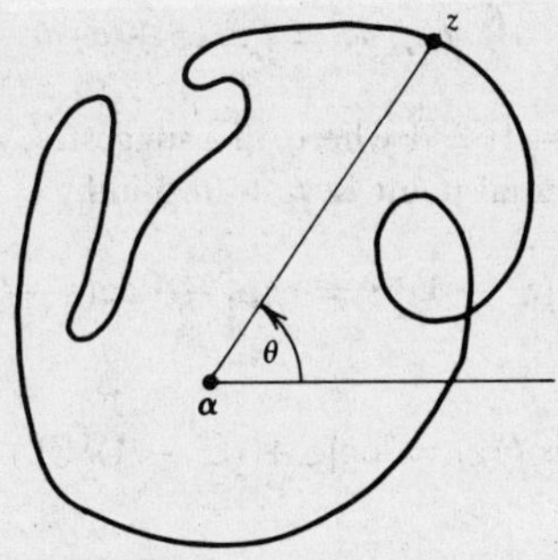

Figure 2-2

$$\frac{1}{2\pi i}\int_C \frac{dz}{z-\alpha} = N \tag{2.15}$$

where $N = N(C,\alpha)$ is an integer[2] called the winding number of C with respect to α. If α is on C, the winding number is not defined.

Problems

1. If C is the circle $|z| = 1$, evaluate

$$\int_C \frac{dz}{z}, \qquad \int_C \frac{dz}{|z|}, \qquad \int_C \frac{|dz|}{z}, \qquad \int_C \frac{dz}{z^2}, \qquad \int_C \frac{dz}{|z^2|}, \qquad \int_C \frac{|dz|}{z^2},$$

2. If C is the circle $|z| = 1$, show by Theorem 2.1 that

$$\left|\int_C \frac{dz}{4+3z}\right| \le 2\pi \qquad \text{and} \qquad \left|\int_C \frac{dz}{4+3z}\right| \le \frac{6}{5}\pi.$$

(In the second case consider the parts of C with $x > 0$ and $x < 0$ separately.)

3. Let C be the line joining the point 1 to the point $2 + 2i$. (a) Get a suitable parametrization of C in the form $z = \alpha + \beta t$, $0 \le t \le 1$, and thus evaluate the first two of the following three expressions:

$$\int_C (z^2 + 2z)\,dz, \qquad \int_C (z^2 + 2z)|dz|, \qquad \int_C |z^2 + 2z||dz|.$$

(b) Estimate the third expression above by Theorem 2.1.

4. Let C_0 be the circle $|z| = \rho$ where ρ is a positive constant. For any complex constant α the translated circle C_α is the circle $|z - \alpha| = \rho$. Illustrate the geometric relation C_0 and C_α by a sketch, parametrize C_0 and C_α as suggested by Example 2.1, and assuming continuity of f, show that

$$\int_{C_\alpha} f(z)\,dz = \int_{C_0} f(z+\alpha)\,dz.$$

5. Let C be the circle $2|z - 1| = 1$ where, in a suggestive notation, the initial point is $\frac{1}{2} - 0i$ and the terminal point is $\frac{1}{2} + 0i$. Find

$$\int_C \frac{dz}{(z^2-1)^{1/2}} \text{ given } (z^2-1)^{1/2} = \exp\left[\frac{1}{2}\text{Log}(z-1) + \frac{1}{2}\text{Log}(z+1)\right].$$

(An antiderivative is $F(z) = \log[z + (z^2 - 1)^{1/2}]$.)

[2] See also Section 1, Problem 11 with $\zeta(a) = \zeta(b)$.

***Problems on invariance**

6. (a) Following the pattern of the text, establish the equivalence of two arcs C_1 and C_2 under a linear change of parameter $t = c_1\tau + c_0$ where $c_1 \neq 0$ and c_0 are real constants. (b) Show that any arc is equivalent, under a linear change of parameter, to an arc on the interval $0 \leq t \leq 1$. (c) Obtain a similar result for contours by parametrizing C_k so that its interval is $k/n \leq t \leq (k+1)/n$. (d) Discuss a definition of $-C$ by the change of parameter $\tau = -t$.
7. If $f = u + iv$ and $dz = dx + i\,dy$, computation suggests that

$$\int_C f(z)\,dz = \int_C [u(x,y)\,dx - v(x,y)\,dy] + i\int_C [u(x,y)\,dy + v(x,y)\,dx]$$

where each of the integrals on the right is a line integral as introduced in calculus. Show that this formula agrees with the definition given in Section 1 (and hence the invariance properties of complex integrals follow from corresponding properties of line integrals.)
8. Let C_1 and C_2 be the respective arcs

$$z = \zeta_1(t), \qquad a \leq t \leq b \qquad \text{and} \qquad z = \zeta_2(\tau), \qquad \alpha \leq \tau \leq \beta.$$

It is said that C_2 arises from C_1 by an admissible change of parameter if there is a differentiable function $\phi(t)$ such that $\alpha = \phi(a)$, $\beta = \phi(b)$ and such that the condition $\tau = \phi(t)$ gives $\alpha \leq \tau \leq \beta$, $\zeta_2(\tau) = \zeta_1(t)$ for $a \leq t \leq b$. In this case show that C_1 and C_2 are equivalent. (If

$$F_1(t) = \int_a^t f[\zeta_1(s)]\zeta_1'(s)\,ds, \qquad F_2(\tau) = \int_\alpha^\tau f[\zeta_2(\sigma)]\zeta_2'(\sigma)\,d\sigma$$

where $\tau = \phi(t)$, it is easily checked that F_1 and F_2 both have the same derivative with respect to t. Since they agree for $t = a$, they agree for $t = b$.)
9. Show that every contour is equivalent to an arc. (See Section 1, Problem 10.)

3. The Cauchy integral theorem; restricted case. The gist of the foregoing discussion is that differentiation and integration of complex-valued functions is accomplished by carrying out corresponding operations on the real and imaginary parts, and that formal calculations suggested by the notation are in fact justified. For example, if C is the curve $z = \zeta(t)$, $a \leq t \leq b$, then

$$\int_C f(z)\,dz = \int_a^b f[\zeta(t)]\zeta'(t)\,dt$$

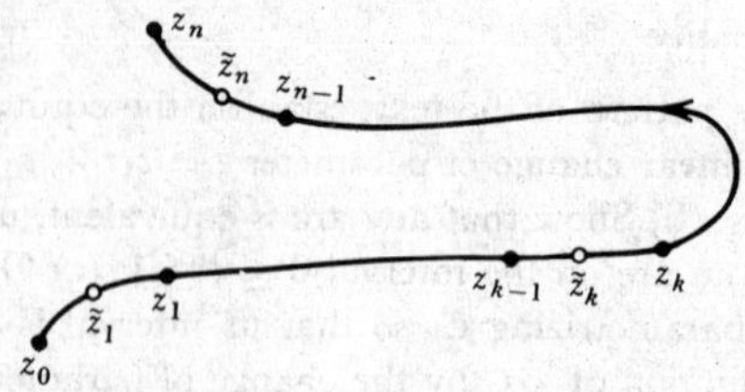

Figure 3-1

provided there is enough continuity. As in real analysis, the integral can also be obtained as a limit of sums of the form

$$\sum_{k=1}^{n} f(\tilde{z}_k)(z_k - z_{k-1})$$

where $z_0, z_1, \ldots, z_n$ are points taken in succession along C and $\tilde{z}_k$ is on C between z_{k-1} and z_k (Figure 3-1). This second formulation makes it obvious that the integral is, to a large extent, independent of the parametric representation of C, and also leads to an important inequality.

Indeed, if $|f(z)| \leq M$ on C, where M is constant, then

$$\left| \sum f(\tilde{z}_k)(z_n - z_{k-1}) \right| \leq \sum M|z_k - z_{k-1}| = M \sum |z_k - z_{k-1}|.$$

Since a straight line is the shortest distance between two points,

$$|z_k - z_{k-1}|$$

does not exceed the length of the arc of C joining z_{k-1} to z_k. It follows that the above sum does not exceed ML where L is the length of C, and passing to the limit gives a similar estimate for the integral. Hence[1]

$$\left| \int_C f(z)\, dz \right| \leq ML,$$

where M is an upper bound for $|f(z)|$ on C and L is the length of C.

[1] Another proof was given in the preceding section.

The Cauchy integral theorem states that in a certain class of domains the integral of an analytic function over a closed contour is zero. The restriction on the domain is that it be simply connected. This concept and the general form of the theorem are considered later. For development of the basic properties of analytic functions a much simpler form of the theorem is sufficient.

As a preliminary theorem a special case of the Cauchy integral theorem is now established for a closed contour C whose trace consists of the three edges of a triangle. Let T denote the closed region consisting of the interior of a triangle and its edges. Let C denote the closed contour formed by the three edges. The direction in which C is traversed is chosen so that if the vector designating this direction along an edge of C is rotated counterclockwise through 90° it points into the interior of the triangle shown in Figure 3-2. This direction on C is called the positive direction.

It will be seen later that the theorem for the triangle is the basis for more general forms of the theorem.

THEOREM 3.1. *Let a closed triangular region T lie in a domain in which a function $f(z)$ is analytic. Let C denote the boundary of T traversed in the positive direction. Then*

$$\int_C f(z)\, dz = 0.$$

Two proofs will be given, one in this section and one in the next. In the first proof the definition of "a function $f(z)$ analytic in a domain D" is restricted so that not only is $f'(z)$ required to exist throughout D but $f'(z)$

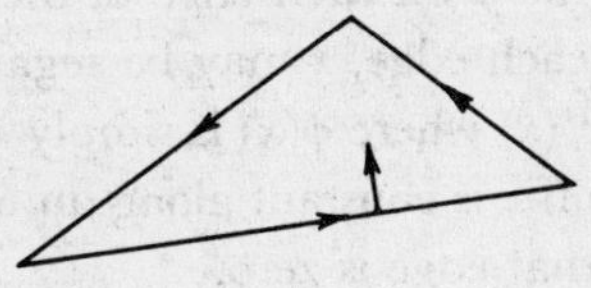

Figure 3-2

is required to be continuous in D. This restricted definition of analyticity is entirely adequate for the theory and applications of analytic functions and can be used for any specific function without any loss whatsoever. This is so because the original definition not requiring continuity leads to the consequence that $f'(z)$ is in fact continuous in D, so that there is no real narrowing of the class of functions under consideration with the restricted definition. The first proof is the one used in nineteenth-century mathematics. It requires a knowledge of the fact that under suitable restrictions a repeated integral is equal to a double integral.

First proof of Theorem 3.1. Consider the triangular region T with boundary C and with vertices α, β and γ in that order as C is traversed in the positive direction. If $f(z) = u(x,y) + i\,v(x,y)$, then since f' exists and is continuous, the same is true for $\partial u/\partial x$, $\partial u/\partial y$, $\partial v/\partial x$ and $\partial v/\partial y$ which satisfy the Cauchy-Riemann equations

$$\frac{\partial u}{\partial x} = \frac{\partial v}{\partial y}, \qquad \frac{\partial u}{\partial y} = -\frac{\partial v}{\partial x}.$$

Furthermore,

$$\int_C f(z)\,dz = \int_C (u\,dx - v\,dy) + i\int_C (u\,dy + v\,dx) \tag{3.1}$$

where, for example,

$$\int_C u\,dx = \left(\int_\alpha^\beta + \int_\beta^\gamma + \int_\gamma^\alpha\right) u(x,y)\,dx$$

and in the above integrals y on each edge of the triangle is represented as a function of x. On each edge, x may be regarded as a parameter so that $x = \phi(x)$ and $y = \psi(x)$ where $\phi(x)$ is simply x itself and ψ is a linear function of x. Of course if x is constant along an edge, that is, the edge is vertical, then $\int u\,dx$ on that edge is zero.

Let the projection of the triangle T on the x axis be a line segment extending from x_1 to x_2. Let a vertical line with abscissa x intersect the triangle in $g_1(x)$ and $g_2(x)$ with $g_1(x) \leq g_2(x)$ as shown in Figure 3-3. Then

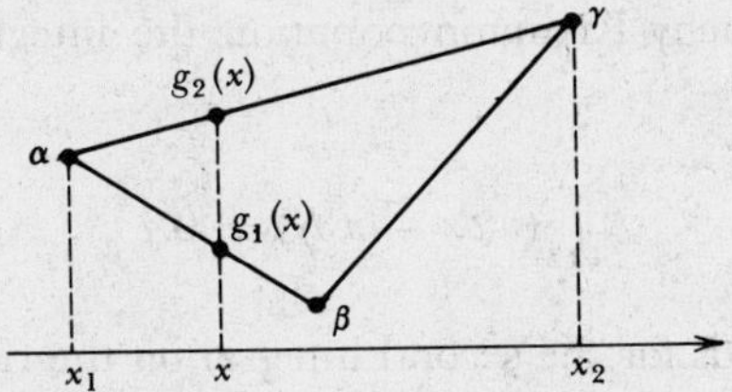

Figure 3-3

$$\int_C u\,dx = \int_{x_1}^{x_2} u(x, g_1(x))\,dx + \int_{x_2}^{x_1} u(x, g_2(x))\,dx$$
$$= \int_{x_2}^{x_1} [u(x, g_2(x)) - u(x, g_1(x))]\,dx.$$

Since

$$u(x, g_2(x)) - u(x, g_1(x)) = \int_{g_1(x)}^{g_2(x)} \frac{\partial u}{\partial y}(x,y)\,dy,$$

it follows that

$$\int_C u\,dx = -\int_{x_1}^{x_2}\left(\int_{g_1(x)}^{g_2(x)} \frac{\partial u}{\partial y}(x,y)\,dy\right)dx.$$

The integral on the right is the repeated integral of $\partial u/\partial y$ over the triangular region T. Replacing the repeated integral by the double integral gives

$$\int_C u\,dx = -\int_T \frac{\partial u}{\partial y}\,dx\,dy.$$

In the same way

$$\int_C v\,dy = \int_T \frac{\partial v}{\partial x}\,dx\,dy.$$

Thus

$$\int_C (u\,dx - v\,dy) = -\int_T \left(\frac{\partial u}{\partial y} + \frac{\partial v}{\partial x}\right)dx\,dy.$$

By the second Cauchy-Riemann equation the integral on the right is zero. Hence

$$\int_C (u\,dx - v\,dy) = 0.$$

A similar result holds for the second integral on the right of (3.1), which proves the theorem.

By means of Theorem 3.1, the existence of an indefinite integral and the Cauchy integral theorem will be proved for functions analytic in a star domain. A domain is called a *star domain* if there is a point z_0 in the domain such that the line segment from z_0 to any point of the domain is contained in it. Thus the interior of the circle or parallelogram is a star domain and z_0 can be taken as any point in these domains. Other examples of star domains are shown in Figure 3-4. The finite z plane, $|z| < \infty$, is a star domain and z_0 can be taken as any point in the plane. When the role of z_0 must be emphasized, it is said that the domain is *star with respect to* z_0. For example, deleting three radial segments from a disk can give a domain that is star with respect to the center, but is star with respect to no other point.

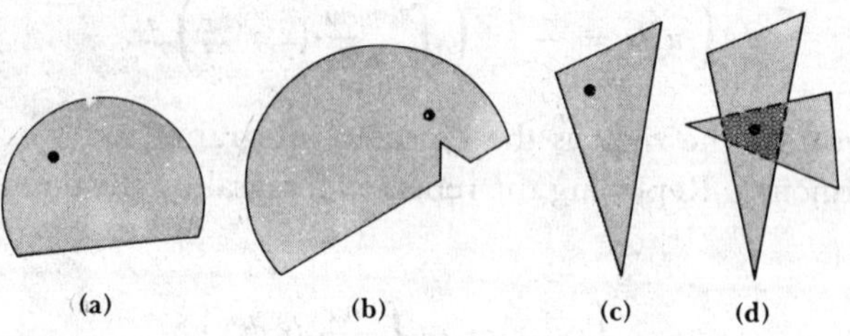

Figure 3-4

We shall establish the following:

THEOREM 3.2. *If $f(z)$ is analytic in a star domain D, then there exists a function $F(z)$ analytic in D such that $F'(z) = f(z)$.*

Proof. Let D be star with respect to z_0, and let z be a point of D. The function F is defined by

$$F(z) = \int_{z_0}^{z} f(\zeta)\,d\zeta \qquad (3.2)$$

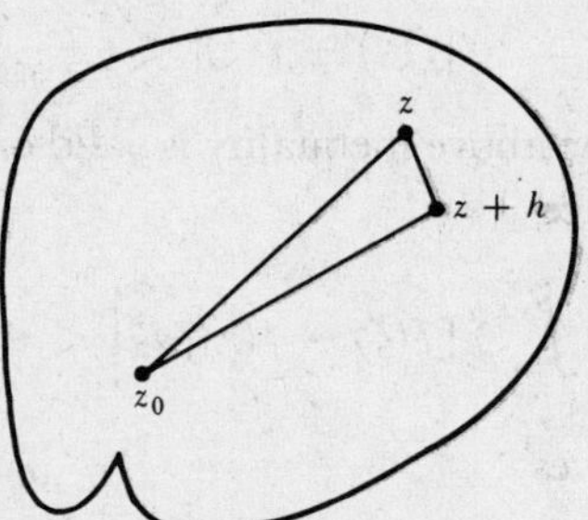

Figure 3-5

where the integration is along the line segment that joins z_0 to z. Since z lies in D there is a neighborhood of z in D. Thus if h is a complex number with $|h|$ sufficiently small, then $z + h$ is in D and the line segment from z to $z + h$ also lies in D, as shown in Figure 3-5.

As any point in the line segment from z to $z + h$ is in D, the line segment from z_0 to any such point must be in D (from the definition of a star domain) and thus the closed triangular region with vertices z_0, z, $z + h$ is in D. From Theorem 3.1

$$\left(\int_{z_0}^{z+h} + \int_{z+h}^{z} + \int_{z}^{z_0} \right) f(\zeta)\, d\zeta = 0 \tag{3.3}$$

or

$$\left(\int_{z_0}^{z+h} - \int_{z_0}^{z} \right) f(\zeta)\, d\zeta = \int_{z}^{z+h} f(\zeta)\, d\zeta.$$

By (3.2) this becomes

$$F(z + h) - F(z) = \int_{z}^{z+h} f(\zeta)\, d\zeta.$$

Therefore, if $h \neq 0$,

$$\frac{F(z + h) - F(z)}{h} - f(z) = \frac{1}{h} \int_{z}^{z+h} [f(\zeta) - f(z)]\, d\zeta. \tag{3.4}$$

Since $f(\zeta)$ is continuous at z, it follows that given any $\epsilon > 0$ there exists a $\delta > 0$ such that

$$|f(\zeta) - f(z)| < \epsilon \qquad \text{if } |\zeta - z| < \delta.$$

Thus when $|h| < \delta$, the above inequality is valid on the line of integration from z to $z + h$. Hence

$$\left| \int_z^{z+h} [f(\zeta) - f(z)]\, d\zeta \right| < \epsilon |h|$$

and therefore (3.4) gives

$$\left| \frac{F(z+h) - F(z)}{h} - f(z) \right| < \epsilon.$$

Since ϵ is an arbitrary positive number this proves that

$$\lim_{h \to 0} \frac{F(z+h) - F(z)}{h} = f(z)$$

or, in other words, that $F'(z) = f(z)$.

In this proof the only use made of the analyticity is to show that $f(z)$ is continuous and that the integral over the triangular contour is 0. Thus the argument has established the following alternative version, which is sometimes more useful:

THEOREM 3.3. *Let $f(z)$ be continuous in a star domain D and let*

$$\int_C f(z)\, dz = 0$$

for every closed triangular contour lying in D. Then there exists a function $F(z)$ analytic in D such that $F'(z) = f(z)$.

Theorem 3.2 leads to the Cauchy integral theorem for a star domain, which reads as follows:

THEOREM 3.4. *If $f(z)$ is analytic in a star domain D and C is a closed contour lying in D, then*

$$\int_C f(z)\, dz = 0.$$

This is an immediate consequence of Theorem 3.2 and (1.10). As shown in Figure 3-6, C can have self-intersections.

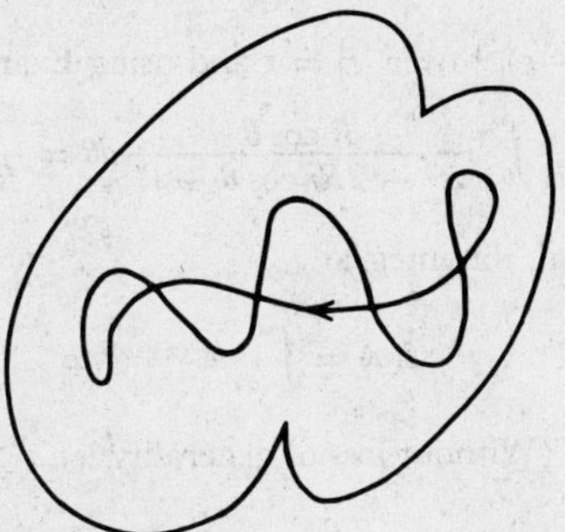

Figure 3-6

Example 3.1. If $0 < r < R$, integrate the function

$$\frac{R+z}{(R-z)z} \equiv \frac{1}{z} + \frac{2}{R-z} \tag{3.5}$$

over the circle $|z| = r$ and deduce that

$$\frac{1}{2\pi}\int_0^{2\pi} \frac{R^2 - r^2}{R^2 - 2Rr\cos\theta + r^2}\,d\theta = 1. \tag{3.6}$$

By Theorem 3.4 the second term on the right of (3.5) integrates to 0. Hence, setting $z = re^{i\theta}$, $dz/z = i\,d\theta$, we get

$$\int_0^{2\pi} \frac{R + re^{i\theta}}{R - re^{i\theta}}\,i\,d\theta = \int_0^{2\pi} i\,d\theta = 2\pi i. \tag{3.7}$$

Since the fraction on the left of (3.7) can be written

$$\frac{(R + re^{i\theta})(R - re^{-i\theta})}{(R - re^{i\theta})(R - re^{-i\theta})} = \frac{R^2 - r^2 + 2iRr\sin\theta}{R^2 - 2Rr\cos\theta + r^2},$$

dividing (3.7) by $2\pi i$ and equating real parts gives (3.6).

Problems

1. By considering $f(z) = 1/z$ in the annulus $1 < |z| < 3$ show that the conclusion of Theorem 3.4 can fail if the region is not assumed to be a star domain. (Take C to be the circle $|z| = 2$.)

2. By integrating $(R - z)^{-1}$ over $|z| = r$ and using Example 3.1, show that

$$\frac{1}{2\pi}\int_0^{2\pi}\frac{R\cos\theta}{R^2 - 2Rr\cos\theta + r^2}\,d\theta = \frac{r}{R^2 - r^2}, \qquad 0 \le r < R.$$

3. Prove that if a is real, the integral

$$I(a) = \int_{-\infty}^{\infty} e^{-(x+ia)^2}\,dx$$

is independent of a. (Without loss of generality let $a > 0$. By Theorem 3.4

$$\int_C e^{-z^2}\,dz = 0$$

where C is the rectangle with vertices at $-b$, b, $b + ia$ and $-b + ia$. Since

$$\left|\int_0^a e^{-(b+iy)^2}\,dy\right| \le e^{-b^2}\int_0^a e^{y^2}\,dy,$$

letting $b \to \infty$ gives $I(a) = I(0)$.)

4. Given that $I(0) = \sqrt{\pi}$ in Problem 3, use the result of that problem to show that

$$\int_{-\infty}^{\infty} e^{-x^2/2}\cos ux\,dx = \sqrt{2\pi}e^{-u^2/2}.$$

5. *Fundamental theorem of algebra.* If $P(z)$ is a nonconstant polynomial, show that $P(z)$ has at least one root. Other proofs will be given. (Let

$$P(z) = a_n z^n + \cdots + a_1 z + a_0 = zQ(z) + a_0$$

where $n \ge 1$, $a_n \ne 0$, and where the polynomial $Q(z)$ is defined by this equation. Then

$$\frac{1}{z} = \frac{P(z)}{zP(z)} = \frac{zQ(z) + a_0}{zP(z)} = \frac{Q(z)}{P(z)} + \frac{a_0}{zP(z)}.$$

If $P(z)$ does not vanish, the first term on the right integrates to 0 by Theorem 3.4 and hence

$$2\pi i = \int_{|z|=R}\frac{a_0}{zP(z)}\,dz.$$

As $|z| = R \to \infty$, clearly $\lim P(z)/a_n z^n = 1$, and hence for large R

$$\left|\frac{P(z)}{a_n z^n}\right| \ge \frac{1}{2} \qquad \text{or} \qquad |P(z)| \ge \frac{1}{2}|a_n||z|^n.$$

The resulting inequality

$$2\pi = \left|\int_{|z|=R}\frac{a_0}{zP(z)}\,dz\right| \le (2\pi R)\frac{2|a_0|}{R|a_n|R^n} = \frac{4\pi|a_0|}{|a_n|R^n}$$

leads to a contradiction as $R \to \infty$.)

6. Let $f(z)$ be analytic for $|z| < R$ and continuous for $|z| \leq R$. If C is the circle $|z| = R$, show that

$$\int_C f(z)\,dz = 0.$$

Outline of solution: For $r < R$

$$\int_C f(z)\,dz = \int_0^{2\pi} f(\mathrm{R}e^{i\theta})\mathrm{R}e^{i\theta} i\,d\theta - \int_0^{2\pi} f(re^{i\theta})re^{i\theta} i\,d\theta$$

because, by Theorem 3.4, the second integral on the right is 0. Since the function $f(z)z$ is continuous in $|z| \leq R$ it is uniformly continuous. Hence, given $\epsilon > 0$, there is a $\delta > 0$ such that

$$|f(Re^{i\theta})Re^{i\theta} - f(re^{i\theta})re^{i\theta}| < \epsilon \qquad \text{for } R - r < \delta.$$

This shows that the magnitude of the above integral does not exceed $2\pi\epsilon$.

7. This problem requires familiarity with the theory of line integrals. Verify that the Cauchy-Riemann equations are the conditions for the integrands in Section 2, Problem 7 to be exact, and thus get an immediate proof of Theorem 3.4 for the case in which $f'(z)$ is continuous.

***4. The Cauchy-Goursat theorem; restricted case.** More than half a century after complex analysis had become a recognized branch of mathematics, it was discovered by Edouard Goursat that the hypothesis of continuity of $f'(z)$ in the foregoing theorems can be dispensed with. In the proof of this, which is given below, the following fact will be required. Let there be given a set of closed intervals

$$a_n \leq x \leq b_n$$

on the line such that each contains the next; that is,

$$a_m \leq b_n, \qquad a_1 \leq a_2 \leq a_3 \leq \cdots \qquad \text{and} \qquad b_1 \geq b_2 \geq b_3 \geq \cdots$$

Furthermore, let $b_n - a_n \to 0$ as $n \to \infty$; that is, let the lengths of the intervals tend to zero. Such a set, shown in Figure 4-1, is called a nested sequence of intervals. Given a nested sequence of intervals, it is a fundamental property of the real number system that there exists a number ξ contained in all the intervals and such that $a_n \to \xi$ and $b_n \to \xi$ as $n \to \infty$.

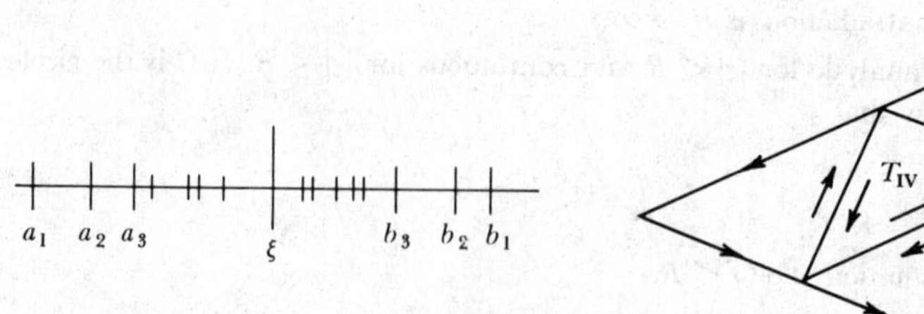

Figure 4-1 **Figure 4-2**

Second proof of Theorem 3.1. In this proof, the definition of analytic is the unrestricted one where f' is not assumed to be continuous. The same triangle T is used as in the last section. Joining the midpoints of the edges of the triangle T by line segments forms four triangles T_{I}, T_{II}, T_{III} and T_{IV} with boundaries C_{I}, C_{II}, C_{III} and C_{IV} traversed in the positive direction as shown in Figure 4-2. Then

$$\int_C f(z)\,dz = \int_{C_{\mathrm{I}}} f(z)\,dz + \cdots + \int_{C_{\mathrm{IV}}} f(z)\,dz \tag{4.1}$$

since the sum of the integrals taken in opposite directions along the edges of T_{IV} is zero. From (4.1) it follows that

$$\left|\int_C f(z)\,dz\right| \leq \left|\int_{C_{\mathrm{I}}} f(z)\,dz\right| + \cdots + \left|\int_{C_{\mathrm{IV}}} f(z)\,dz\right|.$$

Of the four nonnegative quantities appearing on the right above, one must be at least as great as each of the other three. Denote the triangle associated with that one by T_1 and its boundary by C_1. Then there follows

$$\left|\int_C f(z)\,dz\right| \leq 4\left|\int_{C_1} f(z)\,dz\right|.$$

Joining the midpoints of the edges of T_1 forms four triangles. Proceeding with T_1 just as was done with T, we find that one of these four triangles T_2 with boundary C_2 satisfies

$$\left|\int_{C_1} f(z)\,dz\right| \leq 4\left|\int_{C_2} f(z)\,dz\right|.$$

Thus

$$\left|\int_{C} f(z)\,dz\right| \leq 4^2\left|\int_{C_2} f(z)\,dz\right|.$$

Continuing in this way produces a sequence of triangles T, T_1, $T_2, \ldots$, each contained in the preceding one, such that

$$\left|\int_{C} f(z)\,dz\right| \leq 4^n\left|\int_{C_n} f(z)\,dz\right|. \tag{4.2}$$

The horizontal projections on the x axis of T, $T_1, \ldots$ form a nested sequence of intervals with each interval half as long as the interval which precedes it. The vertical projections of T, $T_1, \ldots$ on the y axis also form a nested sequence of intervals. Hence as $n \to \infty$, the respective sequences determine real numbers ξ and η such that all points of T_n converge to $\zeta = \xi + i\eta$ as $n \to \infty$. In other words, given any $\delta > 0$ there is an N depending on δ such that if $n \geq N$, then T_n lies inside the circle $|z - \zeta| < \delta$.

Since $f'(\zeta)$ exists it follows that given any $\epsilon > 0$ there exists a $\delta > 0$ such that

$$\left|\frac{f(z) - f(\zeta)}{z - \zeta} - f'(\zeta)\right| < \epsilon$$

for $0 < |z - \zeta| < \delta$. Or

$$f(z) = f(\zeta) + (z - \zeta)f'(\zeta) + (z - \zeta)\omega$$

where $\omega = \omega(z)$ satisfies $|\omega| < \epsilon$. Choosing n so large that T_n lies inside the circle $|z - \zeta| < \delta$ gives

$$\int_{C_n} f(z)\,dz = f(\zeta)\int_{C_n} dz + f'(\zeta)\int_{C_n} (z - \zeta)\,dz + \int_{C_n} (z - \zeta)\omega(z)\,dz. \tag{4.3}$$

Since C_n is closed,

$$\int_{C_n} dz = z\Big|_{C_n} = 0, \qquad \int_{C_n} z\,dz = \frac{z^2}{2}\Big|_{C_n} = 0.$$

Thus the first two integrals on the right in (4.3) are 0 and

$$\int_{C_n} f(z)\,dz = \int_{C_n} (z - \zeta)\omega(z)\,dz. \tag{4.4}$$

Let the length of C_n be denoted by s_n. Then since ζ is in T_n, $|z - \zeta| \leq s_n$ for z in T_n. Using this and $|\omega| < \epsilon$ in (4.4) gives

$$\left|\int_{C_n} f(z)\,dz\right| \leq \epsilon s_n \int_{C_n} |dz| = \epsilon s_n^2.$$

Since $s_{n+1} = s_n/2$ from the manner in which T_{n+1} is obtained from T_n, there results

$$s_n = \frac{1}{2^n}s$$

where s is the length of C. This gives

$$\left|\int_{C_n} f(z)\,dz\right| \leq \frac{\epsilon s^2}{4^n}$$

and hence, by (4.2),

$$\left|\int_C f(z)\,dz\right| \leq \epsilon s^2. \tag{4.5}$$

Since ϵ is arbitrary, the integral in (4.5) must be 0. Thus Theorem 3.1 has been established with the general definition of analyticity. It follows that the consequences of Theorem 3.1 which were deduced in the preceding section are also valid with the general definition.

Subsequent sections of this chapter have a much higher density of theorems and proofs than heretofore. We have therefore appended some examples and problems that are not specifically associated with the topic of this section, but are chosen so as to illustrate general methods of reasoning about complex integrals and star domains.

Example 4.1. If C is any contour, show that the integral

$$I(\alpha) = \int_C \frac{d\zeta}{\zeta - \alpha}$$

is a continuous function of α at any point α not on C.

Let the distance from α to the nearest point of C be $\delta > 0$, and let $|h| < \delta/2$. Then $I(\alpha + h) - I(\alpha)$ is equal to

$$\int_C \left(\frac{1}{\zeta - \alpha} - \frac{1}{\zeta - \alpha - h} \right) d\zeta = -h \int_C \frac{d\zeta}{(\zeta - \alpha)(\zeta - \alpha - h)}.$$

Since $|\zeta - \alpha| \geq \delta$ and $|\zeta - (\alpha + h)| \geq \delta/2$, the integrand does not exceed $2/\delta^2$ in magnitude, and hence

$$|I(\alpha + h) - I(\alpha)| \leq |h| \frac{2L}{\delta^2}$$

where L is the length of C. This tends to 0 as $h \to 0$.

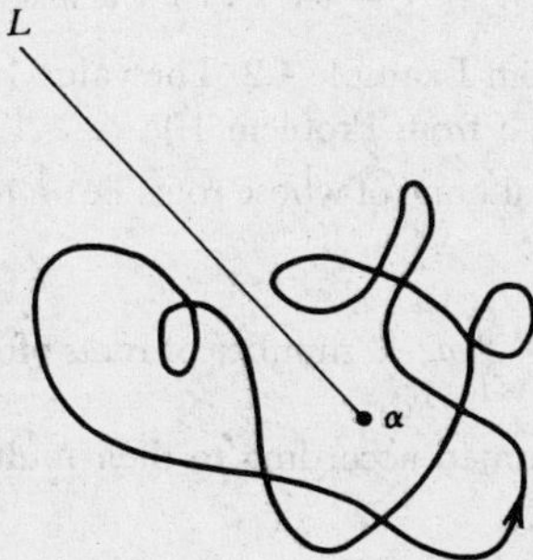

Figure 4-3

Example 4.2. As illustrated in Figure 4-3, let L be a semi-infinite line joining a given point α to ∞ and let C be a closed contour that does not meet L. Show that the integral in Example 4.1 is 0.

The domain D obtained when L is deleted from the complex plane is a star domain in which $1/(z - \alpha)$ has a continuous derivative. Hence the conclusion follows from Theorem 3.4 in the form given in Section 3—the stronger form given here is not needed.

As a consequence of Example 4.2 the integral

$$\int_{z_0}^{z} \frac{dz}{z - \alpha}$$

is independent of the path joining z_0 to z in D and can be used to define a continuous logarithm. This definition is simpler than that in Chapter 2.

Problems

1. Let C be a closed contour and let D be a domain not containing any point of C. Show that the integral $I(\alpha)$ of Example 4.1 is independent of α so long as α is in D. (By Example 2.2, $I(\alpha)/2\pi i$ is integer-valued in D and by Example 4.1 it is continuous in D.)
2. Let C be a closed contour $z = r(\theta)e^{i\theta}$, $0 \leq \theta \leq 2\pi$, where $r(\theta) > 0$. (a) Give an analytic description of the interior domain bounded by C and show that it is star with respect to the origin. (b) Deduce that

$$\frac{1}{2\pi i}\int_C \frac{d\zeta}{\zeta - \alpha} = \begin{cases} 0 \text{ for } \alpha \text{ outside } C \\ 1 \text{ for } \alpha \text{ inside } C. \end{cases}$$

(The value 0 follows from Example 4.2. The value 1 is easily verified for $\alpha = 0$ and follows for general α from Problem 1.)
3. Let $P(z)$ be a polynomial none of whose roots lie on the contour C of Problem 2. Show that

$$\frac{1}{2\pi i}\int_C \frac{P'(z)}{P(z)}\, dz = \text{number of roots of } P(z) \text{ inside } C$$

where the roots are counted according to their multiplicity. (As in Chapter 2, Example 1.3,

$$\frac{P'(z)}{P(z)} = \frac{1}{z - z_1} + \frac{1}{z - z_2} + \cdots + \frac{1}{z - z_n}.$$

In view of Problem 2, the result follows by inspection.)

*** The geometry of star domains**

4. A domain that is star with respect to every one of its points is said to be convex. Show that the half-plane $\text{Im } z > 0$ and the disk $|z| < 1$ are convex.
5. If D_1 and D_2 are domains, the set of points in D_1 and at the same time in D_2 is called the intersection of D_1 and D_2. The set of points in D_1 or D_2 or both is called the union of D_1 and D_2. Show that the intersection of two convex domains is convex but that the union of two convex domains need not be convex.
6. If domains D_1 and D_2 are both star with respect to one and the same point z_0, show that their intersection and their union are star with respect to z_0.
7. Referring to Problems 4–6 show that domains suggested by Figure 3-4(a)–(d) are all star domains.

5. Cauchy's integral formula and its consequences. A remarkable integral formula due to Cauchy shows that the values of an analytic function on the boundary of a disk determine the values at interior points. We denote the center of the disk by α, its radius by $\rho > 0$, and we assume that $f(z)$ is analytic in a larger disk centered at α (see Figure 5-1).

THEOREM 5.1. *Let $f(z)$ be analytic in the disk $|z - \alpha| < R$, let C be the circle $|z - \alpha| = \rho < R$, and let z be any point such that $|z - \alpha| < \rho$. Then*

$$f(z) = \frac{1}{2\pi i} \int_C \frac{f(\zeta)}{\zeta - z} \, d\zeta. \tag{5.1}$$

In agreement with the conventions introduced previously, C is traversed in a counterclockwise direction.

Proof. Regard z as fixed and take ζ as variable of integration. Let z be enclosed in a small circle with z as the center and consider the two closed curves C_1 and C_2 shown in Figure 5-2, where the outer circle is C and the inner one is k. Then

$$\int_{C_1} \frac{f(\zeta)}{\zeta - z} \, d\zeta = 0 \tag{5.2}$$

and similarly for C_2. Indeed C_1 is inside a star domain in the ζ plane in which $f(\zeta)/(\zeta - z)$ is an analytic function of ζ (see Figure 5-3). (That C_1 is in the domain could be shown analytically by giving equations for the boundary of the star domain and C_1.) By Theorem 3.4, (5.2) follows. Adding (5.2) to the corresponding result for C_2 and using the fact that the

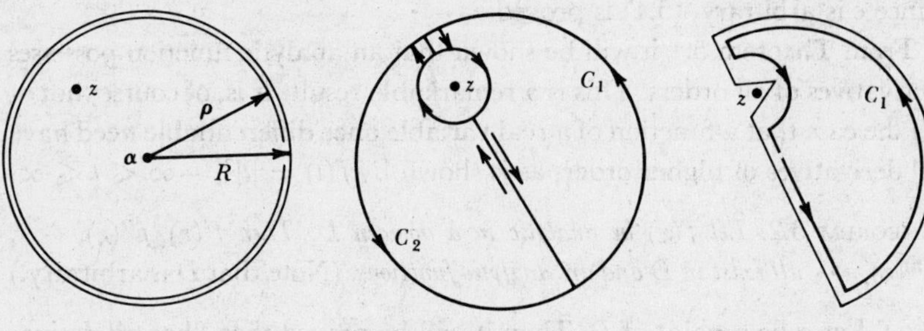

Figure 5-1 **Figure 5-2** **Figure 5-3**

integrals along the straight line segments common to C_1 and C_2 cancel when added, we get

$$\int_C \frac{f(\zeta)}{\zeta - z}\,d\zeta = \int_k \frac{f(\zeta)}{\zeta - z}\,d\zeta$$

where k is also traversed counterclockwise. Thus

$$\int_C \frac{f(\zeta)}{\zeta - z}\,d\zeta = f(z)\int_k \frac{d\zeta}{\zeta - z} + \int_k \frac{f(\zeta) - f(z)}{\zeta - z}\,d\zeta. \tag{5.3}$$

Clearly the first integral on the right is

$$\int_k \frac{d\zeta}{\zeta - z} = \log(\zeta - z)\Big|_k = i\arg(\zeta - z)\Big|_k = 2\pi i. \tag{5.4}$$

For the second integral, since $f(\zeta)$ is continuous at z, given an $\epsilon > 0$ there is a $\delta > 0$ such that

$$|f(\zeta) - f(z)| < \epsilon \qquad \text{for } |\zeta - z| < \delta.$$

If the radius of k is $b < \delta$, then

$$\left|\int_k \frac{f(\zeta) - f(z)}{\zeta - z}\,d\zeta\right| \le \frac{\epsilon}{b}\int_k |d\zeta| = 2\pi\epsilon.$$

Using this and (5.4) in (5.3) gives

$$\left|\int_C \frac{f(\zeta)}{\zeta - z}\,d\zeta - 2\pi i f(z)\right| \le 2\pi\epsilon.$$

Since ϵ is arbitrary, (5.1) is proved.

From Theorem 5.1 it will be shown that an analytic function possesses derivatives of all orders. This is a remarkable result. It is, of course, not at all the case that a function of a real variable once differentiable need have all derivatives of higher order, as is shown by $f(t) = |t|^3$, $-\infty < t < \infty$.

THEOREM 5.2. *Let $f(z)$ be analytic in a domain D. Then $f'(z), f''(z), \cdots, f^{(n)}(z), \cdots$ all exist in D and are analytic functions.* (Note that D is arbitrary.)

Proof. Let α be a point of D. Then it will be proved that f has all derivatives at α. Since α is arbitrary, this will establish the existence of all derivatives in D.

If C is a sufficiently small circle with α as center, then from (5.1) it follows that for any point z inside C

$$f(z) = \frac{1}{2\pi i} \int_C \frac{f(\zeta)}{\zeta - z}\, d\zeta. \tag{5.5}$$

If (5.5) is differentiated formally n times with respect to z, there results

$$f^{(n)}(z) = \frac{n!}{2\pi i} \int_C \frac{f(\zeta)}{(\zeta - z)^{n+1}}\, d\zeta. \tag{5.6}$$

The validity of (5.6) will now be proved.[1]

By (5.5) for z and $z + h$, where h is small enough so that $z + h$ is in the interior of the disk bounded by C,

$$\begin{aligned} \frac{f(z+h) - f(z)}{h} &= \frac{1}{2\pi i}\frac{1}{h} \int_C f(\zeta)\left[\frac{1}{\zeta - z - h} - \frac{1}{\zeta - z}\right] d\zeta \\ &= \frac{1}{2\pi i} \int_C f(\zeta) \frac{1}{(\zeta - z - h)(\zeta - z)}\, d\zeta \\ &= \frac{1}{2\pi i} \int_C \frac{f(\zeta)}{(\zeta - z)^2}\, d\zeta + J \end{aligned} \tag{5.7}$$

where

$$J = \frac{h}{2\pi i} \int_C \frac{f(\zeta)}{(\zeta - z)^2(\zeta - z - h)}\, d\zeta.$$

Since z is inside C, min $|\zeta - z|$ for ζ on C is positive. Denoting this minimum by 2δ, it follows that if $|h| < \delta$ then for ζ on C,

$$|\zeta - z - h| \geq |\zeta - z| - |h| > 2\delta - \delta = \delta.$$

There exists some constant M such that $|f(\zeta)| \leq M$ for ζ on C. (This follows, for example, from the uniform continuity of f on C.) If the radius of C is ρ, then

[1] Another proof is given in Chapter 6, Section 3.

$$|J| \leq \frac{|h|}{2\pi} \frac{M}{(4\delta^2)(\delta)} 2\pi\rho = |h| \frac{M\rho}{4\delta^3}.$$

Letting $|h| \to 0$ we get $|J| \to 0$ and hence (5.7) proves the existence of $f'(z)$. This establishes (5.6) for the case $n = 1$.

The existence of f' was of course assured by the assumption that f is analytic. However, repeating the above procedure starting with (5.6) with $n = 1$ establishes the existence of $f''(z)$ and proves (5.6) for $n = 2$. Thus f' has a derivative f'' and so is itself analytic. This proves that if $f(z)$ is analytic so is $f'(z)$. Applying this to f' instead of to f proves that f'' is analytic. More generally the analyticity of $f^{(n)}$ implies that of $f^{(n+1)}$. Hence, by induction, all derivatives exist and are analytic functions. Since $f^{(n)}(z)$ is analytic,

$$f^{(n)}(z) = \frac{1}{2\pi i} \int_C \frac{f^{(n)}(\zeta)}{\zeta - z} \, d\zeta.$$

Integrating by parts n times establishes (5.6).

If $f(z) = u(x,y) + i\, v(x,y)$, Theorem 5.2 leads to a variety of formulas expressing $f^{(n)}(z)$ in terms of partial derivatives of u and v. Differentiating in a direction parallel to the x axis or to the y axis as in Chapter 2, Section 1, we get

$$f'(z) = u_x + iv_x, \qquad f'(z) = v_y - iu_y. \tag{5.8}$$

Since Theorem 5.2 indicates that $f'(z)$ is analytic, we can now repeat the process starting with $f'(z)$ rather than $f(z)$. The real and imaginary parts of $f'(z)$ (corresponding to u and v in the original discussion) are given by (5.8). The first expression (5.8) leads to

$$f''(z) = u_{xx} + iv_{xx}, \qquad f''(z) = v_{xy} - iu_{xy}$$

and the second expression (5.8) leads to

$$f''(z) = v_{yx} - iu_{yx}, \qquad f''(z) = -u_{yy} - iv_{yy}.$$

Formulas for higher derivatives are obtained similarly.

Since $f'(z)$ is analytic it is certainly continuous and so $\operatorname{Re} f'(z)$ and $\operatorname{Im} f'(z)$ are also continuous. Referring to (5.8) we see that the four partial derivatives of first order are continuous. The eight partial derivatives of second order are likewise continuous, since each of them equals $\pm\text{Re}\, f''(z)$ or $\pm\text{Im}\, f''(z)$. The process is readily extended by induction and gives the following:

THEOREM 5.3. *All partial derivatives of u and v are continuous at any point where $f = u + iv$ is analytic.*

The following inequality is due to Cauchy:

THEOREM 5.4. *In Theorem 5.1 assume that there exists a positive constant M such that $|f(z)| \leq M$ in $|z - \alpha| < R$. Then*

$$|f^{(n)}(\alpha)| \leq \frac{Mn!}{R^n}. \tag{5.9}$$

Proof. From (5.6), with the radius of C equal to ρ and with $z = \alpha$,

$$|f^{(n)}(\alpha)| \leq \frac{n!}{2\pi} \frac{M}{\rho^{n+1}} 2\pi\rho = \frac{Mn!}{\rho^n}.$$

Since this holds for all $\rho < R$, letting $\rho \to R$ gives (5.9).

The following is always called "Liouville's theorem," though it was first established by Cauchy:

THEOREM 5.5 (*Liouville*). *If $f(z)$ is analytic and bounded in the z plane, then $f(z)$ is a constant. (In other words, an entire function cannot be bounded unless it is a constant.)*

Proof. By hypothesis there is a constant M such that $|f(z)| \leq M$. By (5.9), with $n = 1$,

$$|f'(\alpha)| \leq \frac{M}{R}$$

for arbitrary R. Hence, $f'(\alpha) = 0$. Since α is arbitrary, $f'(z) = 0$. From

$$f(z) - f(0) = \int_0^z f'(\zeta)\, d\zeta$$

it follows that $f(z) = f(0)$, which proves the theorem.

We now establish the following:

THEOREM 5.6 (*Morera*). *Let $f(z)$ be continuous in a domain D and let*

$$\int_C f(z)\,dz = 0 \tag{5.10}$$

hold for every triangle C which together with its interior is in D. Then $f(z)$ is analytic in D. (Note that D is arbitrary.)

Proof. Let α be a point in D and let the positive number ρ be so small that the disk G: $|z-\alpha| < \rho$ is in D. Then G is a star domain in which the hypothesis of Theorem 3.3 holds. Thus there is an analytic function $F(z)$ such that $F'(z) = f(z)$ in G. Since $F(z)$ is analytic in G it possesses analytic derivatives of all orders in G and, in particular, $F'(z) = f(z)$ is analytic in G. Since α was any point of D, it follows that $f(z)$ is analytic in D.

Morera's theorem is often an extremely useful device to establish the analyticity of a function, since integration is often easier to justify than differentiation.

Example 5.1. Fundamental theorem of algebra. Show that a nonconstant polynomial $P(z)$ has at least one root. It is easily checked that $|P(z)| \to \infty$ as $|z| \to \infty$ and hence $1/P(z)$ is bounded outside some circle $|z| = R$. If $P(z)$ does not vanish, then $1/P(z)$ is also bounded for $|z| \le R$ since it is continuous, and so $1/P(z)$ is bounded for all z. By Liouville's theorem $1/P(z)$ is constant, and this makes $P(z)$ constant.

Problems

1. If $f(z)$ is an entire function such that $\operatorname{Re} f(z) \le M$ for all z, where M is constant, show that $f(z)$ is constant. (Apply Liouville's theorem to $e^{f(z)}$.)
2. If $f(z)$ is analytic for $|z| \le R$, show that

$$|f^{(n)}(0)| \le \frac{n!}{2\pi R^n} \int_0^{2\pi} |f(\mathrm{Re}^{i\theta})|\, d\theta.$$

3. Let $f(z)$ be an entire function such that $|f(z)| \le M|z|^m$ for large $|z|$, where M is constant and $m \ge 0$ is an integer. Show that $f(z)$ is a polynomial of degree at most m. (Use Cauchy's inequality with $n = m + 1$ to establish

$$|f^{(m+1)}(\alpha)| \le \frac{M(m+1)!(R+|\alpha|)^m}{R^{m+1}}.$$

and thus deduce $f^{(m+1)}(\alpha) = 0$.)

4. Let the entire function $f(z)$ satisfy $|f(z)| \leq |R|\epsilon(R)$ for large $|z| = R$, where $\lim \epsilon(R) = 0$ as $R \to \infty$. Show that $f(z)$ is constant. (Proceed as in the proof of Liouville's theorem.)
5. Let u be harmonic in a domain D and let u have continuous second derivatives. Show that u has derivatives of all orders at every point of D. (In any small disk contained in D construct a harmonic conjugate v by Chapter 2, Section 5.)
6. Let $w = f(z)$ where $f(z)$ is analytic for $|z| < 1$. If $0 \leq r < 1$, show that the image in the w plane of the circle $|z| = r$ has length L such that

$$L \geq 2\pi r|f'(0)|.$$

(The length is

$$L = \int |dw| = r\int_0^{2\pi} |f'(re^{i\theta})|\, d\theta.$$

Estimate the integral by applying Problem 2 to f' instead of f, with $n = 0$ and $R = r$.)
7. Referring to Section 3, Problem 6, show that if in Theorem 5.1 the further assumption is made that $f(z)$ is continuous for $|z - \alpha| \leq R$, then (5.1) is valid for the case where C is the circle $|z - \alpha| = R$.
8. *Integral of Cauchy's type.* Let f be continuous, but not necessarily analytic, on a contour C. Show that the function

$$F(z) = \int_C \frac{f(\zeta)}{\zeta - z}\, d\zeta \qquad \text{satisfies} \qquad F'(z) = \int_C \frac{f(\zeta)}{(\zeta - z)^2}\, d\zeta$$

at any point z not on C. Hence, F is analytic off of C. The conditions on f could be weakened; for example, piecewise continuity would suffice.
9. Let $f(t)$ be continuous for $a \leq t \leq b$. (a) Show that the function

$$F(z) = \int_a^b e^{izt} f(t)\, dt \qquad \text{satisfies} \qquad F'(z) = i\int_a^b e^{izt} t f(t)\, dt$$

at every z and hence is an entire function. (b) Establish a similar result when the path of integration is an arbitrary contour C.
10. Use Morera's theorem to give an independent proof of the analyticity of the functions considered in Problems 8 and 9.

6. Taylor series.

In calculus one encounters Taylor series for e^x, $\sin x$, $\log(1 + x)$, and so forth. It will be shown here that every analytic function can be represented by a Taylor series.

Although the systematic study of series is deferred to Chapter 6, a brief review of terminology is given now. The partial sums of a power series

$$a_0 + a_1(z - \alpha) + a_2(z - \alpha)^2 + \cdots,$$

where the a_j are complex numbers, are defined by

$$s_n(z) = a_0 + a_1(z - \alpha) + \cdots + a_n(z - \alpha)^n.$$

The partial sums form a sequence of polynomials $\{s_n(z)\}, n = 0, 1, 2, \ldots.$

Let a function $f(z)$ and a sequence of functions $\{s_n(z)\}$ be given in a region G of the complex plane. Then the sequence is said to *converge uniformly* to the function in the region if given any $\epsilon > 0$ there exists an integer N, which can depend on ϵ, such that

$$|f(z) - s_n(z)| < \epsilon \tag{6.1}$$

for $n \geq N$ and z in G. The word uniform is used because N in (6.1) depends only on ϵ and not on z, so long as z is in G. The uniform convergence of a sequence $\{s_n(z)\}$ to $f(z)$ in a region G is sometimes stated in the form

$$f(z) = \lim_{n\to\infty} s_n(z) \tag{6.2}$$

uniformly in the region G. In particular, if $s_n(z)$ are the partial sums of a power series, (6.2) is sometimes written in the form

$$f(z) = \sum_0^\infty a_k(z - \alpha)^k$$

uniformly in the region G.

THEOREM 6.1. *Let $f(z)$ be analytic for $|z - \alpha| < R$ and let $\rho < R$. Then*

$$f(z) = \sum_0^\infty \frac{f^{(k)}(\alpha)}{k!}(z - \alpha)^k$$

uniformly in $|z - \alpha| \leq \rho$. That is, $f(z)$ is represented uniformly by its Taylor series in $|z - \alpha| \leq \rho$.

Before giving the proof we mention another form of Theorem 6.1 which is sometimes more convenient. If z is replaced throughout by $z + \alpha$, the equation in the conclusion is

$$f(z+\alpha) = \sum_{0}^{\infty} \frac{f^{(k)}(\alpha)}{k!} z^k \tag{6.3}$$

and the condition $|z - \alpha| \leq \rho$ in Theorem 6.1 becomes $|z| \leq \rho$ in (6.3). Thus Theorem 6.1 is equivalent to the assertion that (6.3) converges uniformly for $|z| \leq \rho$.

Proof. Define $h(z) = f(z + \alpha)$. Since $f(z + \alpha)$ is analytic for

$$|(z + \alpha) - \alpha| < R,$$

evidently $h(z)$ is analytic for $|z| < R$. Let ρ_1 be chosen so that $\rho < \rho_1 < R$ and let C_1 be the circle $|z| = \rho_1$. By the Cauchy integral formula,

$$h(z) = \frac{1}{2\pi i} \int_{C_1} h(\zeta) \frac{1}{\zeta - z} \, d\zeta. \tag{6.4}$$

If n is a positive integer and $w \neq 1$, the identity

$$\frac{1}{1-w} = 1 + w + w^2 + \cdots + w^{n-1} + \frac{w^n}{1-w}$$

is verified by multiplying both sides by $1 - w$. Setting $w = z/\zeta$ gives

$$\frac{1}{\zeta - z} = \frac{1}{\zeta} + \frac{z}{\zeta^2} + \cdots + \frac{z^{n-1}}{\zeta^n} + \frac{z^n}{\zeta^n} \frac{1}{\zeta - z}$$

after division by ζ. Hence (6.4) becomes

$$h(z) = \frac{1}{2\pi i} \int_{C_1} \frac{h(\zeta)}{\zeta} \, d\zeta + \cdots + \frac{z^{n-1}}{2\pi i} \int_{C_1} \frac{h(\zeta)}{\zeta^n} \, d\zeta + z^n h_n(z) \tag{6.5}$$

where

$$h_n(z) = \frac{1}{2\pi i} \int_{C_1} \frac{h(\zeta)}{\zeta^n} \frac{d\zeta}{\zeta - z}. \tag{6.6}$$

By (5.6) with $z = 0$,

$$\frac{1}{2\pi i} \int_{C_1} \frac{h(\zeta)}{\zeta^{k+1}} \, d\zeta = \frac{h^{(k)}(0)}{k!} = \frac{f^{(k)}(\alpha)}{k!} \tag{6.7}$$

where the last equality follows from $h(z) = f(z + \alpha)$. Thus (6.5) is

$$f(z+\alpha) = f(\alpha) + f'(\alpha)z + \cdots + \frac{f^{(n-1)}(\alpha)}{(n-1)!}z^{n-1} + z^n h_n(z). \quad (6.8)$$

According to (6.8) the term $z^n h_n(z)$ is the difference between $f(z+\alpha)$ and the first n terms of its Taylor series (6.3). To prove the theorem it suffices to show that $|z^n h_n(z)|$ is small if n is large; that is, one must show that given $\epsilon > 0$ there exists an N such that

$$|z^n h_n(z)| < \epsilon \quad (6.9)$$

for $n \geq N$ and $|z| \leq \rho$.

Let M_1 be an upper bound for $h(z)$ on $|z| = \rho_1$ and let $|z| \leq \rho$. Then from (6.6), since $|\zeta - z| \geq |\zeta| - |z| \geq \rho_1 - \rho$,

$$|z^n h_n(z)| \leq \rho^n \frac{M_1}{2\pi\rho_1{}^n(\rho_1 - \rho)} 2\pi\rho_1 = \frac{M_1\rho_1}{\rho_1 - \rho}\left(\frac{\rho}{\rho_1}\right)^n. \quad (6.10)$$

Since $\rho < \rho_1$, the right-hand side tends to 0 as $n \to \infty$, and hence, we can find an N such that (6.9) holds. This completes the proof.

The function $h_n(z)$ in (6.8) is analytic for $|z| < R$, as is shown next. Clearly (6.8) defines $h_n(z)$ and shows it to be analytic for $0 < |z| < R$. Thus it remains only to show that $h_n(z)$ is analytic at $z = 0$. But this follows from (6.6) since $h(\zeta)/\zeta^n$ is continuous on C_1 (see Section 5, Problem 8). In fact, (6.7) and (6.6) give $h_n(0) = f^{(n)}(\alpha)/n!$

As noted above, a Taylor series admits two forms. Either we can expand $f(z+\alpha)$ in powers of z or we can expand $f(z)$ in powers of $z - \alpha$. Writing $z - \alpha$ in place of z in (6.8) and defining $g_n(z) = h_n(z - \alpha)$ gives the following:

THEOREM 6.2. *Under the hypothesis of Theorem 6.1*

$$f(z) = f(\alpha) + f'(\alpha)(z-\alpha) + \cdots + \frac{f^{(n-1)}(\alpha)}{(n-1)!}(z-\alpha)^{n-1} + g_n(z)(z-\alpha)^n$$

where $g_n(z)$ is analytic for $|z - \alpha| < R$ and

$$g_n(\alpha) = \frac{f^{(n)}(\alpha)}{n!}. \quad (6.11)$$

Taylor series behave in many respects like polynomials. In illustration of this fact we establish the following:

THEOREM 6.3. *Let $f(z)$ be analytic for $|z - \alpha| < R$. Then the series obtained by differentiating the Taylor series for f termwise converges uniformly to $f'(z)$ in any disk $|z - \alpha| \leq \rho < R$. Furthermore, the differentiated series is actually the Taylor series for $f'(z)$.*

Proof. Since $f(z)$ is analytic for $|z - \alpha| < R$, the same is true of $f'(z)$ by Theorem 5.2. Thus $f'(z)$ has a Taylor series, namely,

$$f'(\alpha) + \frac{f''(\alpha)}{1!}(z - \alpha) + \cdots + \frac{f^{(n)}(\alpha)}{(n-1)!}(z - \alpha)^{n-1} + \cdots.$$

Clearly the same series is obtained by formal differentiation of the Taylor series for f, that is, by differentiating

$$f(\alpha) + f'(\alpha)(z - \alpha) + \frac{f''(\alpha)}{2!}(z - \alpha)^2 + \cdots + \frac{f^{(n)}(\alpha)}{n!}(z - \alpha)^n + \cdots$$

term by term with respect to z. Since the formally differentiated series agrees with the Taylor series, Theorem 6.1 indicates that it converges uniformly to $f'(z)$ in any disk $|z - \alpha| \leq \rho < R$.

Example 6.1. Prove that

$$\cos z = 1 - \frac{z^2}{2!} + \frac{z^4}{4!} - \frac{z^6}{6!} + \cdots \tag{6.12}$$

uniformly in $|z| \leq \rho$ for every ρ.

If $f(z) = \cos z$, the derivatives are $-\sin z$, $-\cos z$, $\sin z$, $\cos z, \ldots$ and hence

$$f(0) = 1, \qquad f'(0) = 0, \qquad f''(0) = -1, \qquad f'''(0) = 0.$$

Since $f^{(iv)}(z) = f(z)$, the sequence repeats, so that (6.12) is the Taylor series about the point $\alpha = 0$. Since $\cos z$ is analytic in $|z| < R$ for every R, Theorem 6.1 shows without further discussion that the convergence is uniform in $|z| \leq \rho$ for each fixed ρ. This is in marked contrast to the situation for real analysis, in which a separate investigation of the remainder must be made.

Example 6.2. Estimate the remainder when the series of Example 6.1 is stopped at the term $z^{2m}/(2m!)$. If $z = x + iy$,

$$2|\cos z| = |e^{iz} + e^{-iz}| \leq e^y + e^{-y} \leq e^{|y|} + 1$$

and hence $2|\cos \zeta| \leq 1 + \exp \rho_1$ when $|\zeta| = \rho_1$. This gives the value M_1 for use in (6.10) and, setting $n = 2m + 1$, we conclude that

$$\left|\cos z - \sum_{k=0}^{m} (-1)^k \frac{z^{2k}}{(2k)!}\right| \leq \frac{1 + e^{\rho_1}}{\rho_1 - \rho} \frac{\rho_1}{2} \left(\frac{\rho}{\rho_1}\right)^{2m+1}$$

for $|z| \leq \rho < \rho_1$. It is not easy to make an optimum choice of ρ_1. However if $\rho_1 = 2\rho$, the bound is $(1 + e^{2\rho})/2^{2m+1}$, which is easy to compute.

Example 6.3. Find $\lim(1 - \cos z)/z^2$ as $z \to 0$. By (6.12) and Theorem 6.2

$$\cos z = 1 - \frac{z^2}{2} + z^3 g(z)$$

where $g(z)$ is entire. Hence for $z \neq 0$

$$\frac{1 - \cos z}{z^2} = \frac{1}{2} - z g(z)$$

and the limit as $z \to 0$ is ½.

Example 6.4. If $f(z) = (1 - \cos z)/z^2$ for $z \neq 0$ and $f(0) = ½$, show that f is an entire function. By the preceding example $f(z) = ½ - zg(z)$ for $z \neq 0$ and also for $z = 0$. Since g is entire, so is f.

Example 6.5. Prove that a Taylor series can be integrated term by term. Specifically, show that if $f(z)$ is analytic for $|z| < R$ and has a Taylor series

$$f(z) = \sum_0^\infty a_n z^n,$$

then

$$\int_0^z f(\zeta)\, d\zeta = \sum_0^\infty \frac{a_n}{n + 1} z^{n+1} \tag{6.13}$$

uniformly for $|z| \le \rho < R$.

If the integral is denoted by $F(z)$, then $F(z)$ is analytic for $|z| < R$ and $F'(z) = f(z)$. Applying Theorem 6.3 to $F(z)$ shows that the Taylor series for $F(z)$ must be (6.13), and the result follows from Theorem 6.1.

Problems

1. Show that, uniformly for $|z| \le \rho < \infty$,

$$e^z = 1 + z + \frac{z^2}{2!} + \frac{z^3}{3!} + \cdots, \qquad \sin z = z - \frac{z^3}{3!} + \frac{z^5}{5!} - \cdots.$$

2. Obtain the Taylor series for $(1 - z)^{-1}$ about the point 0 and deduce

$$\frac{1}{(1-z)^2} = 1 + 2z + 3z^2 + \cdots, \quad \mathrm{Log}(1 - z) = -z - \frac{z^2}{2} - \frac{z^3}{3} - \cdots$$

uniformly in $|z| \le \rho < 1$.

3. Let β be any complex number. Show that if $(1 + z)^\beta$ is taken as $e^{\beta\,\mathrm{Log}(1+z)}$, then uniformly for $|z| \le \rho < 1$

$$(1 + z)^\beta = 1 + \frac{\beta}{1}z + \frac{\beta(\beta - 1)}{1\cdot 2}z^2 + \frac{\beta(\beta-1)(\beta-2)}{1\cdot 2\cdot 3}z^3 + \cdots.$$

This is the binomial theorem. Hint: By induction

$$\frac{d^n}{dz^n} e^{\beta\,\mathrm{Log}(1+z)} = \beta(\beta - 1)\cdots(\beta - n + 1)e^{(\beta-n)\mathrm{Log}(1+z)}.$$

4. Show that if $(\sin z)/z$, $(e^z - 1)/z$, and $(1/z)\mathrm{Log}(1 + z)$ are defined to be 1 at $z = 0$, the resulting functions are analytic near 0. Thus obtain Taylor series for

$$S(z) = \int_0^z \frac{\sin \zeta}{\zeta}\, d\zeta, \qquad E(z) = \int_0^z \frac{e^\zeta - 1}{\zeta}\, d\zeta, \qquad L(z) = \int_0^z \frac{\mathrm{Log}(1 + \zeta)}{\zeta}\, d\zeta.$$

5. After the manner of Example 6.2 estimate the remainder in the series for e^z. According to your estimate, about how many terms are needed to compute e^z within $\pm 10^{-20}$ for $|z| \le 1$? Answer: 70 terms, if $\rho_1 = 2\rho$.

6. If $P(z)$ is a polynomial of degree ≤ 3, find a_n such that

$$\frac{P(z)}{(z^2 + 1)(z - 1)(z - 2)} = a_0 + a_1 z + a_2 z^2 + \cdots$$

for $|z| < 1$. (Use partial fractions; see Chapter 2, Problem 1.1.)

7. Deduce the result of Example 6.5 as in the proof of Theorem 6.3.

8. Let C be the circle $|z - \alpha| = \rho_1$. Referring to (6.6) and to Section 2, Problem 4, show that $h_n(z)$ and $g_n(z) \equiv h_n(z - \alpha)$ of the text satisfy

$$h_n(z) = \frac{1}{2\pi i}\int_C \frac{f(\zeta)}{(\zeta-\alpha)^n}\frac{d\zeta}{\zeta-\alpha-z}, \quad g_n(z) = \frac{1}{2\pi i}\int_C \frac{f(\zeta)}{(\zeta-\alpha)^n}\frac{d\zeta}{\zeta-z}.$$

9. Use Morera's theorem to deduce that $g_n(z)$ in Problem 8 is analytic.
10. Although the dependence is not emphasized by the notation of the text, clearly $h_n(z)$ depends on α; thus $h_n(z) = h_n(z,\alpha)$ and similarly for g_n. Deduce by Morera's theorem and (6.6) that $h_n(z,\alpha)$ is an analytic function of α.
11. Prove that if a power series $\Sigma_0^\infty a_n z^n$ converges uniformly for $|z| \le \rho$ to an analytic function $f(z)$, then it is the Taylor series for f. (Let C be the circle $z = \rho e^{i\theta}$. By (6.1), if k is an integer,

$$\left|\int_C [f(z) - s_n(z)]z^{-k-1}\,dz\right| \le \frac{2\pi\epsilon}{\rho^k}.$$

For $n > k$ this becomes

$$\left|\frac{1}{2\pi i}\int_C \frac{f(z)}{z^{k+1}}\,dz - a_k\right| \le \frac{\epsilon}{\rho^k}.$$

The left side is 0 since ϵ is arbitrary, and (5.6) gives $a_k = f^{(k)}(0)/k!$)

12. Show that Taylor series can be multiplied term by term. Specifically show that if f and g are analytic in $|z| < R$ with Taylor series

$$f(z) = a_0 + a_1z + a_2z^2 + \cdots, \qquad g(z) = b_0 + b_1z + b_2z^2 + \cdots,$$

then the Taylor series for $f(z)g(z)$ is

$$a_0b_0 + (a_0b_1 + a_1b_0)z + \cdots + (a_0b_n + a_1b_{n-1} + \cdots + a_nb_0)z^n + \cdots$$

and converges uniformly to $f(z)g(z)$ in $|z| \le \rho < R$. (It suffices to show that the coefficient of z^n in the latter series is

$$\frac{1}{n!}\frac{d^n}{dz^n}[f(z)g(z)] \qquad \text{at } z = 0.$$

The second conclusion then follows from Theorem 6.1.)

7. Uniqueness and the maximum principle. The Taylor series shows that the values $f(\alpha), f'(\alpha), f''(\alpha), \ldots$ at a point α of a domain D in which f is analytic determine $f(z)$ in a disk $|z-\alpha| < R$ centered at α. Thus if $f(z)$ is known on some infinitely differentiable short arc in $|z-\alpha| < R$, then $f(z)$ is uniquely determined in $|z-\alpha| < R$, since by differentiation of $f(z)$ on the arc its derivatives at a point also are known. This is not implied by the Cauchy integral formula, since there a knowledge of $f(z)$ on the entire boundary of a disk was required to determine f inside.

We now show that $f(z)$ is uniquely determined in the domain D, and not just in a small disk, by knowledge of the derivatives $f^{(k)}(\alpha)$. Results given later show that $f(z)$ is also determined in D by knowledge of the values on any short continuous curve C; the infinite differentiability of C is not needed.

THEOREM 7.1. *Let f and g be analytic in a domain D, and let $f(z) = g(z)$ in a neighborhood of some point α of D. Then $f = g$ in D.*

The proof depends on the following fact of real analysis. Let S be a non-empty set of points on an interval $a \leq t \leq b$. A number t_0 is said to be a least upper bound for S if

(1) Every number t in S satisfies $t \leq t_0$.
(2) For each $\epsilon > 0$, some numbers t in S satisfy $t > t_0 - \epsilon$.

Then the least upper bound of S exists on $a \leq t \leq b$ and is unique.

Proof. Let $F(z) = f(z) - g(z)$ and suppose $f(\beta) \neq g(\beta)$ at some point β of D. Join α to β by a broken line in D as shown in Figure 7-1. The line is denoted by $z = \zeta(t)$, $a \leq t \leq b$, where $\zeta(t)$ is continuous and $\zeta(a) = \alpha$, $\zeta(b) = \beta$. Let S be the set of points on $a \leq t \leq b$ such that F together with all its derivatives vanishes at the point $z = \zeta(t)$, that is, t is in S if and only if

$$F^{(k)}[\zeta(t)] = 0, \qquad k = 0, 1, 2, \ldots.$$

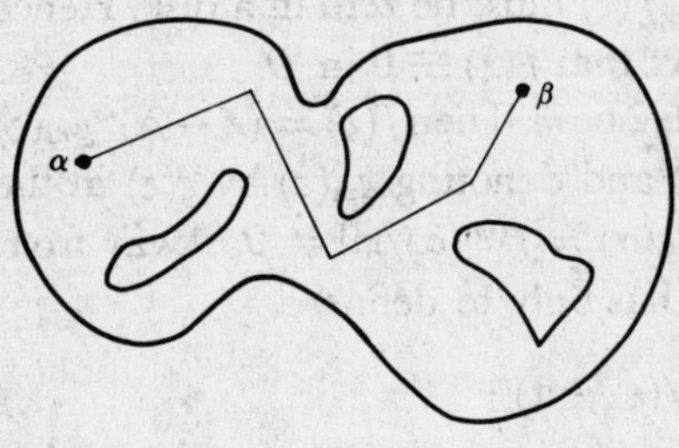

Figure 7-1

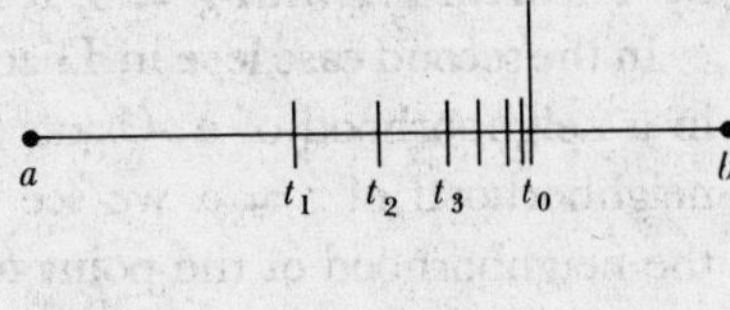

Figure 7-2

The set S is not empty since $t = a$ is in S, and therefore S has a least upper bound t_0. By definition of least upper bound we can find points t_n in S such that $t_n \to t_0$ (Figure 7-2). Since the derivatives are continuous,

$$F^{(k)}[\zeta(t_0)] = \lim_{n\to\infty} F^{(k)}[\zeta(t_n)] = 0$$

where the last equality follows from the fact that each t_n is in S. This shows that F together with all its derivatives vanishes at $z_0 = \zeta(t_0)$, hence t_0 is in S. Since $F(\beta) \neq 0$, we conclude that $t_0 < b$.

The Taylor expansion of $F(z)$ about the point $z_0 = \zeta(t_0)$ shows that $F(z)$ vanishes identically in a neighborhood of z_0. Hence all t for which $|t - t_0|$ is small are in S, and in particular, some t with $t > t_0$ is in S. This contradicts the fact that t_0 was the least upper bound. Thus there is no point β where $f(\beta) \neq g(\beta)$ and the theorem is proved. (An alternative proof can be based on the fact that the set of points in which $f^{(n)}(z) = g^{(n)}(z)$ for all $n \geq 0$ is both open and closed in D and hence is either empty or D itself.)

THEOREM 7.2. *Let $f(z)$ be analytic in a domain D. Let α be a point in D and let $f(\alpha) = 0$. Then either $f(z)$ is identically zero in D or else there exists a positive integer m and a function $g(z)$ analytic in D such that*

$$f(z) = (z - \alpha)^m g(z) \tag{7.1}$$

where $g(\alpha) \neq 0$ and $g(\alpha) = f^{(m)}(\alpha)/m!$

Proof. Either $f^{(k)}(\alpha) = 0$ for all k or else there is an integer $m > 0$ such that $f^{(k)}(\alpha) = 0$ for $k < m$ and $f^{(m)}(\alpha) \neq 0$. In the first case the Taylor series is zero and therefore by Theorem 6.1, $f(z)$ must be zero in a disk. Hence by Theorem 7.1 with $g = 0$, it follows that $f(z) \equiv 0$ in D.

In the second case let n in Theorem 6.2 be m. Then $f(z) = (z - \alpha)^m g_m(z)$ in a neighborhood of α. Using (6.11) and denoting $g_m(z)$ by $g(z)$ in the neighborhood of $z = \alpha$ we see that $g(\alpha) = f^{(m)}(\alpha)/m! \neq 0$. Away from the neighborhood of the point α one has only to define

$$g(z) = f(z)/(z - \alpha)^m$$

to complete the proof of the theorem.

An immediate consequence of Theorem 7.2 is the following:

THEOREM 7.3. *If $f(z)$ is analytic and not identically zero in a domain D, its zeros are isolated; that is, if $f(\alpha) = 0$ there is a $\delta > 0$ such that $f(z) \neq 0$ for $0 < |z - \alpha| < \delta$.*

Proof. By (7.1), $f(z) = (z - \alpha)^m g(z)$ where g is continuous and $g(\alpha) \neq 0$. Because of the continuity of g there is a neighborhood of α in which $g(z) \neq 0$, and the conclusion follows.

The next theorem implies the result stated earlier, to the effect that $f(z)$ is determined in a domain D by knowledge of its values on any short curve contained in D.

THEOREM 7.4. *Let f and g be analytic in a domain D and suppose $f(z_n) = g(z_n)$ on a sequence of distinct points z_n having a limit α in D. Then $f = g$ in D.*

Proof. By continuity the function $F(z) = f(z) - g(z)$ vanishes at α. Since $F(z_n) = 0$ and $z_n \to \alpha$, this zero of F is not isolated, and $F = 0$ by Theorem 7.3.

Although Theorem 7.4 indicates that $f(z)$ is uniquely determined by its values on the sequence $\{z_n\}$, no means of calculation is given. This contrasts with the Cauchy integral formula, which not only shows that f is determined by its values on the boundary of a disk, but provides an effective means of computation. Another difference between Theorem 7.4 and the Cauchy formula is that in Theorem 7.4 it is essential that the limit α be interior to the domain of analyticity. But if f is continuous, f need not be analytic on the boundary for validity of the Cauchy formula.[1] Hence f is determined in the circle by its boundary values even when f is not analytic on the boundary.

We now show that a similar result holds for any bounded domain D, and in fact, that $f(z)$ is determined (apart from an imaginary constant) by knowledge of its real part alone on the boundary. Proof of this hinges on the maximum principle, discussed next.

LEMMA 7.1. *Let $g(\theta)$ be real-valued and continuous on $a \leq \theta \leq b$ and $g(\theta) \leq k$ where k is a constant. If*

$$\frac{1}{b-a}\int_a^b g(\theta)\, d\theta \geq k,$$

[1] See Section 5, Problem 7.

then $g(\theta) = k$.

The lemma states that the average of a continuous function g can be as large as an upper bound of g only if g is everywhere equal to the upper bound.

Proof. If at some point ξ, $g(\xi) < k$, then by the continuity of $g(\theta)$ there are positive numbers ϵ and δ such that

$$g(\theta) \le k - \epsilon \qquad \text{for } \xi - \delta \le \theta \le \xi + \delta.$$

Hence

$$\begin{aligned}\int_a^b g(\theta)\,d\theta &= \left(\int_a^{\xi-\delta} + \int_{\xi-\delta}^{\xi+\delta} + \int_{\xi+\delta}^{b}\right) g(\theta)\,d\theta \\ &\le (\xi - \delta - a)k + 2\delta(k - \epsilon) + (b - \xi - \delta)k \\ &= (b - a)k - 2\delta\epsilon\end{aligned}$$

or

$$\frac{1}{b-a}\int_a^b g(\theta)\,d\theta \le k - \frac{2\delta\epsilon}{b-a},$$

which is contrary to the hypothesis and proves the lemma.

THEOREM 7.5 (*Maximum principle*). *Let $f(z)$ be analytic in a domain D. Then $|f(z)|$ cannot have a maximum anywhere in D unless $f(z)$ is a constant. More precisely, (7.2) implies f is a constant.*

Proof. Let α be a point in D and assume that in some ϵ neighborhood of α

$$|f(\alpha)| \ge |f(z)|, \qquad |z - \alpha| < \epsilon. \tag{7.2}$$

If $f(\alpha) = 0$, then by (7.2) $f(z) = 0$ in a neighborhood of α so $f(z) = 0$ in D, proving the theorem. Now suppose $f(\alpha) \ne 0$. By the Cauchy integral formula, with $\zeta = \alpha + re^{i\theta}$ and $z = \alpha$,

$$f(\alpha) = \frac{1}{2\pi}\int_0^{2\pi} f(\alpha + re^{i\theta})\,d\theta \tag{7.3}$$

for any $r < \epsilon$. If we define

$$F(r,\theta) = \operatorname{Re}\frac{f(\alpha + re^{i\theta})}{f(\alpha)}, \tag{7.4}$$

then F is continuous, and by (7.2) with $z = \alpha + re^{i\theta}$,

$$F(r,\theta) \leq \left|\frac{f(\alpha + re^{i\theta})}{f(\alpha)}\right| \leq 1. \tag{7.5}$$

Dividing (7.3) by $f(\alpha)$ and taking the real part gives

$$1 = \frac{1}{2\pi}\int_0^{2\pi} F(r,\theta)\, d\theta.$$

Since F is continuous, (7.5) and the above equation imply, by Lemma 7.1, that $F(r,\theta) = 1$. By (7.4)

$$\text{Re}\,\frac{f(\alpha + re^{i\theta})}{f(\alpha)} = 1$$

In conjunction with (7.5) this gives

$$\text{Im}\,\frac{f(\alpha + re^{i\theta})}{f(\alpha)} = 0$$

and so

$$\frac{f(\alpha + re^{i\theta})}{f(\alpha)} = 1$$

or, in other words, $f(z) = f(\alpha)$ in a neighborhood of α. By Theorem 7.1 with $g(z) = f(\alpha)$, it follows that $f(z) = f(\alpha)$ in D and the theorem is proved.

THEOREM 7.6 (*Maximum principle*). *Let $f(z)$ be analytic in a bounded region D and let $|f(z)|$ be continuous in the closed region $\overline{D}$. Then $|f(z)|$ assumes its maximum on the boundary of the region.*

Proof. The theorem is trivial if $f(z)$ is a constant. Suppose then $f(z)$ is not a constant. Since $|f(z)|$ is continuous, by a well-known theorem in real variables $|f(z)|$ assumes its maximum somewhere in the closed bounded region $\overline{D}$. By the preceding theorem the maximum cannot be assumed at any interior point and hence must be assumed on the boundary.

THEOREM 7.7. *Let $f(z)$ and $g(z)$ be analytic in the bounded domain D, let* Re $f(z)$ *and* Re $g(z)$ *be continuous in $\overline{D}$, and let* Re $f(z)$ = Re $g(z)$ *on the boundary. Then $f(z) - g(z) = ic$ in D, where c is a real constant.*

A proof is given in Problem 5. The result indicates that f is nearly determined by values of $\operatorname{Re} f(z)$ on the boundary but gives no clue to practical computation. The effective determination of $f(z)$ in D from knowledge of $\operatorname{Re} f(z)$ on the boundary is actually a major problem of pure and applied mathematics. Some methods of attacking this problem are given in subsequent chapters of this book.

Example 7.1. L'Hospital's rule. Let f and g be analytic in a domain D and not identically 0. If $f(\alpha) = g(\alpha) = 0$ at a point α of D, show that

$$\lim_{z\to\alpha} \frac{f(z)}{g(z)} = \lim_{z\to\alpha} \frac{f'(z)}{g'(z)},$$

the value ∞ being allowed in both cases. By Theorem 7.2

$$f(z) = (z - \alpha)^m F(z), \qquad g(z) = (z - \alpha)^n G(z)$$

where $m \geq 1$, $n \geq 1$ and $F(\alpha) \neq 0$, $G(\alpha) \neq 0$. For $z \neq \alpha$ but z near α a short calculation gives

$$\frac{f(z)}{g(z)} = (z - \alpha)^k \frac{F(z)}{G(z)}, \qquad \frac{f'(z)}{g'(z)} = (z - \alpha)^k \frac{mF(z) + (z - \alpha)F'(z)}{nG(z) + (z - \alpha)G'(z)}$$

where $k = m - n$. If $k = 0$, then $m = n$ and the common value of the limits is $F(\alpha)/G(\alpha)$. If $k > 0$, both limits are 0 and if $k < 0$, both limits are ∞.

Problems

Here and elsewhere the letter M introduced without explanation denotes a constant.

1. If two entire functions agree on a segment of the real axis, no matter how short, show that they agree for all z.
2. If f is analytic in a closed bounded region G and $f(z) \neq 0$ in G, show that $|f|$ assumes its minimum value on the boundary of G. (Consider $1/f$.)
3. Use Problem 2 to prove the fundamental theorem of algebra.
4. Let $f(z)$ and $g(z)$ be analytic in a bounded region D and continuous in $\overline{D}$. If $f(z) = g(z)$ on the boundary, show from the maximum principle that the same

equality holds throughout D. Theorem 7.7 is much stronger but is not to be used for this problem.

5. (a) If $h(z)$ is analytic in a domain D and is not constant, show that Re h attains neither a maximum nor a minimum in D. (Consider e^h.) (b) Prove Theorem 7.7 by applying part (a) to $h = f - g$.

6. By l'Hospital's rule, or otherwise, evaluate:

$$\lim_{z\to\pi} \frac{\sin z}{\pi - z}, \quad \lim_{z\to i} \frac{e^{\pi z}+1}{z^2+1}, \quad \lim_{z\to 0} \frac{(1-\cos z)^2}{\sin z + \sinh z - 2z}.$$

7. *Schwarz's lemma.* Let $f(z)$ be analytic for $|z| \le R$, let $f(0) = 0$, and let $|f(z)| \le M$ for $|z| = R$. Prove that

$$|f(z)| < \frac{M}{R}|z| \qquad \text{for } 0 < |z| < R$$

unless $f(z) = cz$ for some constant c. (Theorem 7.2 with $m = 1$ and $\alpha = 0$ gives $f(z) = zg(z)$ where $g(z)$ is analytic for $|z| \le R$. Estimate $|g(z)|$ for $|z| = R$ by the hypothesis and for $|z| < R$ by the maximum principle.)

8. Let $f(z)$ be analytic for $|z| < R$ and let $f(0) = 0$. If $|f(z)| \le M$ for $|z| < R$, show that $|f(z)| \le M|z|/R$ for $|z| < R$. (Apply Problem 7 in $|z| \le \rho < R$ and let $\rho \to R$.)

9. Let $f(z)$ be analytic for $|z| < R$ and let Re $f(z) < A$ where A is constant. If $f(0) = 0$, prove that

$$|f(z)| \le \frac{2A|z|}{R - |z|}, \qquad |z| < R.$$

(Show that $|f(z)/[2A - f(z)]| \le 1$ and use Problem 8.)

10. Let C be a contour of length L and let $f(z)$ be continuous on C. If $|f(z)| \le M$ on C, show that

$$\left|\int_C f(z)\,dz\right| < ML$$

unless $|f(z)| = M$ at every point of C.

11. Explain how the result of Problem 10 can be used instead of Lemma 7.1 to prove Theorem 7.5.

8. Isolated singularities. Let f be analytic in the punctured disk $0 < |z - \alpha| < R$. Thus $f(z)$ is single-valued and has a derivative except possibly at the point α where f need not be defined. It will be shown that there are only three possible ways in which $f(z)$ can behave in the neighborhood of $z = \alpha$. The point $z = \alpha$ is called a singular point of $f(z)$ and f is said to have an isolated singularity at $z = \alpha$. If $f(z)$ can be made an analytic function in the disk $|z - \alpha| < R$ by assigning a value to f at $z = \alpha$, then α is said to be a removable singular point for $f(z)$.

As an example of a removable singular point consider $f(z) = \sin z/z$. At $z = 0$ this quotient is undefined and so $z = 0$ is a singular point of $f(z)$. By Theorem 6.2 applied to the function $\sin z$ with $\alpha = 0$ and $n = 3$

$$\sin z = z + z^3 g(z)$$

where $g(z)$ is analytic for all z. Thus for $z \neq 0$

$$f(z) = \frac{\sin z}{z} = 1 + z^2 g(z).$$

Clearly the function on the right is analytic for all z including $z = 0$. Hence if one defines $f(0) = 1$ and $f(z) = \sin z/z$, $z \neq 0$, then $f(z)$ is analytic at $z = 0$ and so the singular point has been removed by taking $f(0) = 1$.

As another example, let $f(z)$ be analytic in a domain D, let β be a point of D, and define

$$F(z) = \frac{f(z) - f(\beta)}{z - \beta} \qquad (z \neq \beta), \qquad\qquad F(\beta) = f'(\beta). \quad (8.1)$$

Then F is not only continuous in D (as is obvious) but is analytic in D. This follows from Theorem 6.2 with $n = 1$, just as in the above discussion of $(\sin z)/z$. The function defined by the first formula (8.1) has a removable singularity at β, and the singularity is removed by the second formula (8.1). We use this function in the proof of the following:

Theorem 8.1 *(Riemann). Let $f(z)$ be analytic in the punctured disk $0 < |z - \alpha| < R$ and let*

$$\lim_{z \to \alpha} (z - \alpha) f(z) = 0.$$

Then $f(z)$ has a removable singularity at $z = \alpha$. In other words, $f(z)$ can be assigned a value $f(\alpha)$ at α so that the resulting function is analytic in $|z - \alpha| < R$.

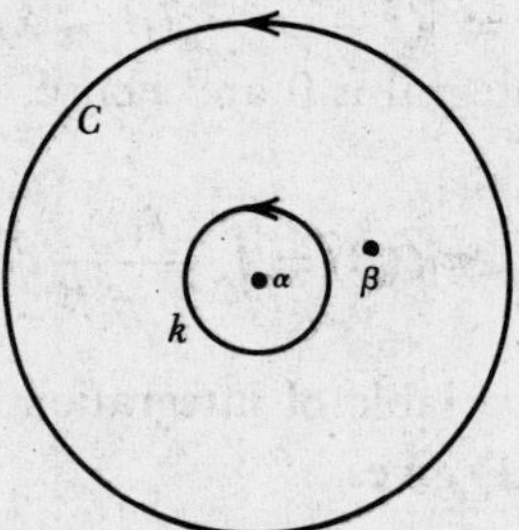

Figure 8-1

Proof. Choose a point β in the punctured disk and define F by (8.1), so that F is analytic in the punctured disk. If C and k are circles as shown in Figure 8-1, then

$$\int_C F(z)\,dz = \int_k F(z)\,dz \tag{8.2}$$

by an argument similar to that used in Section 5 (see Figure 5-3). Substitution of F from (8.1) gives

$$\int_C \frac{f(z)}{z-\beta}\,dz - \int_C \frac{f(\beta)}{z-\beta}\,dz = \int_k \frac{f(z)}{z-\beta}\,dz - \int_k \frac{f(\beta)}{z-\beta}\,dz.$$

By the Cauchy integral formula applied to the constant function $f(\beta)$ the second integral above is $2\pi i f(\beta)$. The fourth integral is 0 since the integrand is analytic in k and hence, rearranging,

$$2\pi i f(\beta) = \int_C \frac{f(z)}{z-\beta}\,dz - \int_k \frac{f(z)}{z-\beta}\,dz. \tag{8.3}$$

The hypothesis $(z-\alpha)f(\alpha) \to 0$ means that if $\epsilon > 0$ there is an $\eta > 0$ such that

$$|(z-\alpha)f(z)| \leq \epsilon \qquad \text{for } |z-\alpha| \leq \eta.$$

Hence if k is the circle $|z-\alpha| = \eta$, it follows that

$$\left|\int_k \frac{f(z)}{z-\beta}\,dz\right| = \left|\int_k \frac{(z-\alpha)f(z)}{(z-\alpha)(z-\beta)}\,dz\right| \leq \frac{2\pi\epsilon}{|\beta-\alpha|-\eta}.$$

We have used $|z - \beta| = |(z - \alpha) - (\beta - \alpha)| \geq |\beta - \alpha| - \eta$. Letting $\epsilon \to 0$ shows that the integral is 0 and hence, by (8.3),

$$2\pi i f(\beta) = \int_C \frac{f(z)}{z - \beta}\, dz. \tag{8.4}$$

We now denote the variable of integration by ζ instead of z and we denote β by z. Thus (8.4) gives

$$f(z) = \frac{1}{2\pi i} \int_C \frac{f(\zeta)}{\zeta - z}\, d\zeta \tag{8.5}$$

where $z \neq \alpha$ and $|z - \alpha| < \rho$. In the proof of Theorem 5.2 it was shown that the integral on the right of (8.5) represents an analytic function for $|z - \alpha| < \rho$. Thus if one defines $f(\alpha)$ as equal to the integral on the right of (8.5) with $z = \alpha$, then $f(z)$ becomes analytic for $|z - \alpha| < \rho$ and hence Theorem 8.1 is proved.

If $f(z)$ is continuous at α, then $\lim(z - \alpha)f(z) = 0$ and Theorem 8.1 applies. The same is true if $|f(z)|$ is bounded; thus $|f(z)| \leq M$ gives

$$|(z - \alpha)f(z)| \leq M\,|z - \alpha|$$

and since the right side tends to 0 the left side must. Still more generally, Theorem 8.1 shows that an analytic function with an isolated singularity at α which is known to satisfy (for example)

$$|f(z)| \leq \frac{M}{|z - \alpha|^{3/4}}$$

in the neighborhood of $z = \alpha$ must in fact be analytic at $z = \alpha$ if one defines $f(\alpha)$ properly.

Because an analytic function is continuous it follows that once it is known that f has a removable singularity at α we need only define $f(\alpha)$ so as to make $f(z)$ continuous at $z = \alpha$ in order to assure that $f(z)$ is analytic at α.

An isolated singularity of $f(z)$ at α is said to be a *pole* if

$$f(z) = \frac{g(z)}{(z - \alpha)^m} \tag{8.6}$$

where $m \geq 1$ is an integer, $g(z)$ is analytic in a neighborhood of α and $g(\alpha) \neq 0$. The terminology "α is a pole of $f(z)$" is also used. The integer m is called the order of the pole. If $m = 1$, the pole is said to be *simple.*

THEOREM 8.2. *A necessary and sufficient condition for an isolated singularity of $f(z)$ at $z = \alpha$ to be a pole is that*

$$\lim_{z \to \alpha} |f(z)| = \infty. \tag{8.7}$$

The meaning of (8.7) is that given any $N > 0$ there is a $\delta > 0$ that depends on N such that

$$|f(z)| > N \qquad \text{for } 0 < |z - \alpha| < \delta.$$

Proof. The necessity of (8.7) is immediate from (8.6) since $g(\alpha) \neq 0$. To prove the sufficiency let $F(z) = 1/f(z)$. Then $F(z)$ has an isolated singularity at $z = \alpha$. Clearly $F(z) \not\equiv 0$. By (8.7) $F(z) \to 0$ as $z \to \alpha$. Hence by Theorem 8.1 $F(z)$ has a removable singularity at α. Since $F(z) \to 0$ as $z \to \alpha$ it follows that if one defines $F(\alpha) = 0$, then $F(z)$ is analytic in a neighborhood of α. Since $F(\alpha) = 0$ it follows from Theorem 7.2 that there exists an integer $m > 0$ such that

$$F(z) = (z - \alpha)^m h(z) \qquad h(\alpha) \neq 0$$

in a neighborhood of α. Since $h(\alpha) \neq 0$, $g(z) = 1/h(z)$ is analytic in a neighborhood of α. Since $f = 1/F$

$$f(z) = \frac{g(z)}{(z - \alpha)^m}, \qquad g(\alpha) \neq 0, \tag{8.8}$$

which proves the theorem.

It follows that if f has a pole of order m at α, then $1/f$ has a zero of order m and conversely.

An isolated singularity of an analytic function which is neither removable nor a pole is called an *essential singularity.* The function $e^{1/z}$ has an essential singularity at $z = 0$. Indeed, since for real positive x, with $\xi = 1/x$, and $m > 0$

$$\lim_{x \to 0} x^m e^{1/x} = \lim_{\xi \to \infty} \frac{e^\xi}{\xi^m} = \infty,$$

it follows that $e^{1/z}$ is not bounded at $z = 0$ nor can $e^{1/z}$ have a pole of order m for any m at $z = 0$. Hence it has an essential singularity.

A remarkable feature of the essential singularity is that a function f in any neighborhood of an essential singularity assumes all values except possibly one. This fact is known as Picard's theorem. For example, given any complex number $\gamma \neq 0$ and any small $\delta > 0$, then it is easy to show that $e^{1/z}$ takes on the value γ an infinite number of times in $0 < |z| < \delta$. Picard's theorem will not be proved here. Instead a weaker theorem will be proved.

THEOREM 8.3 *(Casorati-Weierstrass). If $f(z)$ has an essential singularity at $z = \alpha$ and if γ is any complex number, then in every neighborhood of α, $f(z)$ comes arbitrarily close to γ. More precisely, given γ and any $\epsilon > 0$ and $\delta > 0$, there exists ζ such that $|\zeta - \alpha| < \delta$ and $|f(\zeta) - \gamma| < \epsilon$.*

Proof. Suppose the theorem is false. Then there exists γ, ϵ and δ such that for all z in the punctured disk $0 < |z - \alpha| < \delta$

$$|f(z) - \gamma| \geq \epsilon > 0.$$

Thus in the punctured disk

$$\phi(z) = \frac{1}{f(z) - \gamma}$$

is analytic and is bounded; in fact $|\phi(z)| \leq 1/\epsilon$. Since the singularity of $\phi(z)$ at α is isolated and since $|\phi(z)|$ is bounded, the singularity is removable. Hence by choosing $\phi(\alpha)$ properly, $\phi(z)$ is analytic in the disk $|z - \alpha| < \delta$. Clearly $\phi(z)$ is not identically zero. Since

$$f(z) = \gamma + \frac{1}{\phi(z)}$$

it follows that $f(z)$ is analytic at $z = \alpha$ if $\phi(\alpha) \neq 0$, or $f(z)$ has a pole at α if $\phi(\alpha) = 0$. In either case this contradicts the hypothesis that α is an essential singularity of f and proves the theorem.

It will be recalled that the behavior of a function $f(z)$ at $z = \infty$ is defined by considering the behavior of $f(1/\zeta)$ at the point $\zeta = 0$. For example, $f(z)$ is continuous at $z = \infty$ if $f(1/\zeta)$ is continuous at $\zeta = 0$, or equivalently if $|f(z) - f(\infty)| \to 0$ as $|z| \to \infty$. Let $f(z)$ be a (single-valued) analytic function for $R < |z| < \infty$. Then by using $z = 1/\zeta$ and considering $\zeta = 0$, it follows that the point $z = \infty$ is an isolated singular point. This point may be (1) a removable singularity, (2) a pole, or (3) an essential singularity of f, as described below.

(1) If $|f(z)|$ is bounded as $|z| \to \infty$, or indeed if $|f(z)|/|z| \to 0$ as $|z| \to \infty$, then ∞ is a removable singular point. Thus it is possible to define $f(\infty)$ so as to make $f(z)$ continuous at $z = \infty$ and then $f(z)$ becomes an analytic function at ∞. In the special case in which $|f(z)| \to 0$ as $|z| \to \infty$ we naturally define $f(\infty) = 0$. Then $f(z)$ has a zero at ∞ and, if $f(z) \not\equiv 0$ for $|z| > R$, f has the form

$$f(z) = \frac{h(z)}{z^m}$$

for $|z| > R$ where $m \geq 1$ is an integer, $h(z)$ is analytic for $|z| > R$ (including $z = \infty$) and $h(\infty) \neq 0$. The integer m is called the order of the zero.

(2) If $|f(z)| \to \infty$ as $|z| \to \infty$, then f has a pole at ∞ and is of the form

$$f(z) = z^m h(z)$$

for $|z| > R$ where $m \geq 1$ is an integer, $h(z)$ is analytic for $|z| > R$ (including $z = \infty$) and $h(\infty) \neq 0$. The integer m is called the order of the pole.

(3) If $z = \infty$ is an isolated singular point of f and is not a removable singularity or a pole, then it is an essential singularity. In this case given any complex number γ, any small $\epsilon > 0$ and large N, there exists ζ with $|\zeta| > N$ such that

$$|f(\zeta) - \gamma| < \epsilon.$$

Thus $f(z)$ takes values arbitrarily near any given complex number in every neighborhood of $z = \infty$.

Example 8.1. Discuss the singularities in the extended plane of

$$f(z) = \frac{(z^2 - 1)(z - 2)^3}{(\sin \pi z)^3}.$$

Evidently $f(z)$ is analytic for $|z| < \infty$ except at the points $z = 0, \pm 1,$

$\pm 2, \ldots$ where the denominator vanishes. The zeros of $\sin \pi z$ at these points have order 1, since the derivative $\pi \cos \pi z$ does not vanish there, and so the zeros of $(\sin \pi z)^3$ have order 3. Therefore $f(z)$ has a pole of order 3 at all integral values of z except at $z = -1$, 1 and 2. Since $z^2 - 1 = (z+1)(z-1)$ has zeros of order 1 at -1 and 1, $f(z)$ has a pole of order 2 at -1 and 1. At the point $z = 2$ l'Hospital's rule gives

$$\lim_{z\to 2} \frac{z-2}{\sin \pi z} = \lim_{z\to 2} \frac{1}{\pi \cos \pi z} = \frac{1}{\pi}.$$

Hence $f(z) \to 3/\pi^3$ as $z \to 2$ and the singularity at $z = 2$ is removable.

It remains to consider the point $z = \infty$. Since $f(1/\zeta)$ has a pole at $\zeta = 1/n$ whenever $n > 2$ is an integer, the singularity of $f(1/\zeta)$ at $\zeta = 0$ is not isolated. This shows that the singularity of $f(z)$ at ∞ is not isolated either.

Example 8.2. Let $f(z)$ be analytic for $0 < |z - \alpha| < R$ and have a pole of order m at α. Let $g(z) = (z-\alpha)^m f(z)$. Prove that

$$f(z) = \frac{a_{-m}}{(z-\alpha)^m} + \frac{a_{-m+1}}{(z-\alpha)^{m-1}} + \cdots + \frac{a_{-1}}{(z-\alpha)} + h(z)$$

for $0 < |z - \alpha| < R$, where $h(z)$ is analytic for $|z| < R$ and where

$$a_j = \frac{g^{(m+j)}(\alpha)}{(m+j)!}. \tag{8.9}$$

The series involving a_j with $j < 0$ is called the *principal part* or *singular part* of $f(z)$ at α.

For proof let the singularity of $g(z)$ at α be removed. By Theorem 6.2

$$g(z) = g(\alpha) + g'(\alpha)(z-\alpha) + \cdots + \frac{g^{(m-1)}(\alpha)}{(m-1)!}(z-\alpha)^{m-1} + h(z)(z-\alpha)^m$$

where $h(z)$ has the desired properties. Division by $(z-\alpha)^m$ shows that the principal part of $f(z)$ at α is

$$\frac{g(\alpha)}{0!(z-\alpha)^m} + \frac{g'(\alpha)}{1!(z-\alpha)^{m-1}} + \cdots + \frac{g^{(m-1)}(\alpha)}{(m-1)!(z-\alpha)} \tag{8.10}$$

and this agrees with (8.9)

If we use the full Taylor series for g the conclusion is that

$$f(z) = \sum_{-m}^{\infty} a_j (z - \alpha)^j$$

uniformly in $\eta \leq |z - \alpha| \leq \rho$ for any $\eta > 0$ and $\rho < R$. The coefficients a_j are given by (8.9). However, as was shown in Section 6, they are also given by

$$a_j = \frac{1}{2\pi i} \int_C \frac{g(\zeta)}{(\zeta - \alpha)^{m+j+1}}\, d\zeta = \frac{1}{2\pi i} \int_C \frac{f(\zeta)}{(\zeta - \alpha)^{j+1}}\, d\zeta$$

where C is any circle $|z - \alpha| = c$ with $0 < c < R$. In the following section the same formula is obtained when $f(z)$ has an essential singularity at α.

Problems

1. Classify the singular points in the extended plane:

$$e^z, \quad \frac{\cos z}{z}, \quad \frac{e^z - 1}{z(z-1)}, \quad \frac{z^2 - 1}{z^2 + 1}, \quad \frac{z^5}{z^3 + z}, \quad e^{\cosh z}, \quad \frac{z(z - \pi)^2}{(\sin z)^2}.$$

2. Consider the singularities of $1/f(z)$ in Example 8.1.
3. Prove that a function analytic in the extended plane is a constant.
4. Show that the principal parts of $8z^3(z + 1)^{-1}(z - 1)^{-2}$ at -1 and 1 are, respectively,

$$\frac{-2}{z + 1} \quad \text{and} \quad \frac{4}{(z - 1)^2} + \frac{10}{z - 1}.$$

(Use (8.10) with $g(z) = z^3(z - 1)^{-2}$ and with $g(z) = z^3(z + 1)^{-1}$.)

5. If $f(z)$ has a pole at ∞ its principal part is obtained by putting $\zeta = 1/z$ in the principal part of $f(1/\zeta)$ at $\zeta = 0$. Show that $(z^2 + 1)^2/(z^2 - z)$ has principal part $z^2 + z$ at ∞, and verify by long division.
6. Find constants a_n, b_n and c_n such that

$$\frac{e^z}{z^5} = \sum_{-5}^{\infty} a_n z^n, \qquad \frac{\sin z}{(z - 2\pi)^2} = \sum_{-1}^{\infty} b_n (z - 2\pi)^n, \qquad \frac{z^6}{(1 - z)^3} = \sum_{-3}^{3} c_n (z - 1)^n$$

for $z \neq 0$, $z \neq 2\pi$ and $z \neq 1$, respectively. What are the principal parts of these functions at these points?

7. Verify directly that e^z assumes all values except 0, and that $\sin z$ assumes all values, in every neighborhood of ∞.
8. Prove that if $f(z)$ is analytic at ∞, then $f(z)$ can be represented for large $|z|$ uniformly by a series of form

$$f(z) = \sum_0^\infty \frac{b_n}{z^n}.$$

(Let $\zeta = 1/z$, $g(\zeta) = f(1/\zeta)$. It will be found that $b_n = g^{(n)}(0)/n!$.)
9. Expand $(z-1)/(z+1)$ in powers of $1/z$.

Partial fraction expansion

10. Let $f(z)$ be a rational function, that is, $f = P/Q$ where P and Q are polynomials. Let the poles of f in the finite plane be at $\alpha_1, \alpha_2, \ldots, \alpha_k$ and the singular parts of f at these poles be, respectively, $f_1(z), f_2(z), \ldots, f_k(z)$ where

$$f_j(z) = \frac{A_{j,1}}{(z-\alpha_j)} + \frac{A_{j,2}}{(z-\alpha_j)^2} + \cdots + \frac{A_{j,m}}{(z-\alpha_j)^m} \qquad (m = m_j).$$

Also let the singular part of f at ∞ be a polynomial $f_{k+1}(z)$. Prove that

$$f(z) = f_1(z) + f_2(z) + \cdots + f_{k+1}(z) + C$$

where C is constant. This is the *partial fraction expansion theorem.* The constant C is determined by assigning any convenient value to z; for example, $z = 0$ or $z = \infty$. (Show that $g = f - f_1 - f_2 - \cdots - f_{k+1}$ is a bounded entire function.)
11. Referring to Problems 4, 5 and 10, expand in partial fractions:

$$\frac{8z^3}{(z+1)(z-1)^2}, \qquad \frac{(z^2+1)^2}{z(z-1)}, \qquad \frac{(z^2+1)^4}{z^3}, \qquad \frac{z^3-z}{z^2+4}.$$

12. Find the singular parts at all poles in the extended plane and give the partial fraction expansion:

$$\frac{12}{z^2(z^2+4)}, \qquad \frac{24z^6}{(z-1)^2(z-2)}, \qquad \frac{z^4+1}{z(z^2+1)^2}, \qquad \frac{60z^9+60}{(z-1)^3(z^2+4)^2}.$$

9. Laurent series. In the neighborhood of an isolated singular point, $z = \alpha$, of a function f, the function can be represented as the sum of a function f_1 which is analytic at $z = \alpha$ and a function f_2 which has $z = \alpha$ as its sole singular point in the extended plane. (This has been done in Example 8.2 for the case where $z = \alpha$ is a pole.) The function f_1 has a

Taylor series expansion valid in a circle with α as center and f_2 has a series expansion in powers of $1/(z-\alpha)$ valid except at $z=\alpha$. The sum of the two series is called the Laurent series for f.

These results will now be proved. As already indicated, the case where $z=\alpha$ is a pole has been dealt with in Example 8.2. Hence the present discussion is needed only for the case of an essential singularity.

THEOREM 9.1. *Let $f(z)$ be analytic in the punctured disk $0<|z-\alpha|<R$ with an isolated singularity at α. Then $f(z)$ can be written in one and only one way in the form*

$$f(z)=f_1(z)+f_2(z) \tag{9.1}$$

where $f_1(z)$ is analytic in the disk $|z-\alpha|<R$, $f_2(z)$ is analytic for all $z\neq\alpha$ including $z=\infty$, and $f_2(\infty)=0$.

Proof. Recalling (8.3) and replacing z by ζ and β by z gives

$$f(z)=\frac{1}{2\pi i}\int_C\frac{f(\zeta)}{\zeta-z}\,d\zeta-\frac{1}{2\pi i}\int_k\frac{f(\zeta)}{\zeta-z}\,d\zeta \tag{9.2}$$

where C is the circle $|\zeta-\alpha|=\rho<R$ and k is the circle $|\zeta-\alpha|=\eta>0$. Here η is small enough and ρ large enough so that z lies outside k and inside C, as shown in Figure 9-1. Let

$$f_1(z)=\frac{1}{2\pi i}\int_C\frac{f(\zeta)}{\zeta-z}\,d\zeta \tag{9.3}$$

for $|z-\alpha|<\rho$. For $|z-\alpha|>\eta$, let

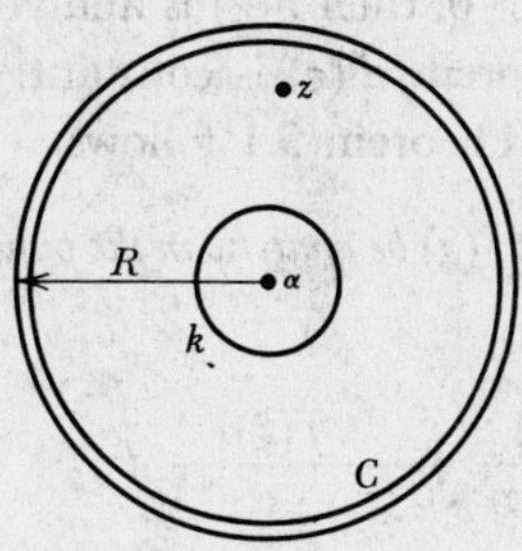

Figure 9-1

$$f_2(z) = -\frac{1}{2\pi i}\int_k \frac{f(\zeta)}{\zeta - z}\,d\zeta. \tag{9.4}$$

Then (9.1) holds. That $f_1(z)$ is analytic for $|z - \alpha| < \rho$ and $f_2(z)$ for $|z - \alpha| > \eta$ follows from the argument used in proving Theorem 5.2.

It is necessary to show that $f_1(z)$ does not change when ρ is increased and similarly that $f_2(z)$ does not change when η is decreased. Holding η fixed, it follows from (9.1), since $f(z)$ and $f_2(z)$ do not depend on ρ, that $f_1(z)$ cannot change as ρ increases so long as $|z - \alpha| < \rho < R$. Similarly, (9.1) shows that $f_2(z)$ does not depend on η so long as $0 < \eta < |z - \alpha|$. Hence $f_1(z)$ is defined by (9.3) for $|z - \alpha| < R$ simply by choosing ρ so that $|z - \alpha| < \rho < R$ and $f_1(z)$ is analytic for $|z - \alpha| < R$.

Similarly, $f_2(z)$ is defined by (9.4) as an analytic function for $|z - \alpha| > 0$ by choosing η so that $|z - \alpha| > \eta > 0$. Since the right side of (9.4) tends to zero as $|z| \to \infty$, it follows that $z = \infty$ is a removable singularity of $f_2(z)$ and if $f_2(\infty) = 0$ then $f_2(z)$ is analytic at $z = \infty$. Hence $f_2(z)$ is analytic for all $z \neq \alpha$ including $z = \infty$.

To show that f_1 and f_2 are uniquely determined let $g_1(z) + g_2(z)$ be another decomposition like that in Theorem 9.1, so that $g_1(z)$ is analytic for $|z - \alpha| < R$, $g_2(z)$ is analytic for $|z - \alpha| > 0$, and $g_2(\infty) = 0$. Then

$$f_1(z) + f_2(z) = g_1(z) + g_2(z), \qquad 0 < |z - \alpha| < R,$$

both sides being equal to f, and thus

$$f_1(z) - g_1(z) = g_2(z) - f_2(z), \qquad 0 < |z - \alpha| < R.$$

If we define $F(z)$ by the left side of this equation for $|z - \alpha| < R$ and by the right side for $|z - \alpha| > 0$, then $F(z)$ is analytic for $0 \leq |z - \alpha| \leq \infty$ and so, by Liouville's theorem, $F(z)$ is constant. The constant is 0 since $g_2(\infty) = f_2(\infty) = 0$ and Theorem 9.1 follows.

THEOREM 9.2 (*Laurent*). *Let $f(z)$ be analytic in the punctured disk $0<|z-\alpha|<R$. Let*

$$a_j = \frac{1}{2\pi i}\int_C \frac{f(z)}{(z - \alpha)^{j+1}}\,dz \qquad -\infty < j < \infty \tag{9.5}$$

where C is the circle $|z - \alpha| = c$ and c is any constant satisfying $0 < c < R$.

Then the Laurent series

$$f(z) = \sum_{-\infty}^{\infty} a_j(z - \alpha)^j \tag{9.6}$$

converges uniformly to $f(z)$ in any closed annulus contained in the punctured disk $0 < |z - \alpha| < R$.

The assertion of uniformity means that given any η and ρ such that $0 < \eta < \rho < R$ and given any $\epsilon > 0$, there exists an N (depending on ϵ, η and ρ, but not on z) such that

$$\left| f(z) - \sum_{-m}^{n} a_j(z - \alpha)^j \right| < \epsilon \tag{9.7}$$

for $n \geq N$, $m \geq N$ and for $\eta \leq |z - \alpha| \leq \rho$.

Before giving the proof we mention that Laurent series are often obtained by judicious use of Taylor series, bypassing the integral formula (9.5). For example, the Laurent series

$$\cos\frac{1}{z} = 1 - \frac{1}{2!z^2} + \frac{1}{4!z^4} - \frac{1}{6!z^6} + \cdots, \qquad 0 < |z|,$$

follows by writing $w = 1/z$ in the Taylor series for $\cos w$. Indirect modes of calculation are justified by the following uniqueness theorem, which asserts that a uniformly convergent expansion of form (9.6) must coincide with the Laurent expansion no matter how it was obtained.

THEOREM 9.3. *Suppose a series of form (9.6) converges uniformly to a continuous function $f(z)$ on the circle $|z - \alpha| = c$. Then the coefficients a_j are necessarily given by (9.5).*

Proof of Theorem 9.3. Let an integer j and a constant $\epsilon > 0$ be given, and choose m and n so that $-m \leq j \leq n$ and also so that (9.7) holds on the circle C: $|z - \alpha| = c$. In this case

$$\left| \frac{1}{2\pi i} \int_C \left[\sum_{-m}^{n} a_j(\zeta - \alpha)^j - f(\zeta) \right] \frac{d\zeta}{(\zeta - \alpha)^{j+1}} \right| \leq \frac{\epsilon}{c^j}.$$

Since

$$\frac{1}{2\pi i}\int_C (\zeta-\alpha)^k\,d\zeta = \begin{cases}1 & \text{for } k=-1\\ 0 & \text{for } k\neq -1\end{cases}$$

where k is any integer, the sum in the above expression integrates to give the single term a_j. Thus the inequality reduces to

$$\left|a_j - \frac{1}{2\pi i}\int_C \frac{f(\zeta)}{(\zeta-\alpha)^{j+1}}\,d\zeta\right| \leq \frac{\epsilon}{c^j}.$$

Since ϵ is arbitrary, and since the left side is independent of ϵ, it follows that the left side is 0. This shows that a_j is given by Laurent's formula (9.5).

Proof of Theorem 9.2. Let $0<\eta<c<\rho<R$ and let $f=f_1+f_2$ be the decomposition given by Theorem 9.1. Since $f_1(z)$ is analytic for $|z-\alpha|<R$ it has a Taylor series,

$$f_1(z) = \sum_0^\infty A_j(z-\alpha)^j$$

uniformly in $|z-\alpha|\leq\rho$. To get a similar representation for $f_2(z)$ define

$$w=\frac{1}{z-\alpha}, \qquad z=\frac{1}{w}+\alpha$$

so that $f_2(z)=f_2(1/w+\alpha)$, which is an entire function of w. Indeed, the only singularity of $f_2(z)$ in the extended plane occurs at $z=\alpha$ and hence the only singularity of $f_2(1/w+\alpha)$ occurs at $1/w+\alpha=\alpha$. This does not happen for any finite w.

By the Taylor series[1]

$$f_2\left(\frac{1}{w}+\alpha\right)=\sum_1^\infty B_j w^j$$

uniformly in $|w|\leq\dfrac{1}{\eta}$ or, equivalently,

$$f_2(z)=\sum_1^\infty B_j(z-\alpha)^{-j}$$

[1] The sum starts at 1 because setting $w=0$ gives $B_0=f_2(\infty)=0$.

uniformly in $|z-\alpha| \geq \eta$. Addition of the two series for $f_1(z)$ and $f_2(z)$ gives a series of the Laurent form which converges uniformly to $f(z)$ in

$$\eta \leq |z-\alpha| \leq \rho.$$

By Theorem 9.3, the series coincides with the Laurent series and Theorem 9.2 follows.

Corresponding to the decomposition $f = f_1 + f_2$ are two parts of the Laurent series, namely,

$$f_1(z) = \sum_0^\infty a_j(z-\alpha)^j, \qquad f_2(z) = \sum_1^\infty \frac{a_{-j}}{(z-\alpha)^j}. \tag{9.8}$$

The function $f_2(z)$ involves negative powers of $z-\alpha$ and is called the *singular part* or *principal part* of $f(z)$ at α. As shown next, there is an intimate connection between the principal part and the nature of the singularity at α.

THEOREM 9.4. *Let $f(z)$ be analytic in the punctured disk $0 < |z-\alpha| < R$, and have Laurent coefficients a_j. Then*

(1) If all $a_j = 0$ for $j < 0$, then $f(z)$ has at worst a removable singularity at α and the Laurent series is the Taylor series for $f(z)$. Conversely, if $f(z)$ has a removable singularity at $z = \alpha$, then all $a_j = 0$ for $j < 0$.

(2) If only a finite number of a_j differ from zero for $j < 0$, then $f(z)$ has a pole at α. Conversely, if $f(z)$ has a pole at α, then only a finite number of a_j for $j < 0$ can differ from zero.

(3) If an infinite number of a_j for $j < 0$ differ from zero, then $f(z)$ has an essential singularity at α. Conversely, if $f(z)$ has an essential singularity, then an infinite number of a_j for $j < 0$ differ from zero.

Proof. Case (1). If all $a_j = 0$ for $j < 0$, then by (9.8) it follows that $f_2(z) = 0$ and hence $f(z) = f_1(z)$ for $0 < |z-\alpha| < R$. Since $f_1(z)$ is analytic for $|z-\alpha| < R$, it follows that $f(z)$ has at worst a removable singularity at $z = \alpha$. Conversely, if $f(z)$ has a removable singularity at $z = \alpha$ it can be made analytic by a proper choice of $f(\alpha)$. Hence from (9.5) and the Cauchy integral theorem it follows that $a_j = 0$ for $j < 0$.

Case (2). In this case let $a_j = 0$ for $j < -m$ and $a_{-m} \neq 0$. Then

$$f(z) = f_1(z) + \sum_{1}^{m} \frac{a_{-j}}{(z-\alpha)^j}.$$

Since $f_1(z)$ is analytic for $|z - \alpha| < R$, this proves that $f(z)$ has a pole of order m at α. The converse statement follows from Example 8.2.

Case (3). Since an isolated singularity of a single-valued function is removable, a pole or an essential singularity, the result for Case (3) follows from Cases (1) and (2).

Example 9.1. Find the Laurent series for $e^{1/z}$ in $0 < |z| \leq \infty$ and thus show for $n = 0, 1, 2, 3, \ldots$ that

$$\frac{1}{\pi}\int_0^{\pi} e^{\cos\theta}\cos(\sin\theta - n\theta)\,d\theta = \frac{1}{n!}. \tag{9.9}$$

Since $1/z$ is analytic for $|z| > 0$ and has a removable singularity at ∞, the same is true of $\exp(1/z)$. Hence $\exp(1/z)$ has a Laurent series without positive exponents. The coefficient of z^j in this series is given by (9.5) as

$$a_j = \frac{1}{2\pi i}\int_C e^{1/\zeta}\zeta^{-j-1}\,d\zeta$$

where C is any circle $|z| = c > 0$. If we take $c = 1$ and $\zeta = e^{i\theta}$, then

$$e^{1/\zeta} = e^{(\cos\theta - i\sin\theta)} = e^{\cos\theta}e^{-i\sin\theta}$$

and the integral is

$$a_j = \frac{1}{2\pi}\int_{-\pi}^{\pi} e^{\cos\theta}e^{-i(\sin\theta + j\theta)}\,d\theta.$$

The imaginary part integrates to 0, since its integrand is an odd function of θ, and so a_j is given by the real part:

$$a_j = \frac{1}{2\pi}\int_{-\pi}^{\pi} e^{\cos\theta}\cos(\sin\theta + j\theta)\,d\theta. \tag{9.10}$$

Direct evaluation of this integral is by no means an easy task. However, since

$$e^w = 1 + w + \frac{w^2}{2!} + \cdots + \frac{w^j}{j!} + \cdots$$

is the Taylor series for e^w, it follows that

$$e^{1/z} = 1 + \frac{1}{z} + \frac{1}{2!z^2} + \cdots + \frac{1}{j!z^j} + \cdots$$

is the Laurent series for $e^{1/z}$. The coefficient a_j can be read off by inspection, and setting $j = -n$ gives (9.9) when we note that the integrand in (9.10) is an even function of θ.

Problems

1. Let $f(z) = A_{-m}z^{-m} + A_{-m+1}z^{-m+1} + \cdots + A_0 + A_1 z + \cdots + A_n z^n$ where m and n are nonnegative integers. Show by direct calculation that the Laurent coefficients in the expansion about $\alpha = 0$ satisfy $a_j = A_j$ for $-m \leq j \leq n$ and $a_j = 0$ otherwise.
2. Applying Problem 1 to $(z + 1/z)^m$ and to $(z - 1/z)^m$, show that

$$\int_0^{2\pi} (\cos\theta)^m e^{ij\theta}\, d\theta = \int_0^{2\pi} (\sin\theta)^m e^{ij\theta}\, d\theta = 0$$

if m and j are integers and $|m| < |j|$. (Set $z = e^{i\theta}$.)
3. What integral formula do you get in Example 9.1 when $c \neq 1$?
4. Let w be a complex number. Show that

$$\exp\left[\frac{w}{2}\left(z - \frac{1}{z}\right)\right] = \sum_{-\infty}^{\infty} J_n(w) z^n$$

for $0 < |z| < \infty$ where

$$J_n(w) = \frac{1}{2\pi}\int_0^{2\pi} e^{iw\sin\theta} e^{-in\theta}\, d\theta = \frac{1}{\pi}\int_0^{\pi} \cos[w\sin\theta - n\theta]\, d\theta.$$

The function $J_n(w)$ is called the Bessel function of order n.
5. By Section 5, Problem 9, the above expression $J_n(w)$ is an entire function of w and its derivatives can be computed by differentiating under the integral sign. Assuming this, show that $J_n(w)$ has a zero of order n at $w = 0$. (See Problem 2.)
6. Let C_0 be the circle $|z| = c$, $0 < c < R$. Show that the Laurent expansion as given by Theorem 9.2 is equivalent to

$$f(z + \alpha) = \sum_{-\infty}^{\infty} a_j z^j, \qquad a_j = \frac{1}{2\pi i}\int_C \frac{f(\alpha + \zeta)}{\zeta^{j+1}}\, d\zeta,$$

where the series converges uniformly in $\eta \le |z| \le \rho$.

7. By use of the contours shown in Figure 9-2 prove that the integrals defining the coefficients a_j in the preceding problem and in Theorem 9.2 are independent of c, $0 < c < R$.

Direct derivation of Laurent series

8. Let $f(z)$ be analytic for $0 < |z - \alpha| < R$ and let $0 < \eta_1 < \eta < \rho < \rho_1 < R$. If $h(z) = f(z + \alpha)$ show by (9.2) that

$$h(z) = \frac{1}{2\pi i} \int_C \frac{h(\zeta)}{\zeta - z}\, d\zeta - \frac{1}{2\pi i} \int_{k_1} \frac{h(\zeta)}{\zeta - z}\, d\zeta$$

where C_1 is the circle $|z| = \rho_1$ and k_1 is the circle $|z| = \eta_1$.

9. In the first integral of Problem 8 write $1/(\zeta - z)$ as

$$\frac{1}{\zeta}\frac{1}{1 - z/\zeta} = \frac{1}{\zeta} + \frac{z}{\zeta^2} + \cdots + \frac{z^{n-1}}{\zeta^n} + \frac{z^n}{\zeta^n}\frac{1}{\zeta - z}$$

and in the second integral write $-1/(\zeta - z)$ as

$$\frac{1}{z}\frac{1}{1 - \zeta/z} = \frac{1}{z} + \frac{\zeta}{z^2} + \cdots + \frac{\zeta^{m-1}}{z^m} + \frac{\zeta^m}{z^m}\frac{1}{z - \zeta}.$$

Following the pattern of the proof of Theorem 6.2 obtain the Laurent expansion in the form given in Problem 6. (Use Problem 7 to replace the circles C_1 and k_1 by C_0 in computing the coefficients a_j.)

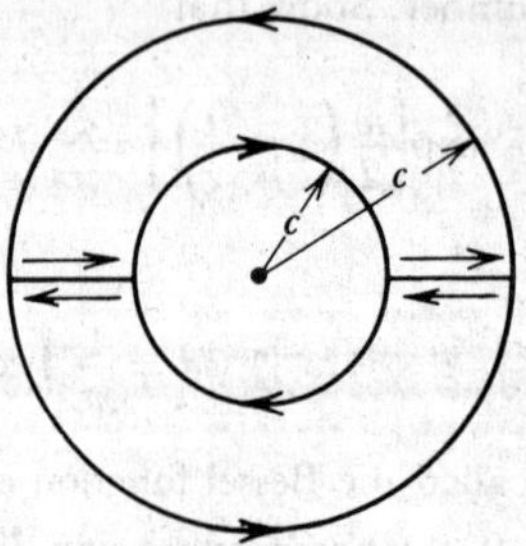

Figure 9-2

10. The definition of analyticity. The statement that $f(z)$ is analytic in a domain D can be interpreted in several ways. Some of these are:

(1) The derivative $f'(z)$ exists in D except perhaps at a finite number of points, where $f(z)$ is required only to be continuous.
(2) $f'(z)$ exists throughout D.
(3) $f'(z)$ is continuous throughout D.
(4) $f(z)$ has continuous derivatives of all orders throughout D.
(5) $f(z)$ can be represented uniformly by its Taylor series in the neighborhood of every point of D.

It follows from the results of this chapter that these five interpretations are equivalent; that is, they all lead to the same class of functions. One is justified, therefore, in using whichever formulation may be most convenient for the purposes at hand.

Additional problems on Chapter 3

1.1. Using $(2\cos\theta)e^{i\theta} = 1 + e^{2i\theta}$, show that

$$\int_0^{\pi} (\cos\theta)^3 \cos 3\theta \, d\theta = \frac{\pi}{8}.$$

1.2. The position of a moving point at time t is given by

$$z = x(t) + iy(t) = r(t)e^{i\theta(t)}$$

where r and θ are polar coordinates with $r \neq 0$. Assuming existence of the needed derivatives, show that

$$z' = (r' + ir\theta')e^{i\theta}, \qquad z'' = \left[r'' - r(\theta')^2 + i\frac{(r^2\theta')'}{r}\right]e^{i\theta}$$

where the primes denote differentiation with respect to t. The first expression gives the complex velocity and the second gives the complex acceleration.

1.3. A force field is said to be central if the complex force vector has the form $F = f(r,\theta)e^{i\theta}$ where f is real. For motion in a central force field show that Newton's law $mz'' = F$ gives $(r^2\theta')' = 0$, hence $r^2\theta' = C$ where C is constant. Deduce that

$$\int_{\theta_0}^{\theta_1} \frac{1}{2} r^2 \, d\theta = \int_{t_0}^{t_1} \frac{1}{2} C \, dt$$

and interpret geometrically. When applied to planetary motion this is Kepler's second law.

2.1. Let C be the circle $|z| = 1$ and let α be a complex number such that $|\alpha| \neq 1$. If

$$I(\alpha) = \int_C \frac{dz}{z - \alpha},$$

show analytically that $I(\alpha) + I(1/\alpha) = 2\pi i$ and interpret geometrically. (Parametrize C in the form $z = e^{i\theta}$ for $I(\alpha)$ and in the form $z = e^{-i\theta}$ for $I(1/\alpha)$. The sum leads to an easy integral.)

2.2. *A formula of Wallis.* If n is a positive integer, evaluate

$$\int_{|Z|=1} \left(z + \frac{1}{z}\right)^{2n} \frac{dz}{z}$$

by the binomial theorem and thus show that

$$\frac{1}{2\pi} \int_0^{2\pi} (2 \cos \theta)^{2n} \, d\theta = \frac{(2n)!}{n!n!}.$$

2.3. Let C be the ellipse $x = a \cos t$, $y = b \sin t$ where $a > 0$, $b > 0$ and t ranges from 0 to 2π. Evaluate the winding number of the origin with respect to C by inspection and thus show that

$$\frac{1}{2\pi} \int_0^{2\pi} \frac{dt}{a^2 \cos^2 t + b^2 \sin^2 t} = \frac{1}{ab}.$$

2.4. This problem assumes familiarity with the theory of fluid flow. As in Chapter 2, Section 6 let the complex velocity $V = p + iq$ of a flow field be given by $f'(z) = p - iq$. If C is an arc $z = \zeta(t)$ with $\zeta'(t) \neq 0$, the fluid-flow velocity, unit tangent and unit normal at a given point of C are, respectively,

$$\mathbf{v} = p\mathbf{i} + q\mathbf{j}, \qquad \mathbf{t} = \frac{dx}{ds}\mathbf{i} + \frac{dy}{ds}\mathbf{j}, \qquad \mathbf{n} = \frac{dy}{ds}\mathbf{i} - \frac{dx}{ds}\mathbf{j}.$$

Referring to Section 2, Problem 7, show that

$$\int_C f'(z)\, dz = \int_C \mathbf{v} \cdot \mathbf{t}\, ds + i \int_C \mathbf{v} \cdot \mathbf{n}\, ds.$$

The first integral on the right is the total circulation along C and the second is the total flow across C.

3.1. Let C be the circle $|\zeta| = c$ where $c > 0$ is constant, and let C^* be the circle $|\zeta| = 1/c$. If $F(z)$ is continuous on C, show that

$$\int_C F(\zeta) \frac{d\zeta}{\zeta} = \int_{C^*} F\left(\frac{1}{\zeta}\right) \frac{d\zeta}{\zeta}.$$

Deduce that the integral is 0 if $\zeta F(\zeta)$ is analytic for $c \leq |\zeta| \leq \infty$.

3.2. Let $F(z)$ be continuous on the closed contour C and suppose $F(z)$ can be uniformly approximated with arbitrary precision by a polynomial; thus for every $\epsilon > 0$ there is a polynomial $P(z)$ such that $|f(z) - P(z)| \leq \epsilon$, for z on C. Prove that the integral of f along C is 0.

3.3. It is desired to approximate $1/z$ on $|z| = 1$ by a function $f(z)$ which is analytic for $|z| \leq 1$. Show that the maximum error is at least 1, that is,

$$\max_{|z|=1} \left| \frac{1}{z} - f(z) \right| \geq 1.$$

(By Cauchy's theorem

$$\int_C \left[\frac{1}{z} - f(z) \right] dz = 2\pi i$$

where C is the circle $|z| = 1$. Estimate the left side.)

4.1. Let C be a closed contour and let α be a point not on C which can be joined to ∞ by a continuous curve that does not intersect C. Show that the winding number of C with respect to α is 0. (The winding number is an integer, is a continuous function of α, and is 0 at ∞.)

4.2. Let $P(z)$ be a polynomial of degree n with roots $z_1, z_2, \ldots, z_n$ none of which lies on a given closed contour C. Show that

$$\frac{1}{2\pi i} \int_C \frac{P'(z)}{P(z)} dz = N(z_1) + N(z_2) + \cdots + N(z_n)$$

where $N(z_k)$ is the winding number of z_k with respect to C.

5.1. If $f = u + iv$ is analytic in a region D, show that uv is harmonic in D but that u^2 need not be.

5.2. A function f has the period p if $f(z + p) = f(z)$ for all z. Prove that if an entire function has the periods 1 and i it must be constant.

5.3. If $f(z)$ is analytic for $|z| \leq 1$, show that

$$\frac{1}{\pi} \int_{|z| \leq 1} f(x + iy)\, dx\, dy = f(0).$$

(Express the integral in polar coordinates and use Cauchy's formula.)

5.4. If $f = u + iv$ is analytic and p is real, show that where $f(z) \neq 0$

$$\frac{\partial^2 |f(z)|^p}{\partial x^2} + \frac{\partial^2 |f(z)|^p}{\partial y^2} = p^2 |f(z)|^{p-2} |f'(z)|^2$$

and where $u(x,y) \neq 0$

$$\frac{\partial^2}{\partial x^2} |u(x,y)|^p + \frac{\partial^2}{\partial y^2} |u(x,y)|^p = p(p-1)|u|^{p-2}|f'(z)|^2.$$

6.1. The Bernoulli numbers B_n are given by the Taylor series

$$\frac{z}{e^z - 1} = \sum_0^\infty \frac{B_n}{n!} z^n, \qquad |z| < 2\pi.$$

Express B_n as a real integral.

6.2. Let $f(z)$ be an entire function which is real on a short segment of the real axis; for example, real for $-\epsilon < x < \epsilon$. Prove that $f(x)$ is real for all real x and that $f(z)$ assumes conjugate values for conjugate values of z.

6.3. Let $f(z)$ be an entire function which is real on the real axis and imaginary on the imaginary axis. Prove f is odd.

6.4. If $P(z)$ is a polynomial and C is the circle $|z - \alpha| = R$, show that

$$\frac{1}{2\pi i}\int_C \overline{P(z)}\,dz = R^2\overline{P'(\alpha)}.$$

(Expand $P(z)$ in powers of $z - \alpha$.)

6.5. Show that the two functions $[(z - \alpha)(z - \beta)]^{1/2}$ are analytic in the plane slit along the line segment from α to β, and that each of these functions has a simple pole at $z = \infty$. Show that for large $|z|$ one of the functions is

$$f(z) = z\exp\left[\frac{1}{2}\operatorname{Log}\left(1 - \frac{\alpha}{z}\right)\left(1 - \frac{\beta}{z}\right)\right]$$

and that the other is the negative of this one. Given the binomial theorem

$$\left(1 - \frac{\alpha}{z}\right)^{1/2} = \exp\left[\frac{1}{2}\operatorname{Log}\left(1 - \frac{\alpha}{z}\right)\right] = 1 - \frac{1}{2}\left(\frac{\alpha}{z}\right) - \frac{1}{8}\left(\frac{\alpha}{z}\right)^2 + \cdots,$$

find the first three terms of the series representing $f(z)$ for large $|z|$.

6.6. Let $f(z)$ be analytic at $z = 0$ and have a zero of order m there. Prove that in any sector $a < \arg z < b$ where $b - a > \pi/m$, Re f changes its sign in the neighborhood of $z = 0$. (By Theorem 6.2, $f(z) = z^m g(z)$, $g(0) = Ae^{i\gamma}$,

$$\operatorname{Re} f(z) = |z|^m |g(z)| \cos(m\theta + \arg g)$$

where $\arg g \to \gamma$ as $|z| \to 0$.)

7.1. If u is harmonic and twice continuously differentiable for $|z| \le 1$ show that $|\text{grad } u|$ attains its maximum over $|z| \le 1$ on the boundary. (The function $g = u_x - iu_y$ is analytic.)

7.2. Let $f(z)$ be analytic for $|z| \le 1$ and let $f(\alpha) = 0$ where $|\alpha| < 1$. If $|f(z)| \le 1$ for $|z| = 1$, prove that

$$|f(z)| \le \left|\frac{z - \alpha}{1 - \bar{\alpha}z}\right|, \qquad |z| \le 1.$$

7.3. For $j = 1, 2, 3, \ldots, n$ let a_j and b_j be positive constants and define

$$f(z) = \sum\left(\frac{a_i}{A}\right)^z\left(\frac{b_i}{B}\right)^{1-z}, \qquad A = \sum a_j, \qquad B = \sum b_j.$$

(a) Show that $f(z)$ is entire, and that on each vertical line $z = x_0 + iy$, $|f(z)|$ assumes its maximum at $y = 0$. (b) Show that the maximum of $|f(z)|$ on the disk $|z - ½| \le ½$ is taken on at $z = 0$ and 1. Deduce

$$\sum a_j^x b_j^{1-x} \le \left(\sum a_j\right)^x \left(\sum b_j\right)^{1-x}, \qquad 0 \le x \le 1.$$

This is known as Hölder's inequality.

7.4. Let $f(z)$ be analytic in a domain D containing the point $z = 0$ and let $|f(1/n)| < 1/2^n$ for $n = 1, 2, 3, \ldots$. Prove $f = 0$ in D.

8.1. Define a single-valued branch $f(z)$ of the many-valued function $(1 + z)^{1/z}$ which is analytic for $0 < |z| < 1$ and real for real z. Remove the singularity of f at 0 by suitably defining $f(0)$ and then calculate $f'(0)$.

8.2. What is the order of the pole of $(\sin z + \sinh z - 2z)^{-2}$ at $z = 0$?

8.3. Referring to Section 5, Problem 3, show that a function which has a pole at $z = \infty$ as its only singularity in the extended plane is a polynomial.

8.4. A certain function $f(z)$ has as sole singularities poles at $0, i, -i$, and ∞. Prove that

$$f(z) = \frac{P(z)}{[z(z^2+1)]^m} \qquad (z \ne 0, i, -i, \infty)$$

where $P(z)$ is a polynomial and m is a suitable integer. (If m is sufficiently large, $z^m(z+i)^m(z-i)^m f(z)$ has removable singularities at 0, i, $-i$ and a pole at ∞.)

8.5. A function $f(z)$ has a finite number of poles as its only singularities in the extended plane. Prove f is rational.

8.6. If $f(z)$ has an isolated singularity at $z = \alpha$, show that $e^{f(z)}$ cannot have a pole there. (Note that $f(z)$ is single-valued.)

8.7. If $f(z)$ has an isolated singularity at α, and $\operatorname{Re} f(z) \le -m \log |z - \alpha|$ for some constant m, show that the singularity is removable.

8.8. Let $f(z)$ be entire and let $f(1) = 2f(0)$. If $\epsilon > 0$, prove there is a point z where $|f(z)| < \epsilon$.

8.9. For $0 < |z| < 1$ let $f(z)$ be analytic and satisfy $|f(z)| \le 4|z|^{1.1}$. Prove that $|f(1/2)| \le 1$.

9.1. By comparing the Laurent expansion of $(z + 1/z)^m$ with the binomial expansion, evaluate

$$\int_0^{2\pi} (\cos\theta)^m \cos n\theta \, d\theta, \qquad m,n \text{ integers.}$$

9.2. Show that

$$\cosh\left(z + \frac{1}{z}\right) = \sum_0^\infty b_n\left(z^n + \frac{1}{z^n}\right)$$

where

$$b_n = \frac{1}{2\pi}\int_0^{2\pi} \cos n\theta \cosh(2\cos\theta)\, d\theta.$$

9.3. Let $f(z)$ be analytic for $0 < |z| < R$, let $|z| = r$, and let

$$0 < \eta \leq r \leq \rho < R.$$

If a_j are the Laurent coefficients of f for expansion about the point $\alpha = 0$, show that

$$\left| f(z) - \sum_{-m}^{n} a_j z^j \right| \leq \left(\frac{r}{\rho}\right)^n \frac{\rho}{\rho - r} M(\rho) + \left(\frac{\eta}{r}\right)^m \frac{\eta}{r - \eta} M(\eta)$$

where $M(\rho)$ and $M(\eta)$ are the mean values of $|f(z)|$ on the circles $|z| = \rho$ and $|z| = \eta$, respectively. (See Section 9, Problem 9.)

9.4. Let $f(z)$ be analytic for $|z| < R$. If $0 < r < R$ and $j \geq 1$, show that the jth Taylor coefficient a_j satisfies

$$a_j = \frac{1}{\pi r^j} \int_0^{2\pi} [\operatorname{Re} f(re^{i\theta})] e^{-ij\theta}\, d\theta.$$

(If C is the circle $|z| = r$, then

$$a_j = \frac{1}{2\pi i} \int_C \frac{f(z)}{z^{j+1}}\, dz, \qquad 0 = \frac{1}{2\pi i r^{2j}} \int_C f(z) z^{j-1}\, dz.$$

Write C as $z = re^{i\theta}$, $0 \leq \theta \leq 2\pi$, and add the conjugate of the second equation to the first.)

9.5. *A theorem of Borel.* For $|z| < 1$ let $f(z)$ be analytic and let $\operatorname{Re} f(z) \geq 0$. If $f(0) = 1$, prove that the jth Taylor coefficient satisfies $|a_j| \leq 2$. (The preceding problem and the condition $f(0) = 1$ give, respectively,

$$|a_j| \leq \frac{1}{\pi r^j} \int_0^{2\pi} |\operatorname{Re} f(re^{i\theta})|\, d\theta, \qquad 1 = \frac{1}{2\pi} \int_0^{2\pi} \operatorname{Re} f(re^{i\theta})\, d\theta$$

for $0 < r < 1$. Estimate a_j by use of $|\operatorname{Re} f| = \operatorname{Re} f$, and let $r \to 1$.)

Chapter 4

Residue theory

When Cauchy's integral theorem is extended to functions having isolated singularities, the value of the integral is not zero, in general, but each singularity contributes a term called the residue. The residue at α depends only on the coefficient of $1/(z - \alpha)$ in the Laurent expansion, not on the other terms. The reason for this is that $1/(z - \alpha)$ has a many-valued integral, $\log(z - \alpha)$, while the other powers of $z - \alpha$ have single-valued integrals and integrate to 0. Since residues can often be computed with ease, the calculus of residues is an effective method for evaluating definite integrals. It is also useful in the study of the location of the zeros and poles of a function, and it leads to some significant integral formulas pertaining to functions defined in a half-plane. On the theoretical side, the calculus of residues is connected with the problem of extending Cauchy's theorem to general regions, and brings complex analysis into contact with nontrivial theorems of plane topology.

***1. Simply connected domains.** For most purposes in pure and applied mathematics the formulation of Cauchy's theorem in the previous chapter is adequate. The restriction to star domains is not a serious impediment because, by introducing further line segments, one can often extend the results to curves which in their original form might not lie in star domains. For example, in Figure 1-1 it is seen that $C = C_1 + C_2$ in the sense that

$$\int_C f(z)\,dz = \int_{C_1} f(z)\,dz + \int_{C_2} f(z)\,dz$$

for every continuous function f. When f is analytic in the domain of Figure 1-1 the integrals over C_1 and C_2 are 0, because C_1 and C_2 lie in

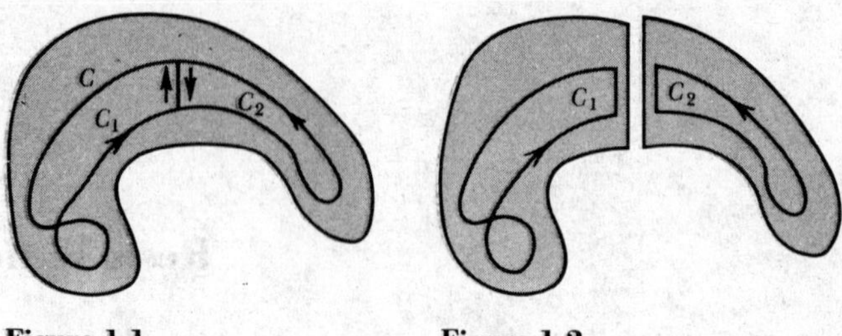

Figure 1-1 **Figure 1-2**

star domains of analyticity, as shown in Figure 1-2. This gives

$$\int_C f(z)\,dz = 0$$

even though the domain of Figure 1-1 is not a star domain. Further illustrations of this device can be found in Chapter 3, Section 5.

The procedure suggested by Figure 1-1 is sufficient for all applications intended here. However it is of considerable theoretical interest to formulate the results in less restrictive form by introducing the notion of a simply connected domain. This may be done in several ways. Here it is done by use of the Jordan curve theorem for simple closed polygons. These terms are now defined.

A broken line (also called a polygonal curve) is a curve

$$z = \zeta(t), \qquad a \leq t \leq b$$

where ζ is continuous and piecewise linear. That is, $[a,b]$ can be divided by points t_k,

$$a = t_0 < t_1 < t_2 < \cdots < t_n = b$$

such that $\zeta(t)$ is a linear function[1] on each interval

$$t_k < t < t_{k+1}.$$

We assume also that ζ is not constant on any of these intervals. The points $z_k = \zeta(t_k)$ are the vertices of the polygonal curve and the straight lines

[1] A *linear function* $\zeta(t)$ has the form $\alpha + \beta t$ where α and β are constant.

given by

$$z = \zeta(t), \qquad t_k < t < t_{k+1}$$

are the edges.

A *simple closed polygon* is a polygonal curve that is closed and does not intersect itself. As explained in Chapter 3 this means that $\zeta(a) = \zeta(b)$ and that, except in this one case,

$$\zeta(t) \neq \zeta(t^*) \qquad \text{for } t \neq t^*$$

It is an important theorem that *a simple closed polygon divides the plane into two domains, one interior to the polygon and the other exterior.* This is an instance of the Jordan curve theorem. The exterior domain contains the point at ∞ in the extended plane, and the interior domain is bounded. All points of the extended plane not on the polygon itself are either in the interior or exterior.

The Jordan curve theorem is by no means obvious for an arbitrary simple closed polygon, which can be an extremely complicated configuration. For simple configurations such as triangles, circles, etc. there is no real problem since the interior and exterior can be described by one or several inequalities. (For example, the interior of the unit circle, $|z| = 1$, is $|z| < 1$.) The Jordan curve theorem for a simple closed polygon is not hard to prove. The proof will not be given here.

A domain is called *simply connected* if the interior of every simple closed polygon in the domain is contained in the domain. An intuitive description is that a simply connected domain is free of holes or cuts in its interior. The disk $|z| < 1$ is simply connected. The annulus $1 < |z| < 2$ is not simply connected (but is doubly connected). The plane or the plane cut along a half-line from a point z_0 to $z = \infty$ is simply connected. The punctured disk $0 < |z| < 1$ is not simply connected since the point $z = 0$ is not in this punctured disk.

The importance of simply connected domains in complex analysis stems, in part, from the following theorem:

THEOREM 1.1. *Let D be a simply connected domain and let $f(z)$ be analytic in D. Then there exists a function $F(z)$ analytic in D such that $F'(z) = f(z)$ in D.*

A proof is outlined later in this section. The function $F(z)$ is of course single-valued and this is in fact the main feature of the theorem. The function $F(z)$ is called an indefinite integral of $f(z)$ and the theorem states that an analytic function in a simply connected domain has an indefinite integral in the domain. Obviously a constant can be added to F and the result will also be an indefinite integral.

Whenever f has the indefinite integral F the formula

$$\int_C f(z)\,dz = F(z)\Big|_C$$

was established in Chapter 3, Section 1 with no restrictions on the domain. The value is 0 when C is closed, and Theorem 1.1 therefore gives:

THEOREM 1.2. *Let D be simply connected and let $f(z)$ be analytic in D. Let C be a closed contour in D. Then*

$$\int_C f(z)\,dz = 0.$$

This is the Cauchy integral theorem. It generalizes Theorem 3.4 of Chapter 3, just as Theorem 1.1 generalizes Theorem 3.2 of Chapter 3.

The condition in Theorems 1.1 and 1.2 that D be simply connected is essential. Consider the function $f(z) = 1/z$. If D is the punctured plane $|z| > 0$, then $1/z$ is analytic in D but D is not simply connected, since $z = 0$ is a boundary point. Let C be the circle $z = e^{it}$, $0 \leq t \leq 2\pi$. Then

$$\int_C \frac{dz}{z} = i\int_0^{2\pi} dt = 2\pi i.$$

Thus the integral of f around the closed contour C in D is not zero. This shows that there can be no single-valued function in D with $1/z$ as its derivative.

The proof of these results requires the concept of a positively oriented polygon. As t increases, a direction is determined on the polygonal curve $z = \zeta(t)$, and a simple closed polygon traversed in a specified direction is said to be *oriented.* The polygon is positively oriented if the vector giving the direction of increasing t along an edge is such that a counterclockwise rotation of $\pi/2$ causes the vector to point into the interior of the polygon. By means of the Jordan curve theorem for polygons it can be shown that

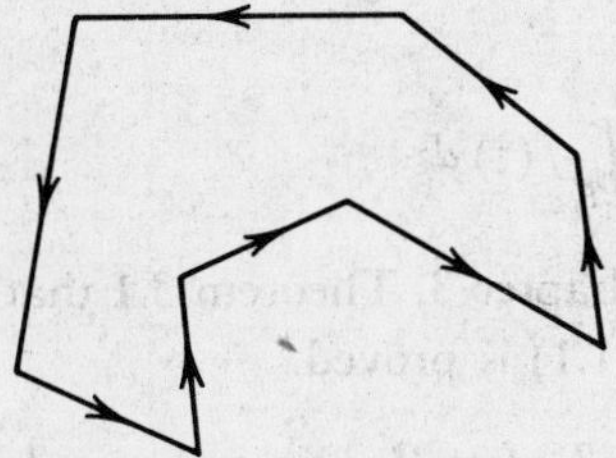
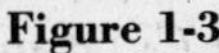

Figure 1-3

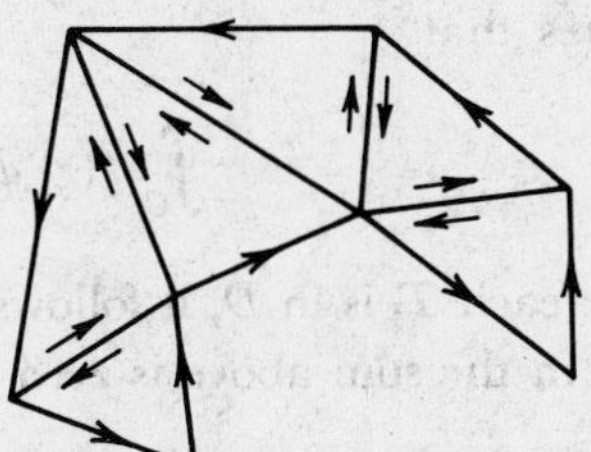

Figure 1-4

if this happens for any edge of an oriented polygon it will happen for every edge. A positively oriented polygon is illustrated in Figure 1-3.

It can be shown by induction that a simple, closed, positively oriented polygon C can be decomposed by diagonals into a set of $n - 2$ positively oriented triangles with the following properties: Each point in the interior of C lies in one of the triangles or on the edge of a triangle, and any point in the interior of a triangle lies in the interior of C. An edge of a triangle which does not lie on an edge of the polygon C is called an interior edge. If all triangles are described in the positive direction, then each interior edge is described twice, once in one direction and once in the opposite direction. Moreover, each edge of the polygon C is described once and in the positive direction of C, as shown in Figure 1-4. The whole configuration is called a *triangulation* of the polygon.

The triangulation of a simple closed polygon leads to Lemma 1.1.

LEMMA 1.1. *Let f be an analytic function in a simply connected domain D. Let C be a simple, closed, oriented polygon in D. Then*

$$\int_C f(z)\,dz = 0. \tag{1.1}$$

Proof. There is no restriction in assuming that C is positively oriented, since the opposite orientation would only change the sign of the left side of (1.1). Hence the proof for the positively oriented case implies both cases. Since D is simply connected, the interior of the polygon C is also in D. If C has n vertices, it can be triangulated by means of $n - 2$ positively oriented triangles $T_1, T_2, \ldots, T_{n-2}$. The definition of "triangulation" ensures that

$$\int_C f(z)\,dz = \sum_{j=1}^{n-2} \int_{T_j} f(z)\,dz.$$

Since each T_j is in D, it follows from Chapter 3, Theorem 3.1 that each term in the sum above is zero. Hence (1.1) is proved.

Proof of Theorem 1.1. Let z_0 be a point in D. Let C_1 be a polygonal curve from z_0 to a point z of D and let C_1 lie entirely in D. Let C_2 be another such polygonal curve from z_0 to z. Then

$$\int_{C_1} f(\zeta)\,d\zeta = \int_{C_2} f(\zeta)\,d\zeta. \tag{1.2}$$

To see this let C_1–C_2 designate the closed polygonal curve which consists of C_1 traversed from z_0 to z followed by C_2 traversed in the reverse direction, that is, from z to z_0. Then (1.2) is equivalent to

$$\int_{C_1-C_2} f(\zeta)\,d\zeta = 0. \tag{1.3}$$

It can be shown by induction that the closed (but not necessarily simple) polygonal curve C_1–C_2 consists of a number of line segments traversed in opposite directions and a number of simple, closed, oriented polygons, as suggested by Figure 1-5. The line segments are traversed once in each direction and thus the integrals over the segments add up to zero. It has been seen in (1.1) that the integral around a simple, closed, oriented polygon is zero. Hence (1.3) follows, and this implies (1.2).

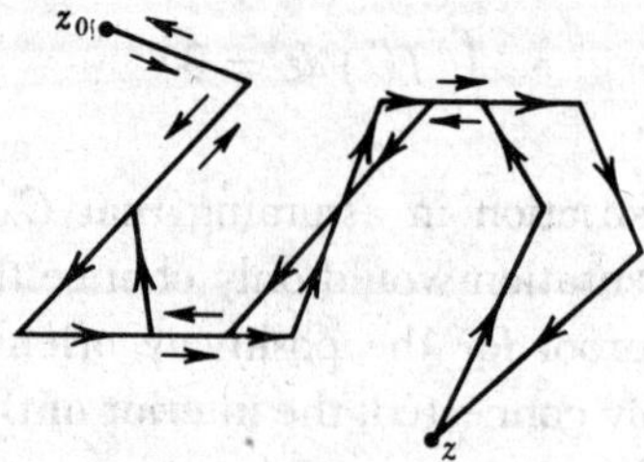

Figure 1-5

If

$$F(z) = \int_{z_0}^{z} f(\zeta)\, d\zeta$$

where the integral is taken along any polygonal curve in D from z_0 to z, then by (1.2) $F(z)$ is independent of the choice of the polygonal curve joining z_0 to z. Hence for small $|h|$

$$F(z+h) - F(z) = \int_{z}^{z+h} f(\zeta)\, d\zeta$$

where the integral is along the line from z to $z+h$. As in the proof of Theorem 3.2 in Chapter 3 it follows that $F'(z) = f(z)$, and this gives Theorem 1.1.

Example 1.1. Cauchy integral formula. Let $f(z)$ be analytic in a simply connected domain containing a closed contour C, and suppose the point a is not on C. Show that

$$\frac{1}{2\pi i}\int_C \frac{f(z)}{z-a}\, dz = Nf(a) \tag{1.4}$$

where $N = N(C,a)$ is the winding number of C with respect to a.

If a is in D, Chapter 3, Section 8 shows that the function

$$\phi(z) = \frac{f(z) - f(a)}{z-a}$$

has a removable singularity at $z = a$, and the singularity is removed by defining $\phi(a) = f'(a)$. That being done, Theorem 1.2 gives

$$\int_C \frac{f(z) - f(a)}{z-a}\, dz = 0$$

or, rearranging,

$$\int_C \frac{f(z)}{z-a}\, dz = f(a)\int_C \frac{dz}{z-a}. \tag{1.5}$$

Since the winding number is

$$N(C,a) = \frac{1}{2\pi i}\int_C \frac{dz}{z-a},$$

by definition, dividing (1.5) by $2\pi i$ gives (1.4).

If a is not in D, then $1/(z-a)$ is analytic in D, Theorem 1.2 shows that both integrals (1.5) are 0, and (1.4) follows again.

In the vast majority of applications C is a simple closed curve that winds just once around the point a. Then $N = 1$ and (1.4) becomes

$$f(a) = \frac{1}{2\pi i}\int_C \frac{f(z)}{z-a}\,dz.$$

Example 1.2. The analytic logarithm. Let $f(z)$ be analytic and free of zeros in a simply connected domain D. Prove that f has an analytic logarithm; that is, there is a function $g(z)$, analytic in D, such that $e^g = f$.

We define $g(z)$ by the formula

$$g(z) = \int_{z_0}^{z} \frac{f'(\zeta)}{f(\zeta)}\,d\zeta + \operatorname{Log} f(z_0)$$

where z_0 is a fixed point of D, z is an arbitrary point of D, and the path is any path lying in D; for instance, a polygonal path from z_0 to z. By Theorem 1.2 the integral is independent of the path. Since

$$\frac{d}{dz}e^{-g}f = e^{-g}f' + fe^{-g}(-g') = 0,$$

$e^{-g}f$ is constant. Setting $z = z_0$ shows that the constant is 1, and hence $f = e^g$.

Problems

1. With D and f as in Example 1.2 show that there are two analytic branches of $\sqrt{f}$ in D, namely $\exp(1/2)g(z)$ and $-\exp(1/2)g(z)$. Higher roots are defined similarly.

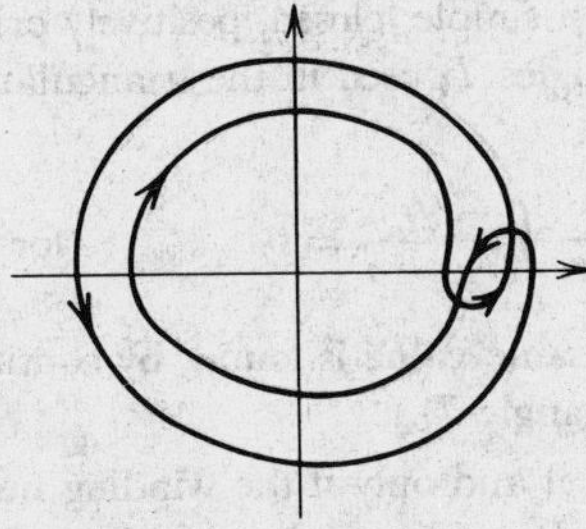

Figure 1-6 Figure 1-7

2. Let C be the overlapping contour shown in Figure 1-6. Introduce a cut as shown in Figure 1-7 and prove

$$\int_C \sqrt{z}\, dz = \int_{C_1} \sqrt{z}\, dz + \int_{C_2} \sqrt{z}\, dz.$$

Since C_1 and C_2 are closed contours and each is in a domain in which $\sqrt{z}$ is analytic, conclude that

$$\int_C \sqrt{z}\, dz = 0.$$

Note that Theorem 1.2 is not directly applicable to C and $\sqrt{z}$.

3. If C is a positively oriented triangle, show by reference to Figure 1-8 that[2]

$$\frac{1}{2\pi i}\int_C \frac{dz}{z-a} = \begin{cases} 0 & \text{for } a \text{ outside } C \\ 1 & \text{for } a \text{ inside } C. \end{cases}$$

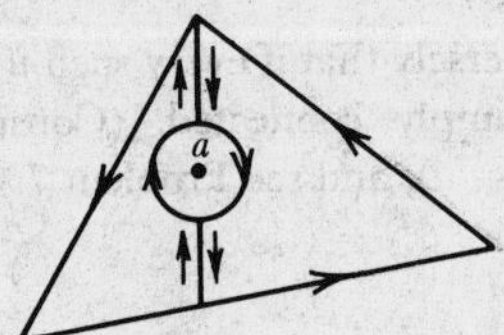

Figure 1-8

[2] Problems 3–6 are closely related to the examples and problems of Chapter 3, Section 4.

4. Obtain the result of Problem 3 when C is a simple, closed, positively oriented polygon. (If a is interior to one of the triangles T_i used in the triangulation of C then Problem 3 gives

$$\frac{1}{2\pi i}\int_{T_i}\frac{dz}{z-a}=1, \qquad \frac{1}{2\pi i}\int_{T_j}\frac{dz}{z-a}=0 \qquad \text{for } j\neq i.$$

The value 1 is obtained by addition. The same value is found, by continuity, when a is in C but on the boundary of a triangle T_i.)

5. Prove: The domain D is simply connected if and only if the winding number satisfies $N(C,a)=0$ for every point a not in D and every closed contour C lying in D. (If D is not simply connected, there is a simple, closed, positively oriented polygon C and a point such that a is in C but not in D. Problem 4 gives $N(C,a)=1$.)

6. Deduce from Problem 5 that every star domain is simply connected.

7. Let D be a domain such that every function f, which is analytic in D and has no zeros in D, has an analytic logarithm in D. Prove D must be simply connected. (If a is not in D then the equation

$$z-a=e^{g(z)}$$

holds for some g, the analytic logarithm of $z-a$. But this gives

$$\int_C\frac{dz}{z-a}=\int_C e^{-g(z)}\,dz=0$$

for every closed contour C in D, and the conclusion follows from Problem 5.)

8. Let $u(x,y)$ be harmonic and twice continuously differentiable in the simply connected domain D. Show that u admits a harmonic conjugate, v, such that $u+iv=f$ is analytic in D. (As in Chapter 2, Section 5, the function $g=u_x-iu_y$ is analytic in D. By integrating g over any convenient path in D construct f such that $f'=g$.)

9. In Problem 8 show conversely that if every such u admits a harmonic conjugate v, then D must be simply connected. (Consider the harmonic function $\operatorname{Log}|z-a|=\operatorname{Re}\log(z-a)$, and see Problem 7.)

2. The residue theorem. For the reader's convenience the gist of the foregoing discussion is summarized here. A simply connected domain is free of holes, as illustrated in Figure 2-1. A typical example is the region bounded by a closed polygonal contour without self-intersections. Such a contour is called a simple closed polygon and the region which it bounds

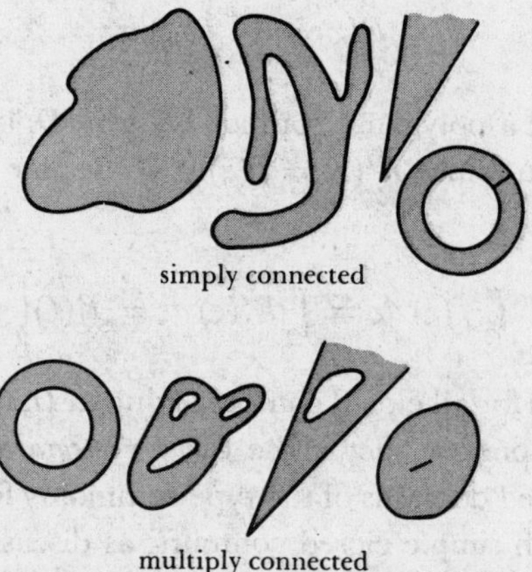

Figure 2-1

is called the interior of the polygon, or a polygonal region. If the boundary of a polygonal region lies in a simply connected domain D then its interior also lies in D; and conversely, if this happens for every polygonal region whose boundary is in D, then D is simply connected. By triangulation (see Figure 2-2) it is found that

$$\int_C f(z)\,dz = 0 \tag{2.1}$$

if C is any simple closed polygon lying in a simply connected region D in which f is analytic. This in turn shows that the integral

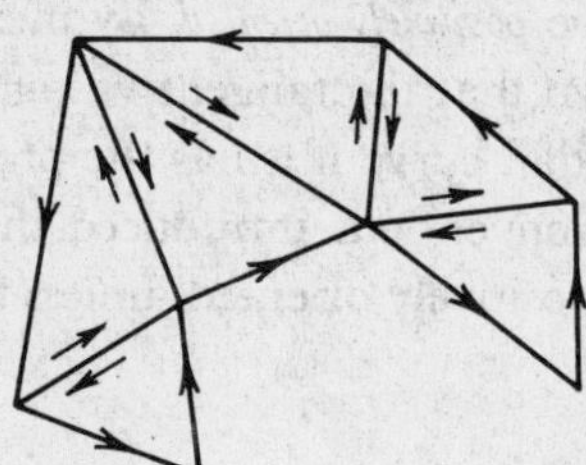

Figure 2-2

$$F(z) = \int_{z_0}^{z} f(z)\, dz,$$

when taken along a polygonal contour lying in D, is independent of the path of integration. Since $F'(z) = f(z)$ we have, for the *arbitrary* contour C in D,

$$\int_C f(z)\, dz = \int_C F'(z)\, dz = F(z)\Big|_C .$$

Hence, (2.1) holds for all closed contours lying in D, not just for polygonal contours. This important fact is the *Cauchy integral theorem.*

Simply connected domains of the type commonly found in applications are associated with simple closed contours, as discussed next. Let

$$z = \zeta(t), \qquad a \leq t \leq b$$

be a contour in the z plane. If $\zeta(a) = \zeta(b)$ and, except for this one case,

$$\zeta(t) \neq \zeta(t^*) \qquad \text{for } t \neq t^*$$

then the contour is called a *simple closed contour* or a *Jordan contour.* According to the Jordan curve theorem, which will not be proved here, a simple closed contour divides the plane into two domains; an interior domain, which is bounded, and an exterior domain. The interior domain is simply connected. Any point not on the contour must lie in one domain or the other. The boundary of each domain is the Jordan contour.

Let t_0 be chosen so that $\zeta(t_0)$ is not an end point of an arc of the Jordan contour and further so that $\zeta'(t_0) \neq 0$. If the vector obtained by rotating $\zeta'(t_0)$, the tangent vector at $\zeta(t_0)$, counterclockwise through an angle of $\pi/2$ points into the interior domain bounded by the Jordan contour, then the contour is said to be *positively oriented.* By means of the Jordan curve theorem it can be shown that the tangent vector has the stated property on every arc of the Jordan curve if it has the property on any arc.

In agreement with conventions introduced in Chapter 3, we assume that a Jordan curve is positively oriented, unless the contrary is explicitly stated. Thus, "Jordan contour," without further qualification, means "positively oriented Jordan contour."

For Jordan contours it is possible to state a form of the Cauchy integral theorem that applies even when $f(z)$ has isolated singularities. This theorem, called the *residue theorem,* is of fundamental importance in complex analysis.

Let $f(z)$ be analytic in the punctured neighborhood of a point α. Let k be a small, positively oriented circle with center at α. Then the *residue* of f at α is defined as

$$\frac{1}{2\pi i}\int_k f(z)\,dz.$$

By Chapter 2, Theorem 9.2, this is the coefficient a_{-1} in the Laurent expansion of f about the point α, and so a_{-1} can also be regarded as the residue of f at α. The residue is sometimes denoted by $\operatorname{Res} f(\alpha)$, read "the residue of f at α."

The following lemma may serve to explain why the coefficient a_{-1} is singled out in preference to other coefficients, and also explains the use of the term "residue":

LEMMA 2.1. *Let $f(z)$ have an isolated singularity at $z = \alpha$ with singular part*

$$f_1(z) = \frac{a_{-1}}{z-\alpha} + \frac{a_{-2}}{(z-\alpha)^2} + \cdots + \frac{a_{-n}}{(z-\alpha)^n} + \cdots$$

and let C be a Jordan contour not passing through the point α. Then

$$\frac{1}{2\pi i}\int_C f_1(z)\,dz = \begin{cases} 0 & \text{for } \alpha \text{ outside C} \\ a_{-1} & \text{for } \alpha \text{ inside C.} \end{cases}$$

Outline of proof. Assume first that α is outside of C. Since the exterior domain of C is connected and contains $z = \infty$, we can join α to ∞ by a simple polygonal line, L, that does not meet C (see Figure 2-3). The domain obtained when L is deleted from the complex plane is simply connected and contains the curve C. Since $f_1(z)$ is analytic for $z \neq \alpha$, it is analytic

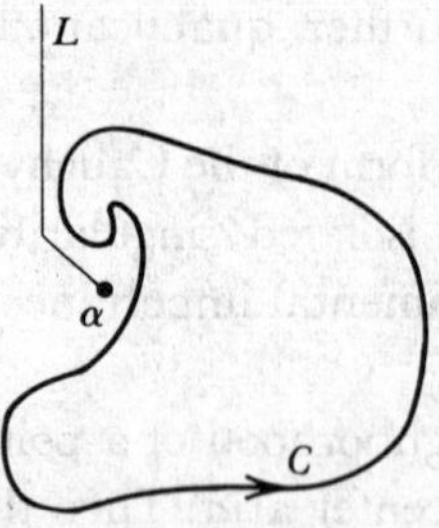

Figure 2-3

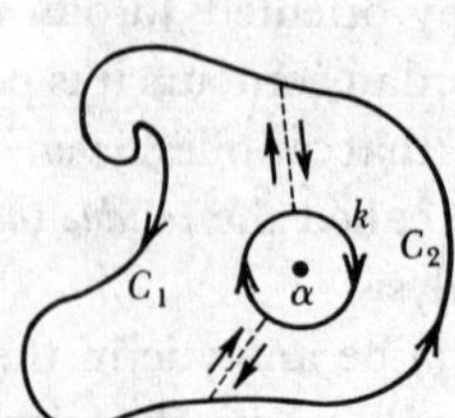

Figure 2-4

in this simply connected domain, and so the integral is 0 by Cauchy's theorem.

Suppose next that α is inside C. Let k be a small circle with α as center and contained in the interior of C. Consider two points on k and from each of them extend a line segment exterior to k but along the radial direction emanating from α, and terminate each segment where it first intersects C as shown in Figure 2-4. In this way, positively oriented Jordan contours C_1 and C_2 are obtained such that

$$\int_C f_1(z)\,dz = \int_{C_1} f_1(z)\,dz + \int_{C_2} f_1(z)\,dz + \int_k f_1(z)\,dz$$

where the last integral is taken counterclockwise around k. By Cauchy's theorem the integrals over C_1 and C_2 are 0, since $f_1(z)$ is analytic for $z \neq \alpha$, and hence the result reduces to the third integral. This gives $2\pi i a_{-1}$ by the definition of a_{-1}.

THEOREM 2.1 (*Cauchy's residue theorem*). *Let $f(z)$ be analytic in a simply connected domain D except for finitely many points α_j at which f may have isolated singularities. Let C be a Jordan contour that lies in D and does not pass through any point α_j. Then*

$$\int_C f(z)\,dz = 2\pi i \sum (\text{residue of } f \text{ at } \alpha_j)$$

where the sum is extended over the points α_j that are inside C.

Proof. Let the principal or singular part of f at α_j be denoted by $f_j(z)$. Then

$f_j(z)$ is analytic in the extended plane except at α_j, and if there are m points α_j,

$$g(z) = f(z) - \sum_1^m f_j(z) \tag{2.2}$$

is analytic in D. By Cauchy's theorem

$$\int_C g(z)\, dz = 0$$

or, rearranging,

$$\int_C f(z)\, dz = \sum_1^m \int_C f_j(z)\, dz.$$

Applying Lemma 2.1 to each term on the right gives Theorem 2.1.

The usefulness of Theorem 2.1 is greatly increased by some formulas for calculating residues. If $f(z)$ has a simple pole at $z = \alpha$, then

$$f(z) = \frac{a_{-1}}{z - \alpha} + g(z)$$

where $g(z)$ is analytic at α. Hence the residue is

$$a_{-1} = \lim_{z \to \infty} (z - \alpha) f(z) \qquad \text{(simple pole).} \tag{2.3}$$

If $f(z) = g(z)/h(z)$ where g and h are analytic at α but $h(z)$ has a simple zero at α, then the residue of f at α is

$$a_{-1} = \frac{g(\alpha)}{h'(\alpha)} \qquad \text{(simple pole).} \tag{2.4}$$

This is so because by l'Hospital's rule

$$\lim (z - \alpha) \frac{g(z)}{h(z)} = \lim g(z) \lim \frac{z - \alpha}{h(z)} = g(\alpha) \frac{1}{h'(\alpha)}.$$

These formulas are valid if f has a simple pole at α or a removable singularity (in which case the residue is 0). If f has a pole of order m at α then

$$f(z) = \frac{a_{-m}}{(z-\alpha)^m} + \frac{a_{-m+1}}{(z-\alpha)^{m-1}} + \cdots + \frac{a_{-1}}{z-\alpha} + g(z)$$

where $g(z)$ is analytic at α. Hence $(z-\alpha)^m f(z)$ is equal to

$$a_{-m} + a_{-m+1}(z-\alpha) + \cdots + a_{-1}(z-\alpha)^{m-1} + (z-\alpha)^m g(z)$$

and thus $(z-\alpha)^m f(z)$ is analytic at α, its value at α being taken as a_{-m}. Differentiating $m-1$ times and then setting $z=\alpha$ gives

$$a_{-1} = \frac{1}{(m-1)!}\frac{d^{m-1}}{dz^{m-1}}[(z-\alpha)^m f(z)]_{z=\alpha}. \tag{2.5}$$

Thus (2.5) is the formula for the residue of $f(z)$ at a pole of order m. (Since one or more of the coefficients $a_{-m}, a_{-m+1}, \ldots$ could be zero, the formula is valid also even if f has a pole of order less than m at α. For simplicity in computation of the right side of (2.5), however, it is best to take m as the actual order of the pole at α and not any larger.)

A function analytic in a domain D except for poles in D is said to be meromorphic in D. The above formulas indicate that when Theorem 2.1 is applied to meromorphic functions, the residues can be determined by differentiation. As a consequence of this fact, the calculus of residues provides an extremely effective method for evaluating definite integrals. Systematic discussion of this technique is deferred to the following sections. However, an illustration of it is given now.

The integral

$$\int_0^{2\pi} f(\cos\theta, \sin\theta)\, d\theta$$

can be recast as a contour integral by setting $z = e^{i\theta}$. When θ increases from 0 to 2π, z traverses the unit circle C counterclockwise. Moreover,

$$\cos\theta = \frac{e^{i\theta}+e^{-i\theta}}{2}, \qquad \sin\theta = \frac{e^{i\theta}-e^{-i\theta}}{2i}, \qquad dz = ie^{i\theta}\, d\theta.$$

Upon recalling that $e^{i\theta} = z$ we see that

$$\cos\theta = \frac{z^2+1}{2z}, \qquad \sin\theta = \frac{z^2-1}{2iz}, \qquad d\theta = \frac{dz}{iz}. \tag{2.6}$$

Hence the given integral becomes

$$\frac{1}{i}\int_C f\left(\frac{z^2+1}{2z}, \frac{z^2-1}{2iz}\right)\frac{dz}{z}.$$

For broad classes of functions f, this can be evaluated by residue theory.

For example, let it be required to evaluate

$$I = \int_0^{2\pi} \frac{d\theta}{a+\cos\theta}, \qquad a > 1.$$

Setting $e^{i\theta} = z$ and using (2.6), we get

$$I = \frac{1}{i}\int_C \frac{2}{z^2+2az+1}\,dz.$$

The denominator of the integrand vanishes at $-a \pm (a^2-1)^{1/2}$. The root $-a-(a^2-1)^{1/2}$ is less than -1 and hence lies outside of the unit circle. Since the product of the two roots is 1, $-a+(a^2-1)^{1/2}$ must lie inside the circle. By (2.4) the residue at this point is the value of $2/(2z+2a)$ at this point, which is $1/(a^2-1)^{1/2}$. Hence

$$I = \frac{2\pi i}{i}\frac{1}{(a^2-1)^{1/2}} = \frac{2\pi}{(a^2-1)^{1/2}}.$$

This result can be generalized to a wider range of values of a. Indeed, let

$$F(w) = \int_0^{2\pi} \frac{d\theta}{w+\cos\theta}$$

where w is any complex number in the w plane slit along the real axis from -1 to 1. (The cut is made to get a region in which the denominator of the integrand does not vanish.) In the slit plane it is readily verified that $F'(w)$ exists and hence $F(w)$ is analytic. The above result shows that the analytic function

$$F(w) - \frac{2\pi}{(w^2-1)^{1/2}}$$

vanishes on the real axis of w to the right of $w = 1$. Since an analytic function cannot vanish on an interval without vanishing in its whole domain of analyticity, this implies that the above function vanishes for all w in the slit plane. Thus in the slit plane

$$\int_0^{2\pi} \frac{d\theta}{w + \cos\theta} = \frac{2\pi}{(w^2 - 1)^{1/2}}. \tag{2.7}$$

As is clear from the derivation, the branch of the square root above must be chosen so as to be positive for $w > 1$.

Example 2.1. Find the residue of $f(z) = e^{-iz}/(1 + z^2)$ at $z = i$, and thus evaluate a class of contour integrals. Here

$$(z - i)f(z) = (z - i)\frac{e^{-iz}}{(z + i)(z - i)} = \frac{e^{-iz}}{z + i}.$$

As $z \to i$ the limit is $e/2i$, which is the residue by (2.3). Alternatively, (2.4) indicates that the residue is the value of $e^{-iz}/2z$ at $z = i$. This gives the same result as before, $e/2i$.

If C is a Jordan contour containing the point i but not $-i$, the residue theorem gives

$$\int_C \frac{e^{-iz}}{z^2 + 1}\,dz = \pi e.$$

Example 2.2. Find the residue of e^z/z^4 at $z = 0$. Dividing the Taylor series for e^z by z^4 gives

$$\frac{e^z}{z^4} = \frac{1}{z^4} + \frac{1}{z^3} + \frac{1}{2!z^2} + \frac{1}{3!z} + \frac{1}{4!} + \cdots.$$

The coefficient of $1/z$ is $1/6$ and this is the residue.

Example 2.3. What is the residue of $f(z) = e^{2z}(z - 1)^{-3}$ at $z = 1$? Here

$$(z - 1)^3 f(z) = e^{2z}.$$

Differentiating twice and setting $z = 1$ gives $4e^2$. By (2.5) with $m = 3$, the residue is $2e^2$.

Example 2.4. Deduce Cauchy's integral formula from the residue theorem.

Let $f(z)$ be analytic in a simply connected domain D containing the

Jordan contour C and let a be inside C. Cauchy's integral formula states that

$$f(a) = \frac{1}{2\pi i} \int_C \frac{f(z)}{z - a}\, dz.$$

To deduce this from Theorem 2.1 observe that $f(z)/(z - a)$ is analytic except at $z = a$, where the residue is $f(a)$ by (2.4).

Problems

1. Show that the residue of e^{1/z^2} at the origin is 0 although the origin is an essential singularity for this function. (Both results follow by setting $w = 1/z^2$ in the Taylor series for e^w.)
2. Find the residues at all singularities in $|z| < \infty$:

$$\frac{1}{z(z-1)}, \qquad \frac{z}{z^4+1}, \qquad \frac{\sin z}{z^2(\pi - z)}, \qquad \frac{ze^{iz}}{(z-\pi)^2}, \qquad \frac{z^3+5}{(z^4-1)(z+1)}.$$

3. If C is the circle $|z| = 1$, show that

$$\int_C \frac{e^{\pi z}}{4z^2+1}\, dz = \pi i, \qquad \int_C \frac{e^z}{z^3}\, dz = \pi i, \qquad \int_C \frac{e^z}{(z^2 + z - 3/4)^2}\, dz = 0.$$

How do these results change if C is the circle $|z| = 2$?

4. (a) If C is the circle $z = e^{i\theta}$, $0 \le \theta \le 2\pi$, show by (2.6) that

$$\int_0^{2\pi} \frac{d\theta}{2 + \sin\theta} = \int_C \frac{2dz}{z^2 + 4iz - 1}.$$

Evaluate the second integral by the residue theorem and deduce that the value of the first integral is $2\pi/\sqrt{3}$. (b) Explain why the integral is unchanged if $\sin\theta$ is replaced by $\cos\theta$, and evaluate by (2.7).

5. If a is complex and $|a| < 1$, show first by (2.6) and then by (2.7) that

$$\int_0^{2\pi} \frac{d\theta}{1 - 2a\cos\theta + a^2} = \frac{2\pi}{1 - a^2}.$$

Obtain the results for $|a| > 1$ by inspection of the above, and check by residues. (In an obvious notation, $I(1/a) = a^2 I(a)$.)

6. Show by the residue theorem that

$$\lim_{R\to\infty} \int_{-R}^{R} \frac{dx}{1 + x^2} = \pi.$$

(If C_R is the semicircular contour $z = \mathrm{R}e^{i\theta}$, $0 \le \theta \le \pi$, consider

$$\int_C \frac{dz}{1+z^2} = \int_{-R}^{R} \frac{dx}{1+x^2} + \int_{C_R} \frac{dz}{1+z^2}.$$

Evaluate the left-hand integral by residues and estimate the right-hand integral by noting that $|1 + z^2| \geq R^2 - 1$.)

7. Plot $I(a)$ vs a for $-\infty < a < \infty$, where

$$I(a) = -i\pi \int_{C_a} \frac{e^z}{z(z^2+\pi^2)} dz$$

and C_a is the boundary of the rectangle $-1 < x < 1$, $a < y < a + 4$.

3. Integrals over the real axis. The calculus of residues gives an effective means for evaluating integrals of the form

$$I = \int_{-\infty}^{\infty} f(x)\, dx. \tag{3.1}$$

An integral such as this is termed improper and can be interpreted in several ways. An appropriate interpretation in complex analysis is to consider that

$$I = \lim_{R\to\infty} \int_{-R}^{R} f(x)\, dx \tag{3.2}$$

provided the limit exists. If the limit does not exist, the integral diverges and no value is assigned to it. The particular definition of the integral given by (3.2) is called the *Cauchy principal value* to distinguish it from other definitions which are more restrictive.

To apply residue calculus to integrals of this type, suppose $f(z)$ has no singularities on the real axis, and only isolated singularities in the upper

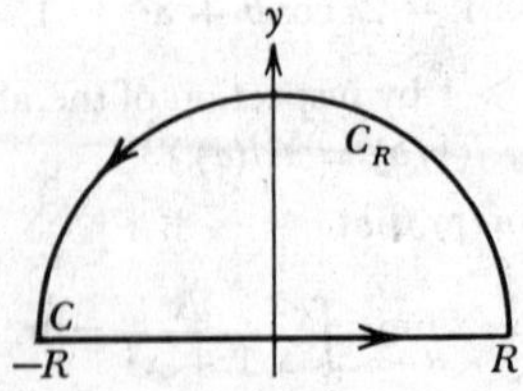

Figure 3-1

half-plane. If C is the semicircular contour shown in Figure 3-1 we have

$$\int_C f(z)\,dz = \int_{-R}^{R} f(x)\,dx + \int_{C_R} f(z)\,dz \tag{3.3}$$

where C_R denotes the curved part of the boundary. The left-hand integral can be evaluated by residues and the right-hand integral can be estimated. For broad classes of functions it is found that letting $R \to \infty$ in (3.3) gives both the existence and the value of the limit (3.2).

Example 3.1. Show that

$$\int_{-\infty}^{\infty} \frac{x^2+3}{(x^2+1)(x^2+4)}\,dx = \frac{5\pi}{6}. \tag{3.4}$$

We consider, instead of (3.4), the integral

$$\int_C \frac{z^2+3}{(z^2+1)(z^2+4)}\,dz = \int_C \frac{z^2+3}{z^4+5z^2+4}\,dz \tag{3.5}$$

where C is the contour shown in Figure 3-1. The first expression in (3.5) shows that the integrand is regular in the upper half-plane except for simple poles at i and $2i$. The second expression and (2.4) show that the residues are given by evaluating

$$\frac{z^2+3}{4z^3+10z}$$

at i and $2i$, respectively. Thus the residues are

$$\frac{-1+3}{-4i+10i} = \frac{1}{3i}, \qquad \frac{-4+3}{-32i+20i} = \frac{1}{12i}.$$

If $R > 2$, then the contour C contains both poles and hence the integral (3.5) satisfies

$$\int_C \frac{z^2+3}{(z^2+1)(z^2+4)}\,dz = 2\pi i\left(\frac{1}{3i} + \frac{1}{12i}\right) = \frac{5\pi}{6}.$$

It remains to estimate the integral on C_R. If $R > 2$, then clearly

$$|z^2 + 3| \le R^2 + 3, \qquad |z^2 + 1| \ge R^2 - 1, \qquad |z^2 + 4| \ge R^2 - 4$$

for $|z| = R$, and so the integral over C_R does not exceed

$$\pi R \frac{R^2 + 3}{(R^2 - 1)(R^2 - 4)} = \pi R \frac{R^2}{R^4} \frac{(1 + 3/R^2)}{(1 - 1/R^2)(1 - 4/R^2)}.$$

This tends to 0 as $R \to \infty$, and letting $R \to \infty$ in (3.3) gives (3.4).

Example 3.2. Show that

$$\int_{-\infty}^{\infty} \frac{\cos x}{x^2 + a^2}\, dx = \frac{\pi}{a} e^{-a}, \qquad a > 0. \quad (3.6)$$

Since $\cos x$ is the real part of e^{ix}, we consider, instead of (3.6), the integral

$$\int_C \frac{e^{iz}}{z^2 + a^2}\, dz$$

where C is the contour of Figure 3-1. The only singularity of the integrand in the upper half-plane is a simple pole at $z = ia$, and the residue at ia is $e^{-a}/2ia$ by (2.4). If $R > a$, this pole is within the contour C, and (3.3) becomes

$$(2\pi i)\frac{e^{-a}}{2ia} = \int_{-R}^{R} \frac{e^{ix}}{x^2 + a^2}\, dx + \int_{C_R} \frac{e^{iz}}{z^2 + a^2}\, dz. \quad (3.7)$$

For $z = x + iy$ with $y \ge 0$ we have

$$|e^{iz}| = |e^{ix - y}| = |e^{ix}| e^{-y} = e^{-y} \le 1. \quad (3.8)$$

This is the reason for considering e^{iz} instead of $\cos z$. The latter function does not admit a useful dominant on C_R, but (3.8) gives

$$\left| \int_{C_R} \frac{e^{iz}}{z^2 + a^2}\, dz \right| \le \pi R \frac{1}{R^2 - a^2} = \frac{\pi R}{R^2} \frac{1}{(1 - a^2/R^2)}$$

for $R > a$. This tends to 0 as $R \to \infty$, and (3.6) follows by equating real parts in (3.7).

Instead of equating real parts one could observe that

$$\int_{-R}^{R} \frac{e^{ix}}{x^2 + a^2}\, dx = \int_{-R}^{R} \frac{\cos x + i \sin x}{x^2 + a^2}\, dx.$$

Since $\sin x$ is odd, that is, satisfies $\sin(-x) = -\sin x$, while $x^2 + a^2$ is even, the term involving $i \sin x$ integrates to 0. This gives (3.6) for all complex a having $\operatorname{Re} a > 0$.

We now give two inequalities which are useful for estimating the integral over C_R.

LEMMA 3.1. *Let $f(z)$ be a rational function such that the degree of the denominator exceeds the degree of the numerator by d. Then there are constants M and R_0 such that*

$$|f(z)| \leq \frac{M}{R^d} \qquad \text{for } |z| = R \geq R_0.$$

Proof. Write $f(z)$ in the form

$$f(z) = \frac{z^n}{z^m} \cdot \frac{a_n + a_{n-1}z^{-1} + \cdots + a_1 z^{1-n} + a_0 z^{-n}}{b_m + b_{m-1}z^{-1} + \cdots + b_1 z^{1-m} + b_0 z^{-m}}$$

where $a_n \neq 0$, $b_m \neq 0$. It is easily checked that the fraction on the right has the limit a_n/b_m as $|z| \to \infty$, and hence is bounded for large $|z|$. This gives Lemma 3.1, and shows that M can be any number larger than $|a_n/b_m|$.

In Example 3.2 the appraisal on C_R uses only $|e^{iz}| \leq 1$. In some cases of interest a more subtle estimate is needed. This is provided by the following lemma:

LEMMA 3.2 (*Jordan's lemma*). *If C_R is the contour $z = Re^{i\theta}$, $0 \leq \theta \leq \pi$, then*

$$\int_{C_R} |e^{iz}||dz| < \pi.$$

Proof. On C_R we have $|e^{iz}| = e^{-y}$, $y = R \sin \theta$, $|dz| = R\, d\theta$. Hence if the integral is denoted by J we can write

$$J = \int_0^{\pi} e^{-R \sin \theta} R\, d\theta = 2\int_0^{\pi/2} e^{-R \sin \theta}\, d\theta \tag{3.9}$$

where the second expression follows from $\sin(\pi - \theta) = \sin \theta$. By the construction suggested in Figure 3-2 it is evident that

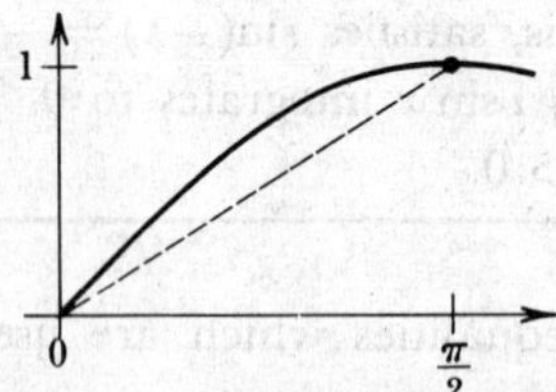

Figure 3-2

$$\sin\theta \geq \frac{2}{\pi}\theta, \qquad 0 \leq \theta \leq \frac{\pi}{2}. \tag{3.10}$$

An analytic proof follows from the fact that $\sin\theta - 2\theta/\pi$ is zero at $\theta = 0$ and at $\theta = \pi/2$, and has a second derivative which is negative for $0 < \theta < \pi/2$. In any case (3.10) is valid and in (3.9) gives

$$J \leq 2\int_0^{\pi/2} e^{-2R\theta/\pi} R\,d\theta = -\pi e^{-2R\theta/\pi}\Big|_0^{\pi/2} < \pi.$$

Example 3.3. If a and b are unequal complex numbers with positive real parts, show that

$$\int_{-\infty}^{\infty} \frac{x^3 e^{ix}}{(x^2 + a^2)(x^2 + b^2)}\,dx = 2\pi i\left(\frac{a^2 e^{-a}}{a^2 - 2b^2} + \frac{b^2 e^{-b}}{b^2 - 2a^2}\right). \tag{3.11}$$

Here the residues in the upper half-plane are the values of

$$\frac{x^3 e^{ix}}{3x^3 + 2(a^2 + b^2)x} = \frac{x^2 e^{ix}}{3x^2 + 2a^2 + 2b^2}$$

at $x = ia$ and $x = ib$, and are found by inspection. The residue theorem gives

$$\int_C \frac{z^3}{(z^2 + a^2)(z^2 + b^2)} e^{iz}\,dz = 2\pi i\left(-\frac{a^2 e^{-a}}{-a^2 + 2b^2} - \frac{b^2 e^{-b}}{-b^2 + 2a^2}\right)$$

which agrees with the desired answer. Moreover, by Lemma 3.1 with $d = 1$,

$$\left|\frac{z^3}{(z^2 + a^2)(z^2 + b^2)}\right| \leq \frac{M}{R}, \qquad |z| = R \geq R_0.$$

Combining this with Lemma 3.2, we see that the integral over C_R does not exceed

$$\int_{C_R} \frac{M}{R} |e^{iz}||dz| < \frac{M}{R}\pi.$$

This tends to 0 and (3.11) follows.

Example 3.4. If $R > 0$, show that

$$\left| \int_{-R}^{R} \frac{\sin x}{x}\, dx - \pi \right| < \frac{\pi}{R}. \tag{3.12}$$

Since a useful estimate for $\sin z$ on C_R is not available, we replace $\sin x$ by e^{ix} and consider

$$\int_C \frac{e^{iz}}{z}\, dz.$$

Now, this integral can be analyzed by methods given later, but it does not fall into the previous pattern because of the singularity at $z = 0$. To fit the previous pattern we subtract 1 from the numerator. Then the integrand has only a removable singularity at $z = 0$, and

$$0 = \int_C \frac{e^{iz} - 1}{z}\, dz = \int_{-R}^{R} \frac{e^{ix} - 1}{x}\, dx + \int_{C_R} \frac{e^{iz} - 1}{z}\, dz.$$

The first equality above follows from the residue theorem and the second from (3.3). Rearranging gives

$$\int_{-R}^{R} \frac{e^{ix} - 1}{x}\, dx = \int_{C_R} \frac{1 - e^{iz}}{z}\, dz = \int_{C_R} \frac{dz}{z} - \int_{C_R} \frac{e^{iz}}{z}\, dz. \tag{3.13}$$

The first integral on the right is $i\pi$ and the second satisfies

$$\left| \int_{C_R} \frac{e^{iz}}{z}\, dz \right| \le \frac{1}{R} \int_{C_R} |e^{iz}||dz| < \frac{\pi}{R}$$

by Lemma 3.2. Equating imaginary parts in (3.13) gives (3.12).

The result indicates that

$$\int_{-\infty}^{\infty} \frac{\sin x}{x}\, dx = \pi$$

and that if this integral is approximated by integrating only from $-R$ to R the error does not exceed π/R. Such estimates are useful in numerical analysis.

Problems

1. Integrate each of the following from $-\infty$ to ∞:

$$\frac{1}{1+x^4},\qquad \frac{1}{1+x^6},\qquad \frac{x^2}{1+x^6},\qquad \frac{1}{(1+x^2)^2},\qquad \frac{x^2}{(1+x^2)^2}.$$

(Answers: $\pi/\sqrt{2}$, $2\pi/3$, $\pi/3$, $\pi/2$, $\pi/2$.)

2. (a) Using the method but not the result of Example 3.2, show that

$$\int_{-\infty}^{\infty}\frac{\cos x}{1+x^2}\,dx=\frac{\pi}{e},\qquad \int_{-\infty}^{\infty}\frac{x\sin x}{1+x^2}\,dx=\frac{\pi}{e},\qquad \int_{-\infty}^{\infty}\frac{\cos x}{(1+x^2)^2}\,dx=\frac{\pi}{e}.$$

(b) Get the third integral above by differentiating[1] (3.6) with respect to a.

3. If a and b are unequal complex numbers with positive real parts, show that

$$\int_{-\infty}^{\infty}\frac{\cos x}{(x^2+a^2)(x^2+b^2)}\,dx=\frac{\pi}{a^2-b^2}\left(\frac{e^{-b}}{b}-\frac{e^{-a}}{a}\right).$$

4. By l'Hospital's rule find the limit of the right-hand member as $b\to a$ in Problem 3. Then determine whether this limit agrees with the value of the integral for $b=a$.

5. If $f(x)$ denotes the integrand in Problem 3 and I denotes the value of the integral, show that

$$\left|\int_{-R}^{R}f(x)\,dx-I\right|\le\frac{\pi}{(R^2-|a|^2)(R^2-|b|^2)},\qquad R>\max(|a|,|b|).$$

(Use Jordan's lemma. It is possible to get an affirmative answer to Problem 4 by the result of this problem, without much calculation.)

6. (a),(b),(c) Formulate and solve analogs of Problems 3, 4 and 5 when the numerator $\cos x$ in Problem 3 is replaced by $x\sin x$.

7. If $0<a<1$, show that

$$\int_{-\infty}^{\infty}\frac{e^{ax}}{e^x+1}\,dx=\frac{\pi}{\sin\pi a},\qquad \int_{-\infty}^{\infty}\frac{\cosh ax}{\cosh x}\,dx=\pi\sec\frac{\pi a}{2}.$$

(Integrate $e^{az}/(e^z+1)$ around a rectangle with vertices at $-R$, R, $R+2\pi i$, $-R+2\pi i$. The second integral can be obtained similarly, or deduced from the first.)

[1] Here and in similar cases elsewhere, the differentiation is justified by theorems on uniform convergence given in Chapter 6.

8. If $0 < c < 1$ and $|\text{Im}\,\alpha| < \pi$, show that

$$\int_{c-i\infty}^{c+i\infty} \frac{e^{\alpha z}}{\sin \pi z}\, dz = \frac{2i}{1 + e^{-\alpha}}.$$

(Integrate around a rectangle with vertices at

$$c - iR, \qquad 1 + c - iR, \qquad 1 + c + iR, \qquad c + iR$$

and let $R \to \infty$. This suggestion also gives a clue to the meaning of the integral.)

4. Improper integrals and principal values. The definition of the integral given in Section 3 can be written in the form

$$⨍_{-\infty}^{\infty} f(x)\, dx = \lim_{R\to\infty} \left(\int_{-R}^{0} f(x)\, dx + \int_{0}^{R} f(x)\, dx \right). \tag{4.1}$$

When this limit exists, the integral is said to be *convergent in the sense of Cauchy* and the value of the limit is the Cauchy principal value. The bar on the integral (4.1) serves to distinguish this definition from another which we want to compare with (4.1), namely,

$$\int_{-\infty}^{\infty} f(x)\, dx = \lim_{R_1\to\infty} \int_{-R_1}^{0} f(x)\, dx + \lim_{R_2\to\infty} \int_{0}^{R_2} f(x)\, dx. \tag{4.2}$$

This formulation is intended when it is said, without further qualification, that the integral is *convergent.*

Convergence of the integral clearly implies convergence (to the same value) in the sense of Cauchy; but the former limit can exist when the latter does not. For example,

$$\int_{-R}^{R} 2x\, dx = x^2 \Big|_{-R}^{R} = 0$$

and hence the limit (4.1) is 0 though neither of the limits (4.2) exists.

The function $f(x) = 2x$ in this example is odd, that is, $f(-x) = -f(x)$. For odd functions the Cauchy principal value (4.1) is 0. On the other hand, if $f(x)$ is even, then Cauchy convergence (4.1) implies convergence. Indeed, for even functions, the equation

$$\tfrac{1}{2} \int_{-R}^{R} f(x)\, dx = \int_{-R}^{0} f(x)\, dx = \int_{0}^{R} f(x)\, dx$$

is verified by inspection, and shows that both limits on the right exist for $R \to \infty$ if the limit on the left exists. Hence (4.2) holds. This remark establishes convergence in Examples 3.1, 3.2, 3.4 and Problems 1–6 of the preceding section.

Integrals with infinite limits are not only ones that are called improper. Other improper integrals are obtained if the interval of integration contains one or more points where the integrand becomes infinite.

Let f be continuous or piecewise continuous on $[a,b]$ except at c, where f may become infinite as shown in Figure 4-1. Then consider

$$\int_a^{c-\epsilon} f(x)\,dx + \int_{c+\eta}^{b} f(x)\,dx. \qquad \epsilon > 0, \eta > 0$$

If the first term tends to a limit as $\epsilon \to 0$ and the second as $\eta \to 0$, then the left side below is defined by the right side[1]

$$\int_a^b f(x)\,dx = \lim_{\epsilon \to 0+} \int_a^{c-\epsilon} f(x)\,dx + \lim_{\eta \to 0+} \int_{c+\eta}^{b} f(x)\,dx \qquad (4.3)$$

and is called an *improper integral.* If these limits fail to exist, then the limit

$$\lim_{\epsilon \to 0+} \left(\int_a^{c-\epsilon} f(x)\,dx + \int_{c+\epsilon}^{b} f(x)\,dx \right)$$

may exist. If this limit exists, it is called the Cauchy *principal value* of the integral and is designated by

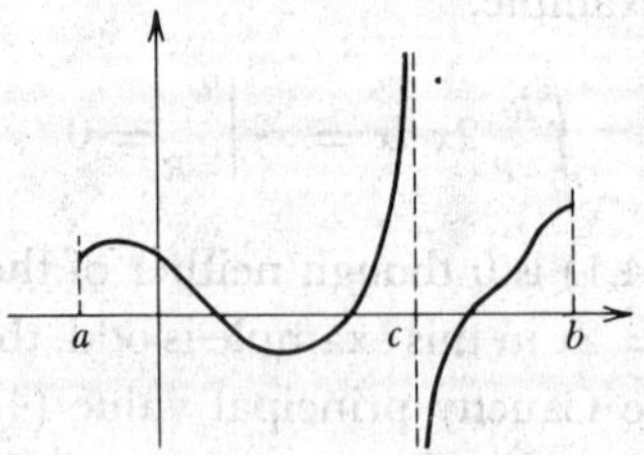

Figure 4-1

[1] The notation $\epsilon \to 0+$ means $\epsilon \to 0$ through positive values, that is, $\epsilon > 0$ throughout the limiting process. Similarly, $\epsilon \to 0-$ means $\epsilon \to 0$ through negative values. The first expression on the right of (4.3) is dropped if $c = a$ and the second is dropped if $c = b$.

$$P\int_a^b f(x)\,dx \qquad \text{or} \qquad ⨍_a^b f(x)\,dx.$$

The same notation is also used for the limit (4.1), taking $a = -\infty$ and $b = \infty$.

As an illustration we show that

$$\int_{-1}^{1} \frac{dx}{x} \quad \text{does not exist, but} \quad ⨍_{-1}^{1} \frac{dx}{x} = 0.$$

Indeed,

$$\int_{\eta}^{1} \frac{dx}{x} = \log\frac{1}{\eta} \to \infty \qquad \text{as } \eta \to 0+.$$

On the other hand,

$$\int_{-1}^{-\epsilon} \frac{dx}{x} + \int_{\epsilon}^{1} \frac{dx}{x} = \log \epsilon + \log\frac{1}{\epsilon} = \log 1 = 0$$

as is also evident geometrically from a graph of $1/x$. In a similar fashion it can be verified that

$$\int_{-\infty}^{\infty} \frac{x\,dx}{1+x^2} \quad \text{does not exist, but} \quad ⨍_{-\infty}^{\infty} \frac{x\,dx}{1+x^2} = 0.$$

The scope of these definitions is extended by addition. Suppose that $f(x)$ is continuous except at a point c where $f(x)$ may become infinite, and suppose the interval of integration is $(-\infty,\infty)$. Choose points a and b such that $-\infty < a < c < b < \infty$. The Cauchy principal value is then

$$⨍_{-\infty}^{\infty} f(x)\,dx = \lim_{R\to\infty}\left(\int_{-R}^{a} + \int_{b}^{R}\right) f(x)\,dx + \lim_{\epsilon\to 0+}\left(\int_{a}^{c-\epsilon} + \int_{c+\epsilon}^{b}\right) f(x)\,dx,$$

provided the limits on the right exist. The expression

$$\left(\int_{-R}^{a} + \int_{b}^{R}\right) f(x)\,dx + \left(\int_{a}^{c-\epsilon} + \int_{c+\epsilon}^{b}\right) f(x)\,dx$$

is equal to

$$\int_{-R}^{c-\epsilon} f(x)\,dx + \int_{c+\epsilon}^{R} f(x)\,dx \tag{4.4}$$

which is independent of a and b. This shows that the Cauchy principal value is also independent of a and b. In most applications one does not bother to introduce intermediate values a and b, but one just considers (4.4) as $R \to \infty$ and $\epsilon \to 0+$ independently. Extension to the case of several points c_i is made analogously, by excluding small intervals $(c_i - \epsilon_i, c_i + \epsilon_i)$ located symmetrically with respect to the points c_i. The ϵ_i tend to $0+$ independently. When the limits defining the improper integral exist, the integral is said to be *convergent.*

Evaluation of Cauchy principal values is facilitated by the following lemma, which pertains to integration over a path of the type illustrated in Figure 4-2.

LEMMA 4.1. *Let $f(z)$ have a simple pole at $z = \alpha$ with residue a_{-1} and let $k = k(\epsilon,\phi)$ be an arc of the circle $|z - \alpha| = \epsilon$ that subtends an angle ϕ at the center. Then*

$$\lim_{\epsilon \to 0+} \int_k f(z)\, dz = i\phi a_{-1}.$$

If we integrate over the whole circle, then $\phi = 2\pi$, and the residue theorem gives $2\pi i a_{-1}$ even without assuming $\epsilon \to 0$. The lemma indicates that integration over a fraction of the circle gives a corresponding fraction of $2\pi i a_{-1}$, provided $\epsilon \to 0$ and the pole is simple.

Proof. We have $f(z) = g(z) + a_{-1}(z - \alpha)^{-1}$ where $g(z)$ is analytic near α; say analytic for $|z - \alpha| \leq \delta$ with $\delta > 0$. If $0 < \epsilon < \delta$, then

$$\int_k f(z)\, dz = \int_k g(z)\, dz + a_{-1} \int_k \frac{dz}{z - \alpha}.$$

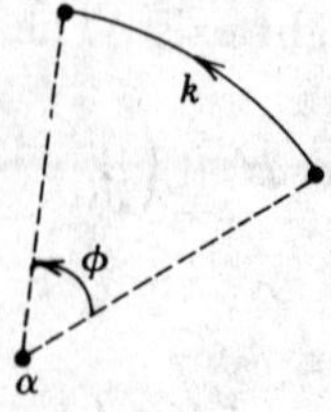

Figure 4-2

The first integral does not exceed $\epsilon\phi M$ where M is a bound for $|g(z)|$ in $|z - \alpha| \le \delta$. Writing $z = \alpha + \epsilon e^{i\theta}$ on k, we see that the second integral is

$$\int_{\theta_0}^{\theta_0+\phi} \frac{\epsilon e^{i\theta} i\, d\theta}{\epsilon e^{i\theta}} = i\phi.$$

This completes the proof.

Example 4.1. If $a > 0$ show that

$$⨍_{-\infty}^{\infty} \frac{\cos x}{a^2 - x^2}\, dx = \pi \frac{\sin a}{a}.$$

Let C be the semicircle indented at $-a$ and a by small semicircles, as shown in Figure 4-3, and let

$$I = \int_C \frac{e^{iz}}{a^2 - z^2}\, dz.$$

We suppose that the radius of the small semicircle at $-a$ is η, while that at a is ϵ and the radius of the large circle is R. There are no singularities of the integrand in C and so $I = 0$. This can be written

$$\left(\int_{-R}^{-a-\eta} + \int_{-a+\eta}^{a-\epsilon} + \int_{a+\epsilon}^{R}\right) \frac{e^{ix}}{a^2 - x^2}\, dx + J_1 + J_2 + J_3 = 0 \quad (4.5)$$

where J_1, J_2, J_3 are the integrals over the semicircles of radii η, ϵ and R, respectively. As in Example 3.2, or more easily by Lemma 3.1, it is seen that

$$\lim_{R\to\infty} J_3 = 0.$$

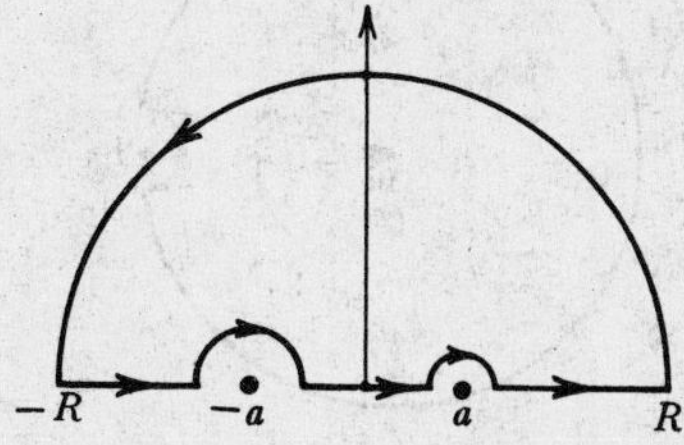

Figure 4-3

To evaluate J_1 and J_2 observe that the residues at $-a$ and a are

$$\frac{e^{-ia}}{2a} \quad \text{and} \quad \frac{e^{ia}}{-2a}$$

respectively. Hence Lemma 4.1 with $\phi = -\pi$ gives

$$\lim_{\eta \to 0+} J_1 = \frac{-\pi i e^{-ia}}{2a}, \qquad \lim_{\epsilon \to 0+} J_2 = \frac{\pi i e^{ia}}{2a}.$$

(The value $\phi = -\pi$ is used instead of $\phi = \pi$ because the small circles are traversed in the negative direction.) Letting $R \to \infty$, $\eta \to 0+$ and $\epsilon \to 0+$ in (4.5), we get

$$⨍_{-\infty}^{\infty} \frac{e^{ix}}{a^2 - x^2}\, dx + \frac{\pi i}{2a}(e^{ia} - e^{-ia}) = 0.$$

Taking real parts gives the desired result.

Example 4.2. Show that

$$\int_{-\pi}^{\pi} \text{Log}|1 - e^{i\theta}|\, d\theta = 0. \tag{4.6}$$

Here we consider

$$\int_C \text{Log}(1 - z)\frac{dz}{z}$$

where C is the circle $|z| = 1$ indented as shown in Figure 4-4. The integrand is single-valued in the plane cut from 1 to ∞, and is real for

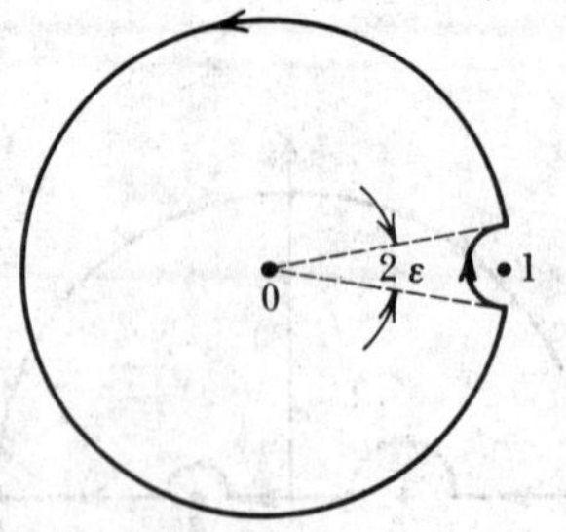

Figure 4-4

$0 < z < 1$. Since the singularity at $z = 0$ is removable, the integral is 0 by Cauchy's theorem. This gives

$$i\int_{\epsilon}^{2\pi-\epsilon} \text{Log}(1 - e^{i\theta})\, d\theta + \int_{C_\epsilon} \text{Log}(1 - z)\frac{dz}{z} = 0 \tag{4.7}$$

where C_ϵ is the small circular arc shown in the figure. Namely, C_ϵ is centered at the point 1 and passes through the two points at which

$$|z| = 1, \quad \text{Arg } z = -\epsilon; \qquad\qquad |z| = 1, \quad \text{Arg } z = \epsilon.$$

The radius ρ of C_ϵ satisfies $\rho < \epsilon$, hence $\rho \to 0$ as $\epsilon \to 0$.

If we write C_ϵ in the form $1 - z = \rho e^{i\phi}$ where $|\phi| \leq \pi$, it is easily checked that

$$|\text{Log}(1 - z)| = |\text{Log } \rho + i\phi| \leq [(\text{Log } \rho)^2 + \pi^2]^{1/2} \leq \sqrt{2}|\text{Log } \rho|$$

provided, in the last inequality, $|\text{Log } \rho| \geq \pi$. Hence the magnitude of the integral over C_ϵ does not exceed

$$2\pi\rho\frac{\sqrt{2}|\text{Log } \rho|}{1 - \rho} \qquad (\rho \leq e^{-\pi}). \tag{4.8}$$

Problem 1 below indicates that $\rho \log \rho \to 0$ as $\rho \to 0$, and hence letting $\epsilon \to 0$ in (4.7) gives (4.6) as a Cauchy principal value. Since the integrand is even, the integral is convergent.

In the next section we discuss integrals in which the path of integration does not avoid the branch cut (as it did here) but follows right along a branch cut.

Problems

1. The Taylor series for e^t implies $e^t > t^n/n!$ for $t > 0$. Derive the first of the following limits by setting $R = e^t$, and get the second from the first:

$$\lim_{R\to\infty} \frac{(\log R)^m}{R} = 0, \qquad \lim_{\rho\to 0+} \rho(\log \rho)^m = 0, \qquad (m = 0, 1, 2, 3, \ldots).$$

2. As in Example 4.1 obtain the formulas

$$\int_{-\infty}^{\infty} \frac{\sin x}{x}\, dx = \pi, \qquad \int_{-\infty}^{\infty} \frac{1 - \cos x}{x^2}\, dx = \pi, \qquad \int_0^{\infty} \frac{\sin \pi x}{x(1 - x^2)}\, dx = \pi.$$

(Note that the third integrand is an even function.)

3. Integrate e^{-z^2} around the boundary of the sector $|z| < R$, $0 < \theta < \pi/4$ and deduce that

$$\int_0^\infty \sin t^2 \, dt = \int_0^\infty \cos t^2 \, dt = \frac{1}{\sqrt{2}} \int_0^\infty e^{-t^2} \, dt.$$

The latter integral is known from real analysis to be $\sqrt{\pi}/2$. (On the curved part of the contour $|e^{-z^2}| = e^{-r^2 \cos 2\theta}$. Show analytically or by inspection of a graph that

$$\cos 2\theta \geq 1 - \frac{4}{\pi}\theta, \qquad 0 \leq \theta \leq \frac{\pi}{4}.$$

This makes it easy to estimate the integral.)

***Absolute convergence**

In these problems all functions are piecewise continuous for $0 \leq x < \infty$. We write $f \in K$ to indicate that f belongs to the class of functions, K, for which the following exists:

$$\lim_{R\to\infty} \int_0^R f(x) \, dx.$$

4. *Linear closure.* Show that $f \in K$ and $g \in K$ implies $\alpha f + \beta g \in K$ for complex constants α, β.
5. *Comparison test.* If $0 \leq f \leq g$, then $g \in K$ implies $f \in K$. (Evidently
$$\int_0^R f(x) \, dx \leq \int_0^R g(x) \, dx \leq \int_0^\infty g(x) \, dx.$$
The expression on the left is bounded and nondecreasing, and hence has a limit.)
6. *An absolutely convergent integral is convergent;* in other words, $|f| \in K$ implies $f \in K$. (Let $f = u + iv$. The inequality $0 \leq |f| + u \leq 2|f|$ shows that $|f| + u \in K$ and hence the same is true of $u = (|f| + u) - |f|$. Likewise $v \in K$ and so $u + iv \in K$.)
7. Deduce $(\cos x)/(1 + x^2) \in K$ by using the inequality
$$\int_1^R \frac{|\cos x|}{1 + x^2} \, dx \leq \int_1^R \frac{1}{x^2} \, dx \leq 1.$$
8. Integrate $(\sin x)/x$ by parts on $1 \leq x \leq R$ and deduce $(\sin x)/x \in K$.
9. Let $g(t)$ be piecewise continuous for $0 < t \leq 1$, let $\epsilon > 0$, and let $R = 1/\epsilon$. Show by setting $x = 1/t$ that
$$\int_\epsilon^1 g(t) \, dt = \int_1^R f(x) \, dx, \qquad \text{where } f(x) = \frac{1}{x^2} g\left(\frac{1}{x}\right).$$
Hence the study of convergence as $\epsilon \to 0+$ falls within the scope of Problems 4–6.

5. Integrands with branch points. As the reader will recall from Chapter 2,

$$\operatorname{Log} z = \operatorname{Log}|z| + i \operatorname{Arg} z \tag{5.1}$$

is single-valued and continuous in the region obtained by deleting 0 and the negative real axis from the z plane. The function $\theta = \operatorname{Arg} z$ is 0 on the positive real axis and is determined at other points by continuity. This gives

$$\lim_{y\to 0+} \operatorname{Arg} z = \pi, \qquad \lim_{y\to 0-} \operatorname{Arg} z = -\pi \qquad x < 0. \tag{5.2}$$

It is as if the cut has two sides, with $\operatorname{Arg} z = \pi$ on the upper side and $\operatorname{Arg} z = -\pi$ on the lower side. For more precise formulation, one can divide the cut plane into two regions D_1, D_2 as shown in Figure 5-1. Each region includes its boundary except for the point $z = 0$, which is not included. Reference to the "upper" or "lower" side of the cut is just an informal way of saying what value is to be assigned in D_1 or D_2, respectively.

Except at the point $z = 0$, D_1 can be extended slightly downward so that, for example, the segment $-R \leq x \leq -\epsilon$ of the real axis would be interior to D_1. Similarly D_2 could be extended upward. In the following examples and problems an extension of this sort makes the path of integration actually interior to the region of analyticity, as required by Cauchy's theorem. The extension is not emphasized because a more efficient method of dealing with this problem is given in Section 9.

The practical importance of these remarks is that integrals along the real axis, that would normally appear to cancel, need not in fact cancel.

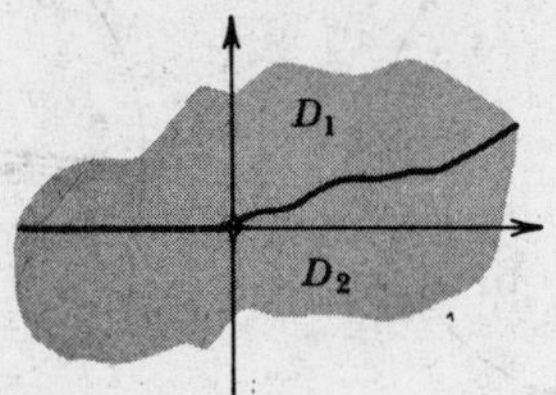

Figure 5-1

For instance

$$\int_{-3}^{-1} \operatorname{Arg} z\, dx + \int_{-1}^{-3} \operatorname{Arg} z\, dx = 4\pi$$

if the path of integration is considered to be in D_1 in the first integral and in D_2 in the second. We now give some examples showing how these ideas are used for the evaluation of integrals.

Example 5.1. If the branch of t^a is chosen so as to be real for $t > 0$, show that

$$\int_0^\infty \frac{t^{a-1}}{1+t}\, dt = \frac{\pi}{\sin \pi a}, \qquad 0 < a < 1.$$

Consider

$$I = \int_C \frac{z^{a-1}}{1-z}\, dz$$

where C is the contour shown in Figure 5-2 with inner radius $\epsilon < 1$ and outer radius $R > 1$. The plane is cut along the negative real axis so that z^{a-1} is single-valued in the cut plane. If $\theta = \operatorname{Arg} z$, we can write $z = re^{i\theta}$ with $r > 0$ and, by definition,

$$z^{a-1} = r^{a-1}e^{i\theta(a-1)}, \qquad (r^{a-1} > 0).$$

By introducing a cut as shown in Figure 5-3 one can express the integral over C as a sum of integrals over C_1 and C_2. The integral over C_1 is 0 because $z^{a-1}/(1-z)$ is analytic in a simply connected domain con-

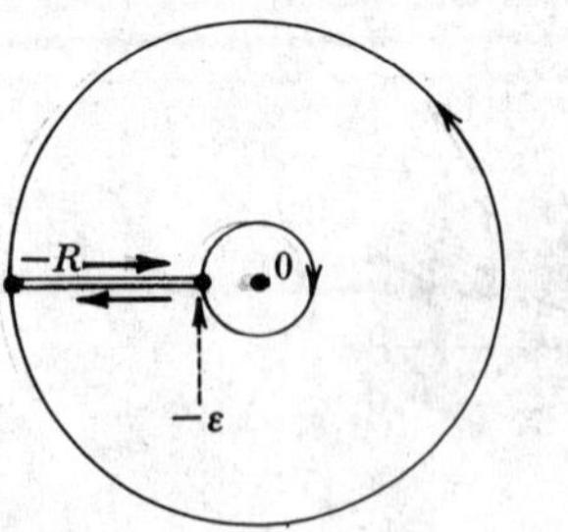

Figure 5-2

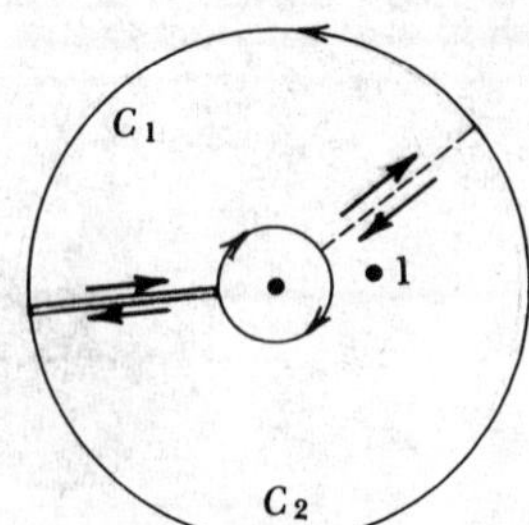

Figure 5-3

taining C_1. However the integral over C_2 has the value $-2\pi i$, since the integrand has a simple pole at $z = 1$ with residue -1. Hence

$$I = -2\pi i.$$

As explained above, $\theta = \pi$ on the upper edge of the cut in Figure 5-2 and $\theta = -\pi$ on the lower edge. Hence

$$I = \int_R^\epsilon \frac{(re^{i\pi})^{a-1}}{1+r} e^{i\pi}\, dr + \int_\epsilon^R \frac{(re^{-i\pi})^{a-1}}{1+r} e^{-i\pi}\, dr + J_1 + J_2$$

where J_1 and J_2 are the integrals over the circles of radius R and ϵ, respectively. In the first integrand we have

$$(re^{i\pi})^{a-1} e^{i\pi} = r^{a-1} e^{i\pi(a-1)} e^{i\pi} = r^{a-1} e^{i\pi a}$$

and similarly in the second. Hence

$$I = \int_\epsilon^R \frac{r^{a-1}}{1+r}\, dr\, (-e^{i\pi a} + e^{-i\pi a}) + J_1 + J_2. \tag{5.3}$$

On the circles of radius R and ϵ, respectively,

$$\left|\frac{z^{a-1}}{1-z}\right| \le \frac{R^{a-1}}{R-1}, \qquad \left|\frac{z^{a-1}}{1-z}\right| \le \frac{\epsilon^{a-1}}{1-\epsilon}$$

and hence

$$|J_1| \le 2\pi R \frac{R^{a-1}}{R-1}, \qquad |J_2| \le 2\pi\epsilon \frac{\epsilon^{a-1}}{1-\epsilon}.$$

Since $a < 1$, the first expression tends to 0 as $R \to \infty$, and since $a > 0$, the second expression tends to 0 as $\epsilon \to 0$. Letting $\epsilon \to 0$ and $R \to \infty$ in (5.3) thus gives

$$\int_0^\infty \frac{r^{a-1}}{1+r}\, dr\, (e^{-i\pi a} - e^{i\pi a}) = -2\pi i$$

when we recall $I = -2\pi i$. Since $e^{-i\pi a} - e^{i\pi a} = -2i \sin \pi a$, this gives the desired result.

Example 5.2. With the usual real-variable meaning for $\log x$ show that

$$\int_0^\infty \frac{\log^2 x}{1+x^2}\,dx = \frac{\pi^3}{8}, \qquad \int_0^\infty \frac{\log x}{1+x^2}\,dx = 0. \tag{5.4}$$

Let

$$I = \int_C \frac{\log^2 z}{1+z^2}\,dz$$

where C is the indented semicircle of Figure 5-4 with inner radius $\epsilon < 1$ and outer radius $R > 1$. Let $\log z = \operatorname{Log} z$, the principal value, on C. Then the integrand of I is analytic except for a simple pole at $z = i$ inside C and hence

$$I = 2\pi i \frac{\operatorname{Log}^2 i}{2i} = \pi\left(\frac{i\pi}{2}\right)^2 = \frac{-\pi^3}{4}. \tag{5.5}$$

On the negative real axis $z = re^{i\pi}$. Hence

$$I = \int_\epsilon^R \frac{\log^2 x}{1+x^2}dx + \int_R^\epsilon \frac{(\operatorname{Log} re^{i\pi})^2}{1+r^2}e^{i\pi}\,dr + J_1 + J_2$$

where J_1 and J_2 are the integrals over the parts of C where $|z| = R$ and $|z| = \epsilon$, respectively. Since $\operatorname{Log} re^{i\pi} = \operatorname{Log} r + i\pi$, we have

$$(\operatorname{Log} re^{i\pi})^2 = \operatorname{Log}^2 r + 2i\pi \operatorname{Log} r - \pi^2$$

and setting $r = x$ in the second integral above thus gives

$$I = \int_\epsilon^R \frac{2\log^2 x + 2\pi i \log x - \pi^2}{1+x^2}\,dx + J_1 + J_2. \tag{5.6}$$

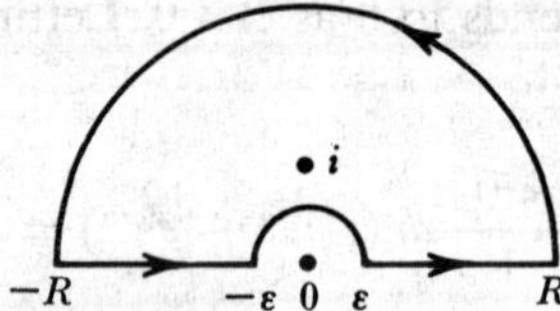

Figure 5-4

On J_1 we have $z = Re^{i\theta}$ with $0 \leq \theta \leq \pi$ and so the integrand satisfies

$$\left|\frac{(\log z)^2}{1 + z^2}\right| = \frac{|\log R + i\theta|^2}{|1 + z^2|} \leq \frac{\log^2 R + \pi^2}{R^2 - 1}.$$

This gives

$$|J_1| \leq \pi R \frac{\log^2 R + \pi^2}{R^2 - 1}.$$

Since $(\log R)^2/R \to 0$ as $R \to \infty$, it follows that $J_1 \to 0$ as $R \to \infty$. A similar calculation with ϵ replacing R shows that $J_2 \to 0$ as $\epsilon \to 0$. Letting $R \to \infty$ and $\epsilon \to 0$ in (5.6) therefore gives

$$\frac{-\pi^3}{4} = \int_0^\infty \frac{2\log^2 x + 2\pi i \log x}{1 + x^2}\, dx - \pi^2 \int_0^\infty \frac{dx}{1 + x^2}$$

when we recall that $I = -\pi^3/4$ by (5.5). The second integral is $\pi/2$, and (5.4) follows by equating real and imaginary parts.

As a check, note that the substitution $x = e^t$ gives the second result (5.4) by inspection.

Problems

In these and subsequent problems use the branch which is customary in real analysis.

1. Show that the result of Example 5.1 remains valid if $a = p + iq$ is complex, $0 < p < 1$. By equating real and imaginary parts evaluate

$$\int_0^\infty \frac{t^{p-1}}{1 + t} \cos(q \log t)\, dt, \qquad \int_0^\infty \frac{t^{p-1}}{1 + t} \sin(q \log t)\, dt.$$

(Use $\sin \pi(p + iq) = \sin \pi p \cosh \pi q + i \cos \pi p \sinh \pi q$.)

2. For $-1 < a < 1$ show that

$$\int_0^\infty \frac{x^a}{1 + x^2}\, dx = \frac{\pi}{2} \sec \frac{\pi a}{2}, \qquad \int_0^\infty \frac{x^a}{(1 + x)^2}\, dx = \pi a \csc \pi a.$$

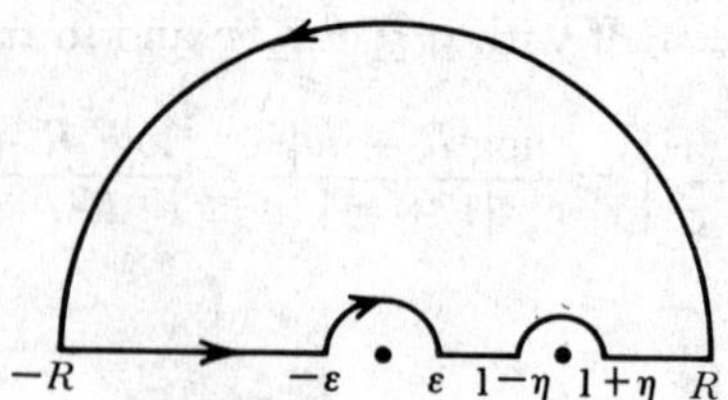

Figure 5-5

3. Do Example 5.1 by use of the contour shown in Figure 5-5.
4. By means of a single contour integral derive both the following:

$$\frac{1}{\pi}\int_0^\infty \frac{x^{1/4}}{x^2+x+1}\,dx = 1-\left(\frac{1}{3}\right)^{1/2}, \qquad \frac{1}{\pi}\int_0^\infty \frac{x^{1/4}}{x^2-x+1}\,dx = \left(\frac{2}{3}\right)^{1/2}.$$

5. Using $f(z) = z^{1/2}(\log z)/(1+z)^2$ show that

$$\int_0^\infty \frac{x^{1/2}\log x}{(1+x)^2}\,dx = \pi, \qquad \int_0^\infty \frac{x^{1/2}}{(1+x)^2}\,dx = \frac{\pi}{2}.$$

6. Integrate from 0 to ∞ as in Example 5.2:

$$\frac{(\log x)^2}{x^3+1}, \qquad \frac{\log x}{x^4+1}, \qquad \frac{\log x}{(x^2+1)^2}, \qquad \frac{\log x}{x^2-1}.$$

(Answers: $3\pi^2\sqrt{2}/64$, $-\pi^2\sqrt{2}/16$, $-\pi/4$, $\pi^2/4$. The last integral requires an indentation in the contour at $x = 1$.)

7. Set $t = x^3$ or $t = x^4$ in the first two integrals of Problem 6 and evaluate by differentiating the result of Example 5.1 with respect to the parameter a.
8. Integrate by parts on (ϵ, R), and thus show that

$$\int_0^\infty \frac{\log(1+x^2)}{x^{1+a}}\,dx = \frac{\pi}{a}\csc\frac{\pi a}{2}.$$

6. Principle of the argument; Rouché's theorem. As stated in Chapter 3, a function is meromorphic in a region if its only singularities there are poles. The residue theorem can be used to get information about the number of zeros of an analytic function, or about the number of zeros minus the number of poles for meromorphic functions.[1]

THEOREM 6.1. *Let $f(z)$ be a meromorphic function in a simply connected domain D containing a Jordan contour C. Suppose f has no zeros or poles on C. Let N be the*

[1] It is possible to incorporate both cases into one by considering a pole to be a zero of negative multiplicity.

number of zeros and P the number of poles of f in the interior of C, where a multiple zero or pole is counted according to its multiplicity. Then

$$\frac{1}{2\pi i}\int_C \frac{f'(z)}{f(z)}\,dz = N - P. \tag{6.1}$$

Proof. The only possible singularities of f'/f inside C are at the zeros or poles of f. If $z = \alpha$ is a zero of f of order m, then

$$f(z) = (z - \alpha)^m h(z)$$

where h is analytic in the neighborhood of α and $h(\alpha) \neq 0$. Hence

$$\frac{f'(z)}{f(z)} = \frac{m}{z - \alpha} + \frac{h'(z)}{h(z)}$$

in the neighborhood of α. Thus the residue of f'/f at α is m. Similarly at a pole, $z = \beta$, of f of order n the residue is $-n$. Applying Theorem 2.1 to the left side of (6.1) yields (6.1) and completes the proof.

If $w = f(z)$, formal calculation suggests that $dw = f'(z)\,dz$ and

$$\frac{1}{2\pi i}\int_C \frac{f'(z)}{f(z)}\,dz = \frac{1}{2\pi i}\int_{C^*} \frac{dw}{w} \tag{6.2}$$

where C^* is the image of C under f, that is, the curve onto which C is mapped by f. The latter expression is the winding number of C^* with respect to the origin. We are thus led to surmise that *$N - P$ is the number of times the image curve C^* winds around the origin in the w plane.* This conclusion can be justified.

Indeed, let the equation of C be $z = \zeta(t)$, $a \leq t \leq b$. The equation of C^* in the w plane is then

$$w = f[\zeta(t)], \qquad a \leq t \leq b.$$

Since $f(z)$ does not vanish on C, the curve C^* does not pass through the origin and it is possible to define a continuous logarithm

$$L(t) = \log f[\zeta(t)]$$

on C^*. On any smooth portion of C two uses of the chain rule give

$$L'(t) = \frac{f'[\zeta(t)]}{f[\zeta(t)]}\zeta'(t)$$

and hence, by the fundamental theorem of calculus,

$$\int_a^b \frac{f'[\zeta(t)]}{f[\zeta(t)]}\zeta'(t)\,dt = \log f[\zeta(t)]\Big|_a^b.$$

This is equivalent to

$$\int_C \frac{f'(z)}{f(z)}\,dz = \log f(z)\Big|_C = \log|f(z)|\Big|_C + i\arg f(z)\Big|_C.$$

The first term on the right is 0, since $\log|f(z)|$ is single-valued and C is closed. Dividing by $2\pi i$ and using Theorem 6.1, we get

$$N - P = \frac{1}{2\pi}\arg f(z)\Big|_C. \tag{6.3}$$

This formulation of the content of Theorem 6.1 is known as the *principle of the argument.*

For a geometrical interpretation, let a line be drawn in the w plane from the origin to the point $w = f(z)$ as shown in Figure 6-1. Then $\arg f(z)$ is the angle which this line makes with a fixed direction, and thus (6.3) describes the number of times the point $w = f(z)$ winds around the point $w = 0$ when z traverses C.

THEOREM 6.2 (*Rouché*). *Let $f(z)$ and $g(z)$ be analytic in a simply connected domain D containing a Jordan contour C. Let $|f(z)| > |g(z)|$ on C. Then $f(z)$ and $f(z) + g(z)$ have the same number of zeros inside C.*

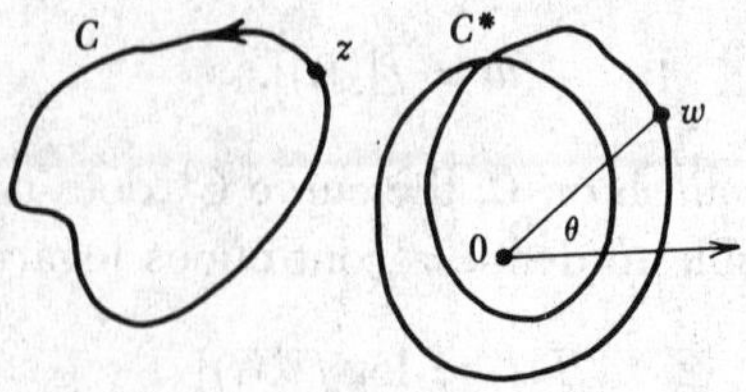

Figure 6-1

First proof. Since $|f| > |g|$ on C, it follows that $f \neq 0$ on C. Also $|f + g| \geq |f| - |g| > 0$ on C. Thus the meromorphic function

$$F(z) = \frac{f(z) + g(z)}{f(z)}$$

has no zeros or poles on C. By the principle of the argument for F,

$$N - P = \frac{1}{2\pi} \arg F(z)\Big|_C . \tag{6.4}$$

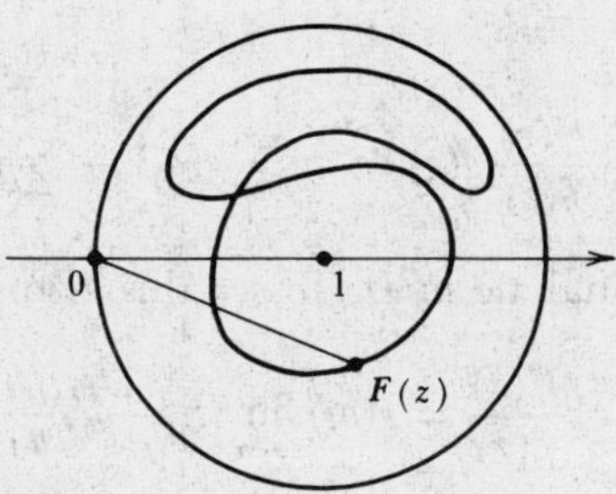

Figure 6-2

Since $|g/f| < 1$ on C, we have $|F(z) - 1| < 1$ when z is on C. Thus the point $F(z)$ lies inside the circle with center at 1 and radius 1, as shown in Figure 6-2. It follows that one of the determinations of $\arg F(z)$ satisfies $|\arg F(z)| < \pi/2$ for z on C, and hence

$$\arg F(z)\Big|_C = 0.$$

(This is true because the only other possibilities are $2\pi k$ for some integer k, and $k \neq 0$ violates the condition $|\arg F(z)| < \pi/2$.) Equation (6.4) now gives $N = P$. Since the poles of F are the zeros of f and the zeros of F are the zeros of $f + g$, the equation $N = P$ indicates that the number of zeros of $f + g$ and of f are equal.

Second proof. For $0 \leq \lambda \leq 1$ let

$$N(\lambda) = \frac{1}{2\pi i} \int_C \frac{f'(z) + \lambda g'(z)}{f(z) + \lambda g(z)}\, dz.$$

Since $|f + \lambda g| \geq |f| - \lambda|g| \geq |f| - |g|$, the denominator does not vanish on C and in fact has a positive lower bound δ independent of λ. From this it follows easily that $N(\lambda)$ is continuous. By Theorem 6.1 applied to $f + \lambda g$ the expression $N(\lambda)$ is an integer, and being both integer-valued and continuous, it must be constant. Thus $N(0) = N(1)$. However $N(0)$ is the number of zeros of f by Theorem 6.1, and $N(1)$ is the number of zeros of $f + g$. This completes the proof.

There is an extension of Theorem 6.1 which is useful in the summation of series. Suppose in Theorem 6.1 that the zeros of f are simple and occur at the points a_i, while the poles of f are simple and occur at the points b_i. Then

$$\frac{1}{2\pi i}\int_C \frac{f'(z)}{f(z)} g(z)\, dz = \sum g(a_i) - \sum g(b_i) \tag{6.5}$$

for every function g analytic in D. To see this, compute the residue

$$\lim_{z\to a_i} (z - a_i)g(z)\frac{f'(z)}{f(z)} = g(a_i) \lim_{z\to a_i} (z - a_i)\frac{f'(z)}{f(z)} = g(a_i).$$

The second equality above follows as in the proof of Theorem 6.1. Similarly the residue at b_i is $-g(b_i)$, and hence, the residue theorem gives (6.5). Equation (6.5) remains valid for multiple zeros and poles if the terms in the sums on the right are repeated a corresponding number of times. The special case $g(z) = 1$ gives Theorem 6.1 again.

A useful choice in (6.5) is $f(z) = \sin \pi z$. This function has no poles, the zeros are simple, and they occur at the integers. Since

$$\frac{f'(z)}{f(z)} = \frac{\pi \cos \pi z}{\sin \pi z} = \pi \cot \pi z$$

it follows that

$$\frac{1}{2\pi i}\int_C g(z)\pi \cot \pi z\, dz = \sum g(n) \quad (g \text{ analytic}) \tag{6.6}$$

where the sum is over the integers n that are within C. In Chapter 6 the integral in (6.6) is used to obtain a variety of significant infinite-series expansions.

In conclusion we mention another formulation of the principle of the argument, due in the main to Cauchy, which is often convenient. Let $f = u + iv$ in Theorem 6.1, so that $\theta = \arg f(z)$ satisfies

$$\cot \theta = \frac{u(x,y)}{v(x,y)}$$

for $z = x + iy$ on C. We begin at some point z_0 of C where $\cot \theta \neq 0$. As z traverses C the function $\cot \theta$ changes sign, sometimes passing through $|\cot \theta| = \infty$, and sometimes through $\cot \theta = 0$. We consider the latter only. Let S^+ denote the number of changes of sign from $+$ to $-$ as $\cot \theta$ passes through the value 0, and let S^- denote the number of changes of sign from $-$ to $+$ as $\cot \theta$ passes through the value 0. Cauchy's theorem states, under the hypothesis of Theorem 6.1, that

$$N - P = (1/2)(S^+ - S^-) \tag{6.7}$$

provided S^+ and S^- are finite.

To see this, consider the graph of $\cot \theta$ vs θ. It is easily checked that increasing θ by π adds 1 to S^+ and decreasing θ by π adds 1 to S^-, provided we start at any point θ_0 where $\cot \theta_0 \neq 0$. The effect of multiple crossings of values θ where $\cot \theta = 0$ is canceled in the subtraction $S^+ - S^-$. Thus the net gain of θ in multiples of 2π is given by the right-hand side of (6.7), and (6.7) is equivalent to (6.3).

The discussion shows that $S^+ - S^-$ gives an estimate for θ as a multiple of π even when only part of C is traversed. This estimate is useful for practical computation.

Example 6.1. Find the number of zeros of

$$P(z) = z^4 + z^3 + 5z^2 + 2z + 4$$

in the first quadrant, $x \geq 0$, $y \geq 0$. The principle of the argument will

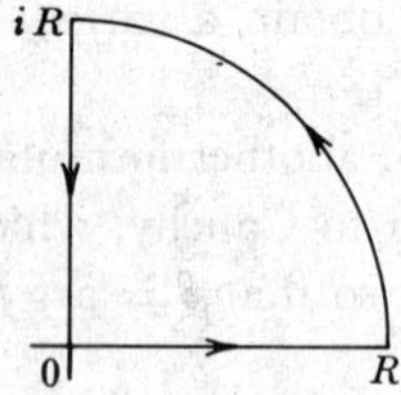

Figure 6-3

be used on a contour C shown in Figure 6-3 where R is large. On $y = 0$, $0 \leq x \leq R$, clearly $P(x) > 0$ and hence

$$\arg P(x) \Big|_0^R = 0;$$

indeed, we can choose $\arg P(x)$ so that $\arg P(x) = 0$ on $0 \leq x \leq R$.

On $z = Re^{i\theta}$, $0 \leq \theta \leq \pi/2$, we have $P(z) = R^4 e^{i4\theta}(1 + \omega)$ where $|\omega| < 2/R$ for large R. Hence $\arg[P(Re^{i\theta})] = 4\theta + \arg(1 + \omega)$. Thus for large R

$$\arg P(Re^{i\theta}) \Big|_{\theta=0}^{\theta=\pi/2} = 2\pi + \delta \tag{6.8}$$

where $\delta \to 0$ as $R \to \infty$.

Finally on $x = 0$, $0 < y < R$

$$\begin{aligned} P(iy) &= (y^4 - 5y^2 + 4) - i(y^3 - 2y) \\ &= (y^2 - 4)(y^2 - 1) - iy(y^2 - 2). \end{aligned}$$

As y decreases from R to 0, $\operatorname{Re} P(iy)$ changes its sign at $y = 2$ and at $y = 1$ while $\operatorname{Im} P(iy)$ changes its sign at $y = \sqrt{2}$. Hence $\operatorname{Re} P(iy) > 0$ and $\operatorname{Im} P(iy) < 0$ for $y > 2$ so that $\arg P(iy)$ is in the fourth quadrant for $2 < y < R$. For $\sqrt{2} < y < 2$, $\arg P(iy)$ is in the third quadrant; for $1 < y < \sqrt{2}$, $\arg P(iy)$ is in the second quadrant; and for $0 < y < 1$, $\arg P(iy)$ is in the first quadrant. Thus as y decreases from R to 0, $\arg P(iy)$ decreases from a value close to 2π to the value 0 at $y = 0$, or

$$\arg P(iy) \Big|_{y=R}^{y=0} = -2\pi + \delta_1$$

where $\delta_1 \to 0$ as $R \to \infty$. Hence

$$\arg P(z)\Big|_C = 0 + 2\pi + \delta - 2\pi + \delta_1.$$

Since C is closed, the change in $\arg P(z)$ must be a multiple of 2π and hence must be zero. Thus $P(z)$ has no zeros in the first quadrant. (Since $P(x)$ is real, this implies that there are also no zeros in the fourth quadrant.)

Problems

1. Since $64 > 17$, the equation $2z^5 + 8z - 1 = 0$ obviously has no roots in $|z| \geq 2$. Confirm this by using Rouché's theorem to show that there are 5 roots in $|z| < 2$. (Take $f = 2z^5$, $g = 8z - 1$.)
2. Show that the above equation has just one root in $|z| < 1$, and that this root is real and positive. (Take $f = 8z - 1$, $g = 2z^5$. There is at least one positive root because $2x^5 + 8x - 1$ has opposite signs at $x = 0$ and $x = 1$.)
3. Show that the above equation has no roots on $|z| = 1$ and hence there are exactly 4 roots in the ring $1 < |z| < 2$.
4. Show that $S^+ = 0$ and $S^- = 2$ for $P(iy)$ in Example 6.1 as y decreases from R to 0.
5. As in Example 6.1, or by (6.7), show that each of the following has just one root in the first quadrant:

$$z^3 - z^2 + 2 = 0, \qquad z^4 + z^2 = 2z - 6, \qquad z^4 + z^3 = 2z^2 - 2z - 4.$$

6. Show that $e^z = 2z + 1$ has exactly one root in $|z| < 1$. (Take

$$f(z) = -2z, \qquad g(z) = e^z - 1 = \int_0^z e^\zeta \, d\zeta.$$

By the latter expression, $|g(z)| \leq e - 1$ in $|z| \leq 1$.)

7. Deduce the fundamental theorem of algebra from Rouché's theorem. (Let $f = a_n z^n$ and $g = a_{n-1}z^{n-1} + \cdots + a_0$ in a circle of large radius R.)
8. *Maximum principle.* Let $g(z)$ be analytic in and on the Jordan contour C. If $|g(z)| \leq m$ on C, where m is constant, deduce from Rouché's theorem that the same inequality holds inside C. (If $|g(z_0)| > m$, Rouché's theorem indicates that

$$-g(z_0) \qquad \text{and} \qquad -g(z_0) + g(z)$$

have the same number of zeros inside C.)

9. Prove that $f(z) = \frac{1}{2}(z + z^{-1})$ assumes every complex value $\alpha = a + ib$,

except for the line segment $b = 0$, $-1 \le a \le 1$, exactly once in the circle $|z| < 1$. (Show that with C the unit circle $z = e^{i\theta}$, $f(e^{i\theta})$ is real and $|f(e^{i\theta})| \le 1$ so that $\arg [f(z) - \alpha]\Big|_C = 0$. Since $f(z) - \alpha$ has exactly one pole in $|z| < 1$, it must have exactly one zero.)

10. Let a_k be the complex zeros of a polynomial $P(z)$, $P(0) \neq 0$. By integrating $z^m P'(z)/P(z)$ for suitable m show that

$$\sum \frac{1}{a_k} = -\frac{P'(0)}{P(0)}, \qquad \sum \frac{1}{a_k^2} = \left(\frac{P'(0)}{P(0)}\right)^2 - \frac{P''(0)}{P(0)}.$$

11. Let $f(z)$ be analytic in and on a simple closed contour C and not zero on C. If $\operatorname{Re} f = 0$ at just $2n$ points of C, show by (6.7) that $f(z)$ has at most n zeros inside C.

12. Let $P(z) = a_0 + a_1 z + \cdots + a_n z^n$ where $0 < a_0 < a_1 < \cdots < a_n$. By applying Enestrom's theorem (p. 42) to $z^n P(1/z)$, show that all n zeros of $P(z)$ lie in $|z| < 1$. Deduce from Problem 11 above that the equation

$$a_0 + a_1 \cos \theta + a_2 \cos 2\theta + \cdots + a_n \cos n\theta = 0$$

has at least $2n$ distinct roots on $0 \le \theta < 2\pi$.

***7. Formulas of Poisson, Hilbert and Bromwich.** Let $f(z)$ be analytic on the imaginary axis and in the right half-plane, and let C be the semicircular contour shown in Figure 7-1. Thus C consists of the segment $z = i\omega$, $-R \le \omega \le R$, together with an arc C_R given by

$$z = Re^{i\theta}, \qquad -\pi/2 \le \theta \le \pi/2.$$

If z is within this contour, the Cauchy integral formula and Cauchy

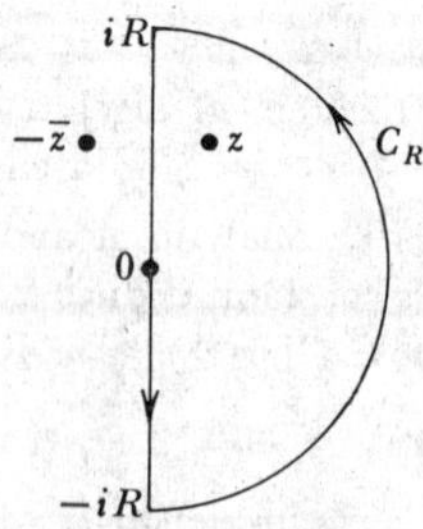

Figure 7-1

integral theorem give, respectively,

$$f(z) = \frac{1}{2\pi i}\int_C \frac{f(\zeta)}{\zeta - z}\,d\zeta, \qquad 0 = \frac{1}{2\pi i}\int_C \frac{f(\zeta)}{\zeta + \bar{z}}\,d\zeta.$$

The second integral is 0 because $-\bar{z}$ is the point obtained by reflecting z in the imaginary axis and so is outside C when z is inside C. Upon multiplying the second integral by a constant, α, and subtracting from the first, we get

$$f(z) = \frac{1}{2\pi i}\int_C \frac{(1-\alpha)\zeta + (\bar{z} + \alpha z)}{(\zeta - z)(\zeta + \bar{z})} f(\zeta)\,d\zeta. \tag{7.1}$$

To estimate the integral over C_R, let $M(R)$ be the maximum of $|f(z)|$ on C_R,

$$M(R) = \max|f(Re^{i\theta})|, \qquad -\pi/2 \leq \theta \leq \pi/2.$$

For given z, the numerator of (7.1) is constant when $\alpha = 1$ and has the order of magnitude R when $\alpha \neq 1$. The denominator has the order of magnitude R^2 and hence the fraction has the order $1/R^2$ or $1/R$, in the two cases respectively. Since the length of C_R is πR, it suffices to assume that

$$\lim_{R\to\infty} \frac{M(R)}{R} = 0 \quad (\alpha = 1), \qquad \lim_{R\to\infty} M(R) = 0 \quad (\alpha \neq 1).$$

These conditions ensure that the integral over C_R tends to 0 as $R \to \infty$. When they hold, the integral for $R \to \infty$ reduces to an integral over the imaginary axis, and setting $\zeta = i\omega$ gives

$$f(z) = \frac{1}{2\pi}\int_{-\infty}^{\infty} \frac{i\omega(1-\alpha) + \bar{z} + \alpha z}{(z - i\omega)(\bar{z} + i\omega)} f(i\omega)\,d\omega. \tag{7.2}$$

The denominator is $|z - i\omega|^2$ and is so written in the sequel.

If $\alpha = 1$, the result (7.2) becomes

$$f(z) = \frac{1}{\pi}\int_{-\infty}^{\infty} \frac{x}{|z - i\omega|^2} f(i\omega)\,d\omega, \tag{7.3}$$

which is called the *Poisson formula for a half-plane.* Other forms of the

Poisson formula are obtained by separating into real and imaginary parts. We set

$$f(z) = u(x,y) + iv(x,y)$$

as usual, and also write $u(0,y) = u(y)$, $v(0,y) = v(y)$. Then

$$f(i\omega) = u(\omega) + iv(\omega),$$

and the Poisson formula gives two real equations of the same structure,

$$u(x,y) = \frac{1}{\pi}\int_{-\infty}^{\infty} \frac{x}{|z - i\omega|^2} u(\omega)\, d\omega, \qquad v(x,y) = \frac{1}{\pi}\int_{-\infty}^{\infty} \frac{x}{|z - i\omega|^2} v(\omega)\, d\omega.$$

Each of these relations is also referred to as the Poisson formula for a half-plane. They have been derived here under the assumption that $f(z)$ is analytic for $x \geq 0$ and that $M(R)/R \to 0$. For instance, the choice $f(z) = 1$ satisfies these conditions and gives

$$1 = \frac{1}{\pi}\int_{-\infty}^{\infty} \frac{x}{|z - i\omega|^2}\, d\omega, \qquad x > 0. \quad (7.4)$$

Instead of $\alpha = 1$ we now suppose that $\alpha = -1$. In this case (7.2) is

$$f(z) = \frac{1}{\pi}\int_{-\infty}^{\infty} \frac{i(\omega - y)}{|z - i\omega|^2} f(i\omega)\, d\omega$$

and separating into real and imaginary parts gives

$$u(x,y) = \frac{1}{\pi}\int_{-\infty}^{\infty} \frac{y - \omega}{|z - i\omega|^2} v(\omega)\, d\omega, \qquad v(x,y) = \frac{1}{\pi}\int_{-\infty}^{\infty} \frac{\omega - y}{|z - i\omega|^2} u(\omega)\, d\omega.$$

These are the *conjugate Poisson formulas.* They were derived here under the assumption that $f(z)$ is analytic for $x \geq 0$ and that $M(R) \to 0$.

In Chapter 3, Section 7 it was shown that an analytic function $f(z)$ is generally determined to within an additive constant by knowledge of its real part alone on the boundary, but no formula was mentioned. For the case of a half-plane this defect has now been remedied. The Poisson formulas give an explicit evaluation of $f(z)$ in $x > 0$ when either the real part or the imaginary part of $f(z)$ is known for $x = 0$. There is no additive constant here, because of the condition $M(R) \to 0$.

If we start with $u(\omega)$ on the imaginary axis we can, in principle, determine $f(z)$ from Poisson's formulas and then recover $v(\omega)$ as the boundary value of $\operatorname{Im} f(z)$. It is natural to inquire whether the determination of $v(\omega)$ from $u(\omega)$ could be carried out more directly by a formula that just involves values of u on the imaginary axis. Such a determination is possible and leads to an important operation known as the Hilbert transform.

Indeed, if $z = iy$ in the conjugate Poisson formulas, the denominator is $(y - \omega)^2$ and the formulas suggest that

$$u(y) = \frac{1}{\pi} \int_{-\infty}^{\infty} \frac{v(\omega)}{y - \omega}\, d\omega, \qquad v(y) = -\frac{1}{\pi} \int_{-\infty}^{\infty} \frac{u(\omega)}{y - \omega}\, d\omega.$$

A function u related to v in this fashion is said to be the *Hilbert transform* of v. Although the method by which we obtained these formulas does not prove their validity, they are verified by making a small semicircular indentation around the point $\omega = iy$ as in Section 4. It suffices to take half the residue at iy in accordance with Lemma 4.1, and to note that the integral over C_R tends to 0 because $M(R) \to 0$. Details of this calculation are not difficult and are left to the reader.

The Hilbert transform has an importance far beyond its connection with Cauchy's formula as sketched above, and has led to the solution of outstanding problems in the theory of Fourier series and harmonic analysis. The Hilbert transform is also the basis for some significant relations in the theory of electrical circuits called the Bode equations. It turns out that the response function of an electrical network is, usually, a rational function which has a zero at ∞ and has all its poles in the left half-plane. Such a function satisfies the conditions of the foregoing derivation, and the relation between $u(\omega)$ and $v(\omega)$ given by the Hilbert transform represents a severe restriction upon the type of response that is actually realizable.

The values $\alpha = 1$ and $\alpha = -1$ are not the only values of technical interest. The choice $\alpha = 0$ leads to an inversion formula in the theory of the Laplace transform and is discussed now.

Setting $\alpha = 0$ in (7.2) gives

$$f(z) = \frac{1}{2\pi} \int_{-\infty}^{\infty} \frac{f(i\omega)}{z - i\omega}\, d\omega.$$

To agree with the notation customary in transform theory, we write F instead of f and s instead of z. Thus

$$F(s) = \frac{1}{2\pi}\int_{-\infty}^{\infty}\frac{F(i\omega)}{s - i\omega}\,d\omega. \tag{7.5}$$

This has been derived here under the hypothesis that $F(s)$ is analytic for $\operatorname{Re} s \geq 0$ and that $M(R) \to 0$ for F.

If $f(t)$ is a given piecewise continuous function of t, the *Laplace transform* is

$$F(s) = \int_0^{\infty} e^{-st} f(t)\,dt \equiv \mathbf{L}f.$$

It is not difficult to show that $F(s)$ is analytic for $\operatorname{Re} s > a$ provided

$$|f(t)| \leq Ae^{at}, \qquad A, a \text{ constant} \tag{7.6}$$

This follows from results on uniform convergence given in Chapter 6.

An important problem in applications is to recover the function $f(t)$ from its transform $F(s)$; that is, given F, it is required to find f such that $\mathbf{L}f = F$. We shall show that this problem is often solved by means of the formula

$$f(t) = \frac{1}{2\pi}\int_{-\infty}^{\infty} e^{i\omega t} F(i\omega)\,d\omega. \tag{7.7}$$

If $f(t)$ is given by (7.7), its Laplace transform is

$$\mathbf{L}f = \frac{1}{2\pi}\int_0^{\infty}\left(e^{-st}\int_{-\infty}^{\infty} e^{i\omega t} F(i\omega)\,d\omega\right)dt$$

provided the integrals exist. A sufficient condition for existence of the integrals is that $\operatorname{Re} s > 0$ and that

$$\int_{-\infty}^{\infty} |F(i\omega)|\,d\omega$$

converge. In this case the integral is absolutely convergent and it is known from real analysis that the order of the integration can be changed. Thus

$$\mathbf{L}f = \frac{1}{2\pi}\int_{-\infty}^{\infty}\left(\int_{0}^{\infty} e^{(i\omega - s)t}\,dt\right)F(i\omega)\,d\omega. \tag{7.8}$$

For Re $s > 0$ the inner integral is

$$\int_0^{\infty} e^{(i\omega - s)t}\,dt = \lim_{R\to\infty} \frac{e^{(i\omega-s)t}}{i\omega - s}\bigg|_0^R = \frac{1}{s - i\omega}$$

and (7.8) therefore agrees with the right-hand side of (7.5). If $F(s)$ satisfies the conditions for (7.5) we conclude that $\mathbf{L}f = F$.

The above analysis generally requires $a < 0$ in (7.6) so that $F(s)$ is analytic for Re $s \geq 0$. To allow $a \geq 0$, let $c > a$ and consider $e^{-ct}f(t)$. This function satisfies (7.6) with a negative exponent and its Laplace transform is

$$\int_0^{\infty} e^{-st}e^{-ct}f(t)\,dt = \int_0^{\infty} e^{-(s+c)t}f(t)\,dt.$$

Thus $\mathbf{L}[e^{-ct}f(t)] = F(s + c)$. If the foregoing analysis can be applied to $e^{-ct}f(t)$ and $F(s + c)$, the formula (7.7) is

$$e^{-ct}f(t) = \frac{1}{2\pi}\int_{-\infty}^{\infty} e^{i\omega t}F(c + i\omega)\,d\omega.$$

Transferring the factor e^{-ct} to the other side gives

$$f(t) = \frac{1}{2\pi}\int_{-\infty}^{\infty} e^{(c+i\omega)t}F(c + i\omega)\,d\omega$$

which is usually written in the form

$$f(t) = \frac{1}{2\pi i}\int_{c-i\infty}^{c+i\infty} e^{st}F(s)\,ds. \tag{7.9}$$

The formula (7.9) is the *Bromwich integral* for inversion of the Laplace transform. It is actually valid at every point of continuity of $f(t)$ provided $f(t)$ is piecewise continuous and (7.6) holds with $a < c$. However, (7.9) was not established in that degree of generality here.

Example 7.1. If $u(\omega)$ is piecewise continuous for $-R \leq \omega \leq R$ and is 0 for $|\omega| > R$, show that the Poisson and conjugate Poisson formulas define harmonic functions when $x \neq 0$. Note that analyticity of $u(\omega)$ is not assumed.

Since $u(\omega) = 0$ for $|\omega| > R$, the formulas in question are

$$\frac{1}{\pi}\int_{-R}^{R} \frac{x}{|z - i\omega|^2} u(\omega)\, d\omega, \qquad \frac{1}{\pi}\int_{-R}^{R} \frac{\omega - y}{|z - i\omega|^2} u(\omega)\, d\omega.$$

If the first expression is denoted by $u(x,y)$ and the second by $v(x,y)$, then

$$u(x,y) + iv(x,y) = \frac{1}{\pi}\int_{-R}^{R} \frac{x + i(\omega - y)}{|z - i\omega|^2} u(\omega)\, d\omega.$$

The numerator is $\bar{z} + i\omega$ and the denominator is $(z - i\omega)(\bar{z} + i\omega)$. Hence

$$u(x,y) + iv(x,y) = \frac{1}{\pi}\int_{-R}^{R} \frac{u(\omega)}{z - i\omega}\, d\omega. \tag{7.10}$$

By an argument virtually identical with that given in Chapter 3, Section 5 this is an analytic function of z for $x \neq 0$. Hence its real and imaginary parts are harmonic.

Example 7.2. Find a function which is harmonic for $x > 0$ and assumes the values π and 0 on the imaginary axis for $|y| < R$ and $|y| > R$, respectively.

The choice $u(\omega) = \pi$ in (7.10) gives

$$\int_{-R}^{R} \frac{d\omega}{z - i\omega} = \frac{\log(z - i\omega)}{-i}\bigg|_{-R}^{R} = i \log \frac{z - iR}{z + iR}$$

with any convenient branch of the logarithm. The real part is

$$u(x,y) = \operatorname{Arg}(z + iR) - \operatorname{Arg}(z - iR)$$

and is harmonic for $x > 0$ by the result of Example 7.1. Geometrically $u(x,y)$ equals the angle subtended by the segment $-R \leq y \leq R$ at the point z as shown in Figure 7-2, and hence the relations $u(0,y) = \pi$ and $u(0,y) = 0$ for $|y| < R$ and $|y| > R$, respectively, are verified by inspection.

The fact that $u(x,y)$ agrees with $u(y)$ on the imaginary axis is not a coincidence. In Problems 6–9 it is shown that this happens at every point of continuity of $u(y)$ provided $u(y)$ is piecewise continuous and

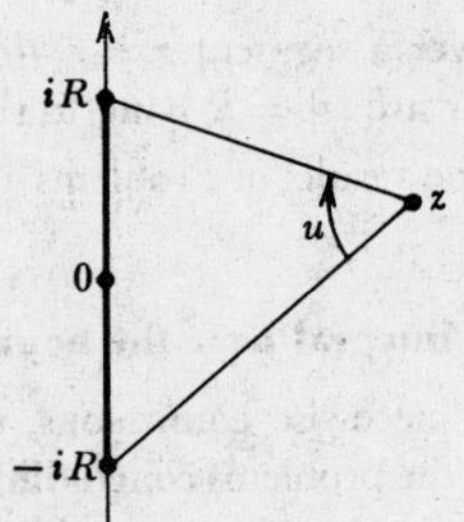

Figure 7-2

$$\int_{-\infty}^{\infty} \frac{|u(\omega)|}{1+\omega^2}\, d\omega \text{ converges.} \tag{7.11}$$

In the same circumstances $u(x,y)$ is harmonic for $x > 0$. This follows from results on uniform convergence given in Chapter 6.

The problem of finding a harmonic function with given boundary values is called the Dirichlet problem and is of great importance in pure and applied mathematics. According to what has just been said, *the Poisson formula solves the Dirichlet problem for a half-plane.* Further discussion of the Dirichlet problem is given in the next chapter.

Problems

1. Apply the Poisson formula to e^{-z} and show that

$$e^{-x}\cos y = \frac{1}{\pi}\int_{-\infty}^{\infty} \frac{x\cos\omega}{x^2+(y-\omega)^2}\, d\omega, \qquad e^{-x}\sin y = \frac{1}{\pi}\int_{-\infty}^{\infty} \frac{x\sin\omega}{x^2+(y-\omega)^2}\, d\omega.$$

2. Apply the Hilbert transform to $1/(z+a)$ where $a > 0$ and get

$$\frac{\pi a}{a^2+y^2} = \fint_{-\infty}^{\infty} \frac{\omega\, d\omega}{(\omega^2+a^2)(\omega-y)}, \qquad \frac{\pi y}{a^2+y^2} = \fint_{-\infty}^{\infty} \frac{a\, d\omega}{(\omega^2+a^2)(y-\omega)}.$$

3. (a) By Poisson's integral get a function $u(x,y)$ which is harmonic for $x > 0$ and assumes the values 0 and 2π on the negative and positive imaginary axis, respectively. (b) Verify that your function actually has these properties.
4. By use of the Bromwich integral with $c > 0$ find functions whose Laplace transforms are

$$\frac{a}{s^2+a^2}, \quad \frac{a}{s(s^2+a^2)}, \quad \frac{1}{(s+a)^2}, \quad \frac{1}{s^8}, \quad \frac{s}{s^2+a^2}.$$

(Consider the integral over a segment $z = c + i\omega$, $-R_0 \leq \omega \leq R_0$, and close this contour by a circular arc $|z| = R$ lying to the left of the line $x = c$.)

5. (a), (b) *Review.* Obtain the results of Problems 1 and 2 by direct evaluation of the integrals.

*The behavior of Poisson's integral near the boundary

In these problems $u(\omega)$ is piecewise continuous, $u(x,y)$ is given by the Poisson formula, and (7.11) holds. By the problems concluding Section 4 the Poisson integral is convergent for $x > 0$. This is taken for granted here.

6. Using (7.4) get

$$u(x,y) - u(y) = \frac{x}{\pi}\int_{-\infty}^{\infty} \frac{u(\omega) - u(y)}{x^2 + (y - \omega)^2}\, d\omega.$$

7. If y is a point of continuity of $u(\omega)$, given any $\epsilon > 0$ we can choose $\delta > 0$ so that $|u(\omega) - u(y)| \leq \epsilon$ for $|\omega - y| \leq \delta$. That being done, let

$$J_1 = \frac{x}{\pi}\int_{y-\delta}^{y+\delta} \frac{u(\omega) - u(y)}{x^2 + (y - \omega)^2}\, d\omega.$$

Deduce from (7.4) that

$$|J_1| \leq \frac{x}{\pi}\int_{y-\delta}^{y+\delta} \frac{\epsilon}{x^2 + (y - \omega)^2}\, d\omega \leq \frac{x}{\pi}\int_{-\infty}^{\infty} \frac{\epsilon}{x^2 + (y - \omega)^2} = \epsilon.$$

8. If $|y - \omega| \geq \delta$, show that there is a positive constant η, depending only on δ and y, such that $(\omega - y)^2/(1 + \omega^2) \geq \eta$, $-\infty < \omega < \infty$.

9. By the preceding problem $x^2 + (y - \omega)^2 \geq \eta(1 + \omega^2)$ in the integrals over $(-\infty, y - \delta)$ and $(y + \delta, \infty)$ in Problem 6. If J_2 denotes the sum of these integrals, deduce that

$$|J_2| \leq \frac{x}{\pi}\int_{|y-\omega|>\delta} \frac{|u(\omega)| + |u(y)|}{\eta(1 + \omega^2)}\, d\omega \leq \frac{x}{\pi\eta}\int_{-\infty}^{\infty} \frac{|u(\omega)| + |u(y)|}{1 + \omega^2}\, d\omega,$$

and hence $|J_2| < \epsilon$ if x is small. Thus $|u(x,y) - u(y)| = |J_1 + J_2| < 2\epsilon$, which gives

$$\lim_{x \to 0+} u(x,y) = u(y).$$

*8. The residue at infinity.

Let $f(z)$ have an isolated singular point at $z = \infty$. The *residue at infinity* is defined by

$$\operatorname{Res} f(\infty) = -\frac{1}{2\pi i}\int_C f(z)\, dz \tag{8.1}$$

where C is any circle $|z| = R$ so large that the only singularity of f in $|z| \geq R$ is the point $z = \infty$. The minus sign is used because the point $z = \infty$, at which the residue is being computed, is exterior to the circle, and the circle is negatively oriented with respect to its exterior.

The residue at ∞ can also be defined to be $-a_{-1}$, where a_{-1} is the coefficient of $1/z$ in the Laurent expansion of $f(z)$ at ∞. Hence $-\text{Res} f(\infty)$ is the coefficient of ζ in the expansion of $f(1/\zeta)$ near 0, and it is the coefficient of $1/\zeta$ in the expansion of

$$F(\zeta) = \frac{1}{\zeta^2} f\left(\frac{1}{\zeta}\right)$$

near 0. It follows that $\text{Res} f(\infty) = -\text{Res}\, F(0)$. The change of variable $z = 1/\zeta$ shows that

$$\int_{|\zeta|=1/R} f\left(\frac{1}{\zeta}\right) \frac{d\zeta}{\zeta^2} = \int_{|z|=R} f(z)\, dz$$

and so the integral in (8.1) agrees with the integral for $\text{Res}\, F(0)$. Hence, the two definitions of $\text{Res} f(\infty)$ are consistent, and the value of the integral (8.1) does not depend on R.

Familiar formulas for $\text{Res}\, F(0)$ give corresponding formulas for $\text{Res} f(\infty)$. As an illustration, $\text{Res}\, F(0) = -\lim_{\zeta \to 0} \zeta F(\zeta)$ if F has a simple pole at $\zeta = 0$, and hence

$$\text{Res} f(\infty) = -\lim_{z\to\infty} z f(z), \qquad \text{if } f(\infty) = 0. \quad (8.2)$$

The same result is obtained by inspection of the Laurent series, since the condition $f(\infty) = 0$ means that $a_n = 0$ for $n \geq 0$. In that case

$$z f(z) = a_{-1} + \frac{a_{-2}}{z} + \frac{a_{-3}}{z^2} + \cdots$$

and $z f(z) \to a_{-1}$ as $z \to \infty$.

It is a remarkable fact that *for every rational function, the sum of all the residues in the extended plane is 0.* This is a special case of the following:

THEOREM 8.1. *If the singularities of $f(z)$ in $|z| \leq \infty$ are isolated and are finite in number, the sum of all the residues is 0.*

Proof. Let C be a circle $|z| = R$ large enough to contain all finite singularities[1] of f, and consider

$$I = \frac{1}{2\pi i}\int_C f(z)\,dz.$$

By the residue theorem I equals the sum of all the residues in $|z| < \infty$, and I also equals $-\operatorname{Res} f(\infty)$ by definition. This completes the proof.

In the integrals considered in Sections 3 and 7, the limit of the integral over the semicircular contour C_R was 0. Sometimes the limit of the integral over C_R exists but is not 0. Evaluation of the limit in such cases is facilitated by the following lemma:

LEMMA 8.1. *Let $f(z)$ be regular at $z = \infty$ and satisfy $f(\infty) = 0$. Let C_R be the contour $z = Re^{i\theta}$, $\theta_0 \le \theta \le \pi + \theta_0$. Then*

$$\lim_{R\to\infty}\int_{C_R} f(z)\,dz = -\pi i \operatorname{Res} f(\infty).$$

If C_R were the whole circle $|z| = R$, we would get $-2\pi i \operatorname{Res} f(\infty)$. The lemma indicates that integration over half the circle gives half this result provided $R \to \infty$ and $f(\infty) = 0$. Lemma 8.1 is proved by applying Lemma 4.1 to $F(\zeta)$ with $\phi = \pi$.

Example 8.1. Show that, as a Cauchy principal value,

$$\int_{-\infty}^{\infty} \frac{5x^3}{1 + x + x^2 + x^3 + x^4}\,dx = -2\pi\left(\sin\frac{\pi}{5} + \sin\frac{2\pi}{5}\right).$$

With the same contour C as in Section 3 consider

$$I = \int_C \frac{5z^3}{1 + z + z^2 + z^3 + z^4}\,dz.$$

Multiplication of the numerator and denominator by $z - 1$ shows that the integrand is

$$f(z) = \frac{5z^3(z-1)}{z^5 - 1}.$$

[1] The term "finite singularity" does not mean that $f(z)$ is finite but means that the singular point is in $|z| < \infty$.

The zeros of the denominator in the upper half-plane are simple and are at the points

$$\omega_1 = e^{2\pi i/5}, \qquad \omega_2 = e^{4\pi i/5}.$$

The corresponding residues are the values of

$$\frac{5z^3(z-1)}{5z^4} = 1 - \frac{1}{z}$$

at ω_1 and ω_2, and the residue at ∞ is

$$-\lim_{z\to\infty} \frac{5z^4(z-1)}{z^5-1} = -\lim_{z\to\infty} \frac{5(1-z^{-4})}{1-z^{-5}} = -5.$$

By the residue theorem

$$I = \int_{-R}^{R} f(x)\,dx + \int_{C_R} f(z)\,dz = 2\pi i\left(1 - \frac{1}{\omega_1} + 1 - \frac{1}{\omega_2}\right).$$

Letting $R \to \infty$ and using Lemma 8.1 with $\operatorname{Res} f(\infty) = -5$, we get

$$\int_{-\infty}^{\infty} f(x)\,dx + 5\pi i = 2\pi i(2 - e^{-2\pi i/5} - e^{-4\pi i/5}).$$

The desired result follows by equating real parts.

As a check, it may be observed that equating imaginary parts gives

$$1 + 2\cos\frac{2\pi}{5} = 2\cos\frac{\pi}{5}$$

which is true because the sum of the roots of $z^5 - 1 = 0$ is 0.

Some important classes of functions have branch points in $|z| < \infty$ but only an isolated singularity at $z = \infty$. The following theorem is useful in the study of functions of this kind:

Theorem 8.2. *Let C be a Jordan contour oriented in the clockwise direction, that is, oriented positively with respect to its exterior domain. Let $f(z)$ be analytic on C and also analytic in the exterior domain except possibly for finitely many isolated singularities α_i. Then*

$$\int_C f(z)\,dz = 2\pi i \sum \operatorname{Res} f(\alpha_i)$$

where the sum includes the term $\operatorname{Res} f(\infty)$.

Outline of proof. Let z_0 be in the interior domain of C and let K be a circle $|z| = R$ large enough to contain C and all finite singularities α_i in its interior. Choose two rays through z_0 that do not meet any α_i. Let them meet K in the two points P_1, P_2. From P_i extend line segments toward z_0 but terminate each segment when it first meets C as shown in Figure 8-1. In this way we get two Jordan contours C_1 and C_2, not meeting any α_i, such that

$$\int_C f(z)\,dz + \int_K f(z)\,dz = \int_{C_1} f(z)\,dz + \int_{C_2} f(z)\,dz. \tag{8.3}$$

Applying the residue theorem to each integral on the right of (8.3) we see that the two integrals together give

$$2\pi i \sum \operatorname{Res} f(\alpha_i)$$

where the sum is over all finite singularities α_i. Since the integral over K is $-2\pi i \operatorname{Res} f(\infty)$ by definition, (8.3) is equivalent to the desired conclusion.

To use Theorem 8.2 we have to pick a single-valued branch in the

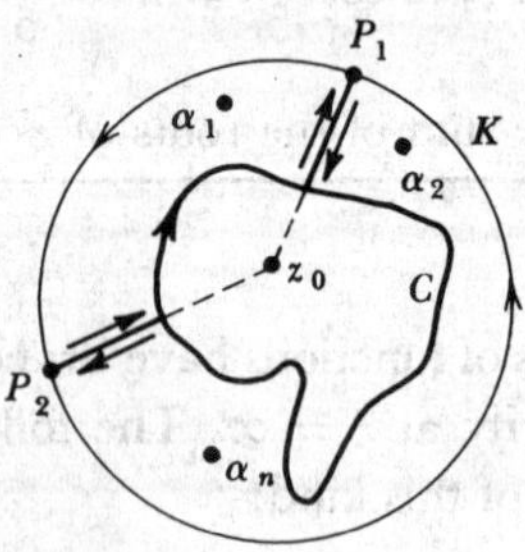

Figure 8-1

neighborhood of ∞. This process is discussed now, insofar as it pertains to the function

$$\sqrt{1 - z^2}.$$

(Other cases are similar.) The region in which we propose to get a single-valued branch is the plane $|z| < \infty$ minus the segment $-1 \leq x \leq 1$ of the real axis, and is referred to as the *cut plane.* We choose the branch that is positive on the positive imaginary axis, so that

$$\sqrt{1 + y^2} > 0 \qquad \text{for } y > 0. \quad (8.4)$$

Hence, the desired branch is positive on the upper edge of the cut.

For $z \neq 1$

$$\sqrt{1 - z^2} = -iz\sqrt{1 - 1/z^2} \tag{8.5}$$

where the minus sign is chosen because of (8.4). By the binomial theorem

$$\sqrt{1 - \zeta} = 1 - c_1\zeta - c_2\zeta^2 - \cdots - c_n\zeta^n - \cdots, \qquad |\zeta| < 1 \quad (8.6)$$

where $-c_n$ is the nth binomial coefficient for exponent $\frac{1}{2}$, and we use this branch with $\zeta = 1/z^2$ in (8.5). Since (8.6) is analytic for $|\zeta| < 1$, the function (8.5) is analytic for $|z| > 1$. Thus, the singularity at ∞ is isolated.

One can easily pick a branch in the upper half-plane which is positive on the y axis, and a branch in the lower half-plane which is negative on the y axis. (The latter choice is dictated by (8.5).) In view of the overlapping of the three regions

$$|z| > 1, \qquad \text{Im } z > 0, \qquad \text{Im } z < 0,$$

the construction gives a single-valued branch in the cut plane.

The problem of constructing a branch near ∞ can also be approached geometrically. Consider, instead of (8.5), the function

$$\sqrt{z^2 - 1} = \sqrt{(z + 1)(z - 1)}.$$

This function is i times the function (8.5). The reversal of sign is intro-

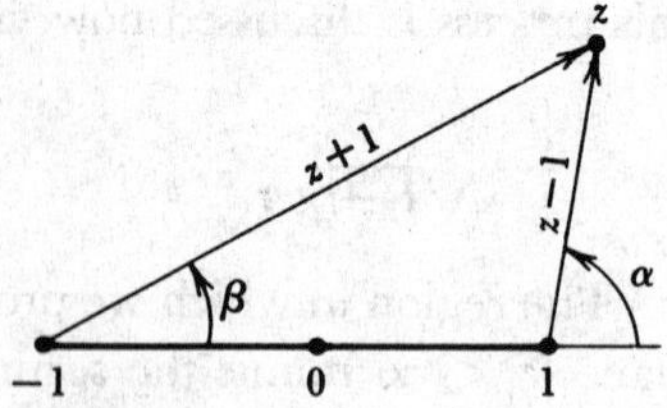

Figure 8-2

duced so that the function will behave like $\sqrt{z^2} = z$ near ∞, and this equation is used to select the branch at ∞. If we define

$$\alpha = \arg(z - 1), \qquad \beta = \arg(z + 1)$$

then these angles have, among other values, the values illustrated geometrically in Figure 8-2. A cut from -1 to 1 forces both α and β to change by the same multiple of 2π as z traverses a closed curve. Hence

$$\tfrac{1}{2}(\alpha + \beta)$$

changes by a multiple of 2π, and $\sqrt{z^2 - 1}$ returns to its original value.

It is often advantageous to use the geometric method and the method of infinite series together. The geometric method gives insight into the general behavior of the branch selected, and the series facilitates computation of $\operatorname{Res} f(\infty)$.

Example 8.2. With the positive branch of the square root show that

$$\int_{-1}^{1} \frac{\sqrt{1 - x^2}}{1 + x^2}\, dx = \pi(\sqrt{2} - 1).$$

Consider

$$I = \int_C \frac{\sqrt{1 - z^2}}{1 + z^2}\, dz$$

where C is the contour illustrated in Figure 8-3 and $\sqrt{1 - z^2}$ is the function discussed above. Since the integrand has opposite signs on the upper

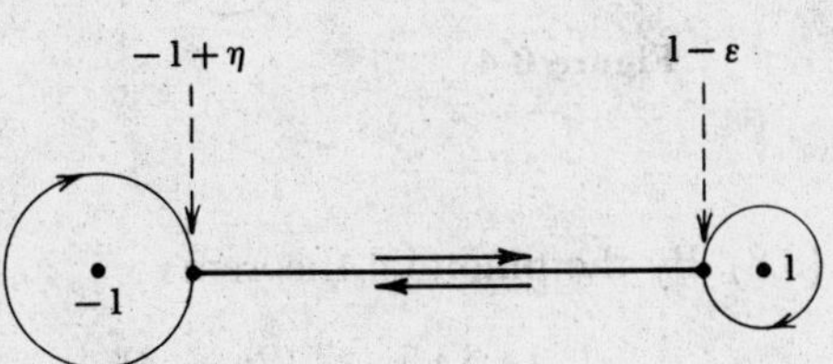

Figure 8-3

and lower edge of the cut, and since the two edges are traversed in opposite directions,

$$I = 2\int_{-1+\epsilon}^{1-\eta} \frac{\sqrt{1-x^2}}{1+x^2}\,dx + J_1 + J_2 \tag{8.7}$$

where J_1 and J_2 are the integrals over the small circles of radius ϵ and η, respectively. Evaluating $\sqrt{1-z^2}/2z$ at i and $-i$, we see that the residues at i and $-i$ are

$$\frac{\sqrt{2}}{2i} \quad \text{and} \quad \frac{-\sqrt{2}}{-2i}$$

respectively. The residue at ∞ is given by (8.5) as

$$-\lim_{z\to\infty} z\frac{-iz}{1+z^2}\sqrt{1-1/z^2} = i.$$

Hence by Theorem 8.2

$$I = 2\pi i\left(\frac{\sqrt{2}}{2i} + \frac{\sqrt{2}}{2i} + i\right) = 2\pi(\sqrt{2}-1). \tag{8.8}$$

It is easily checked that $J_i \to 0$ as $\epsilon \to 0$ and $\eta \to 0$, and hence (8.7) and (8.8) give the desired result.

Example 8.3. With the positive branch of the square root show that

$$\int_0^1 x^{n-2}\sqrt{x(1-x)}\,dx = \pi c_n, \qquad n = 2, 3, 4, \ldots,$$

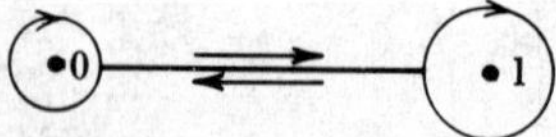

Figure 8-4

where c_n is as in (8.6). By the binomial theorem

$$c_n = \frac{1}{2n}\frac{1\cdot 3\cdot 5\cdots(2n-3)}{2\cdot 4\cdot 6\cdots(2n-2)}$$

but we shall not need this expression in the proof.

Consider

$$I = \int_C z^{n-2}\sqrt{z(1-z)}\,dz$$

where C is the contour in Figure 8-4 and where $\sqrt{z(1-z)}$ is the single-valued branch in the plane cut along the real axis from 0 to 1 which is positive on the upper edge of the cut. By following a small circle around $z = 0$ it is seen that $\sqrt{z}$ is negative on the lower edge of the cut and is a positive multiple of i on the negative real axis. The latter property shows that the appropriate formula for large $|z|$ is

$$\sqrt{z(1-z)} = -iz\sqrt{1-1/z},$$

where the radical is given by (8.6) with $\zeta = 1/z$. Thus the integrand is

$$z^{n-2}\sqrt{z(1-z)} = -iz^{n-1}\left(1 - \frac{c_1}{z} - \cdots - \frac{c_n}{z^n} - \cdots\right)$$

and the coefficient of $1/z$ is ic_n by inspection. The residue at ∞ is $-ic_n$, and since there are no finite singularities, $I = 2\pi c_n$. The rest of the proof is similar to that in Example 8.2.

Problems

1. Let $f(z) = 1/z$ for $0 < |z| < \infty$ and let $f(\infty) = 0$. Show that $f(z)$ is analytic at $z = \infty$ but that $\operatorname{Res} f(\infty) = -1$.

2. Find the residue at ∞:

$$\frac{z^3}{z^4-1},\quad \left(z+\frac{2}{z}\right)^4,\quad \frac{e^z}{z},\quad \left(z^2+\frac{1}{z^2}\right)\sin z.$$

3. Let $P(z)$ be a polynomial of degree $n \geq 1$ and let $f(z) = P'(z)/P(z)$. Prove by (8.2) that $-\text{Res} f(\infty) = n$ and deduce the fundamental theorem of algebra from Theorem 8.1.

4. Show that

$$\int_{-\infty}^{\infty} \frac{2x}{1+x+x^2}\,dx = -\frac{2\pi}{\sqrt{3}},\qquad \int_{-\infty}^{\infty} \frac{x^3}{(x^2+1)(x^2+2x+2)}\,dx = -\frac{4\pi}{5}.$$

5. Deduce the results of Examples 8.2 and 8.3 by considering

$$\int_C \sqrt{z^2-1}\,dz \qquad \text{and} \qquad \int_C z^{n-2}\sqrt{z(z-1)}\,dz.$$

(The geometric construction of Figure 8-2 indicates that

$$\sqrt{x^2-1} = i\sqrt{1-x^2} \qquad \text{or} \qquad \sqrt{x^2-1} = -i\sqrt{1-x^2}$$

on the upper and lower sides of the cut, respectively, where $\sqrt{1-x^2}$ is the positive branch.)

6. Evaluate

$$\int_0^1 \frac{x^2\,dx}{(1+x^2)\sqrt{1-x^2}},\quad \int_0^1 \frac{x^4\,dx}{\sqrt{x(1-x)}},\quad \int_0^1 \frac{x^4\,dx}{(1+x^2)\sqrt{1-x^2}}.$$

7. The function $f(z) = \sqrt[3]{z^2(1-z)}$ has a single-valued branch in the plane cut from $x = 0$ to $x = 1$, because $\text{Arg}[z^2(1-z)]$ increases by 6π when z makes a circuit around the cut. Let $f(z)$ denote the branch which is positive on the upper edge of the cut. If a small circle is traversed in the negative direction around $z = 1$ note that $\arg(1-z)$ diminishes by 2π. Conclude that if $f(x)$ is the value on the upper edge of the cut, $f(x)\exp(-2\pi i/3)$ is the value on the lower edge. Similarly deduce that $f(x)$ has the form $g(x)\exp(2\pi i/3)$ for $x < 0$ where $g(x) > 0$. Hence for large $|z|$

$$f(z) = -z(1-1/z)^{1/3}e^{2\pi i/3}.$$

Using these results show that

$$\int_0^1 \frac{dx}{\sqrt[3]{x^2(1-x)}} = 2\pi\frac{\sqrt{3}}{3},\qquad \int_0^1 \sqrt[3]{x^2(1-x)}\,dx = 2\pi\frac{\sqrt{3}}{27}.$$

8. *Review.* Do Example 8.3 by letting $x = \sin^2\theta$ and using the method of Section 2.

9. In the analysis of Section 7 suppose $f(z)$ has an isolated singularity at ∞, and suppose the conditions on $M(R)$ are replaced by the weaker conditions

$$\lim_{\zeta\to\infty} \frac{f(\zeta)}{\zeta^2} = 0 \quad (\alpha = 1), \qquad \lim_{\zeta\to\infty} \frac{f(\zeta)}{\zeta} = 0 \quad (\alpha \neq 1).$$

How must the formulas of Section 7 be modified? (If $F(\zeta)$ denotes the integrand in (7.1), it is easily checked that Res $F(\infty)$ is

$$-(z + \bar{z}) \lim_{\zeta\to\infty} \frac{f(\zeta)}{\zeta} \qquad \text{or} \qquad (1 - \alpha) \lim_{\zeta\to\infty} f(\zeta)$$

for $\alpha = 1$ or $\alpha \neq 1$, respectively. Use Lemma 8.1.)

*9. Other forms of the residue theorem.

We now give a version of the residue theorem that involves the winding number and is an important aid in the study of many-valued functions. As the reader will recall, the winding number is

$$N(C,\alpha) = \frac{1}{2\pi i} \int_C \frac{dz}{z - \alpha} = \frac{1}{2\pi i} \log(z - \alpha)\Big|_C = \frac{\theta}{2\pi}\Big|_C$$

where C is any closed contour not passing through the point α. Geometrically $N(C,\alpha)$ represents the number of times C winds around α and its value can often be determined by inspection.

LEMMA 9.1. *Let $f(z)$ have an isolated singularity at $z = \alpha$ with singular part*

$$f_1(z) = \frac{a_{-1}}{z - \alpha} + \frac{a_{-2}}{(z - \alpha)^2} + \cdots + \frac{a_{-n}}{(z - \alpha)^n} + \cdots$$

and let C be any closed contour not passing through the point α. Then

$$\frac{1}{2\pi i} \int_C f_1(z)\, dz = a_{-1} N(C,\alpha).$$

Proof. Write $f_1(z)$ in the form

$$f_1(z) = \frac{a_{-1}}{z - \alpha} + s_n(z) + r_n(z) \tag{9.1}$$

where

$$s_n(z) = \frac{a_{-2}}{(z-\alpha)^2} + \cdots + \frac{a_{-n}}{(z-\alpha)^n}$$

and where $r_n(z)$ is defined by the equation. Evidently $s_n(z)$ is the derivative of the single-valued function

$$F(z) = -\frac{a_{-2}}{z-\alpha} - \cdots - \frac{a_{-n}}{(n-1)(z-\alpha)^{n-1}}$$

and integrates to 0 by (2.1). Integrating (9.1) therefore gives

$$\int_C f_1(z)\, dz = a_{-1} \int_C \frac{dz}{z-\alpha} + \int_C r_n(z)\, dz. \tag{9.2}$$

If ϵ is any positive number, the uniform convergence established in Chapter 3, Section 6 shows that $|r_n(z)| < \epsilon$ on C, for sufficiently large n. In this case (9.2) gives

$$\left| \int_C f_1(z)\, dz - a_{-1} \int_C \frac{dz}{z-\alpha} \right| \leq \epsilon L \tag{9.3}$$

where L is the length of C. Since ϵ is arbitrary, while the left side of (9.3) is independent of ϵ, the left side must be 0. This completes the proof.

THEOREM 9.1. *Let $f(z)$ be analytic in a simply connected domain D except at points $\alpha_1, \alpha_2, \ldots, \alpha_n$ where $f(z)$ has isolated singularities. Let C be a closed contour lying in D and not passing through any point α_j. Then*

$$\int_C f(z)\, dz = 2\pi i \sum N(C,\alpha_j) \operatorname{Res} f(\alpha_j).$$

This follows from Lemma 9.1 in the same way that Theorem 2.1 follows from Lemma 2.1.

To illustrate the use of Theorem 9.1 let

$$w = \int_0^z \frac{d\zeta}{1+\zeta^2} \tag{9.4}$$

where the integral is along some contour C_1 that does not pass through

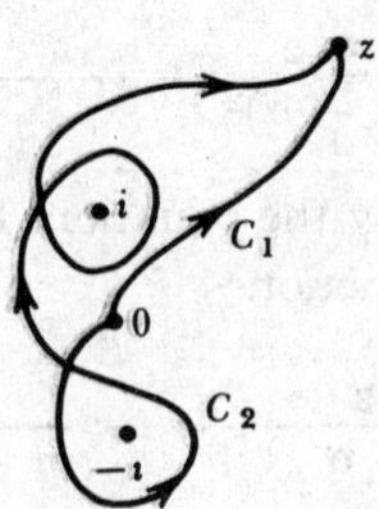

Figure 9-1

the point i or $-i$. Let C_2 be another such contour (Figure 9-1). Traversing C_1 from 0 to z and traversing C_2 in the opposite direction gives a closed contour $C = C_1 - C_2$. By Theorem 9.1

$$\int_C \frac{d\zeta}{1+\zeta^2} = \pi N(C,i) - \pi N(C,-i),$$

since the residues of the integrand at i and $-i$ are $1/(2i)$ and $1/(-2i)$, respectively. Thus

$$\int_{C_1} \frac{d\zeta}{1+\zeta^2} = \int_{C_2} \frac{d\zeta}{1+\zeta^2} + N\pi$$

where N is an integer. This shows that the possible values of w in (9.4) differ by a multiple of π. Actually $w = \tan^{-1}z$, and our result agrees with the familiar fact that the inverse function $z = \tan w$ has period π.

Similar remarks apply to other integrals. If w is defined by

$$w = \int_{z_0}^{z} f(\zeta)\, d\zeta \tag{9.5}$$

then w depends not only on z, in general, but also on the path of integration. Theorem 9.1 indicates that the different values are all obtained from a single value by adding an expression of the form

$$2\pi i(N_1R_1 + N_2R_2 + \cdots + N_kR_k) \tag{9.6}$$

where the N_j are integers and R_j are the residues at the singular points. The numbers $2\pi iR_j$ are often referred to as *periods.* They are not periods of the integral (9.5), however, but of the inverse function.

The same line of thought extends to integrands with branch points. For example, consider

$$w = \int_0^z \frac{d\zeta}{\sqrt{1-\zeta^2}}$$

in the plane cut from -1 to 1. The integral around the cut is

$$2\int_{-1}^{1} \frac{dx}{\sqrt{1-x^2}} = 2\pi.$$

If w is computed by use of two paths C_1 and C_2, it can be shown that the corresponding values w_1 and w_2 satisfy $w_1 - w_2 = 2\pi N$ where N is the number of times the path $C = C_1 - C_2$ winds around the cut (Figure 9-2). This agrees with the fact that $w = \sin^{-1} z$ has values that differ by a multiple of 2π, and with the fact that the inverse function $z = \sin w$ has period 2π.

When there are several branch cuts or singularities the difference in the integrals for two paths is given by (9.6) where R_j is the integral around the jth branch cut or singularity. See Figure 9-3.

In Figure 9-3, and in several previous cases, what has really been done is to use Cauchy's integral theorem in a multiply connected domain. An intuitive discussion of this theorem will now be given. The idea is simple and is easily justified in applications. However the general form is not required elsewhere in this book. The situation considered in Figure 2-4 is sufficient to treat seemingly more complicated cases, by the device of subtracting the singular parts f_j and using linearity. By this device the topological niceties needed for rigorous treatment of the general case have been avoided here.

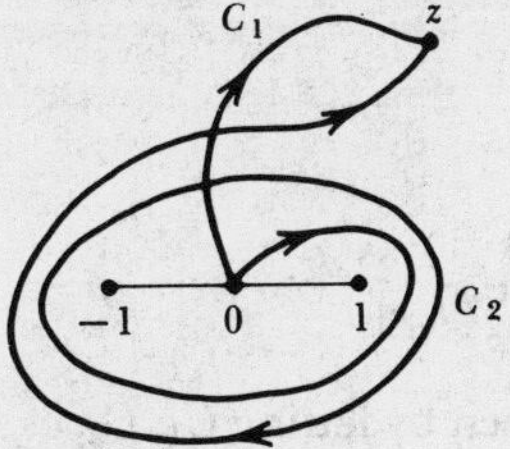

Figure 9-2

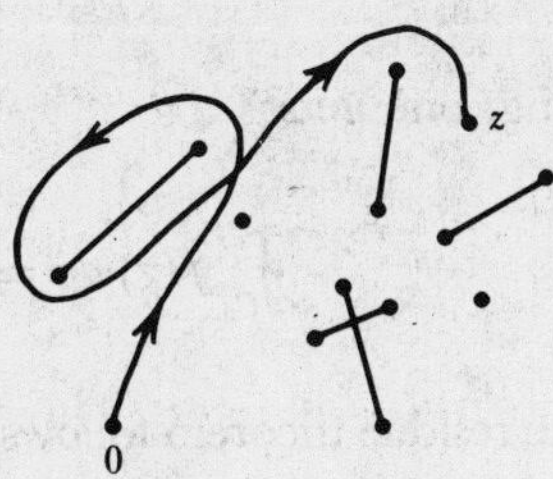

Figure 9-3

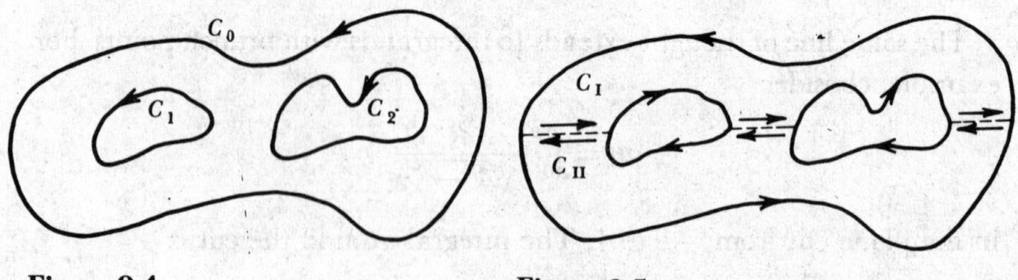

Figure 9-4 **Figure 9-5**

Let C_0 be a positively oriented Jordan contour. Let $C_1, C_2, \ldots, C_n$ each be a Jordan contour lying in the interior of C_0. Let $C_1, C_2, \ldots, C_n$ be disjoint, that is, have no points in common, and let no C_j lie inside C_k for $j = 1, 2, \ldots, n$ and $k = 1, 2, \ldots, n$. Consider the domain D bounded by $C_0, C_1, \ldots, C_n$. Let $f(z)$ be analytic in a larger domain containing $\overline{D}$, the closure of D. Then

$$\int_{C_0} f(z)\,dz = \sum_{j=1}^{n} \int_{C_j} f(z)\,dz \tag{9.7}$$

where each C_j is positively oriented with respect to its interior domain, as shown in Figure 9-4.

For the special case shown in Figure 9-4 the theorem is proved by making the cuts shown in Figure 9-5. Let C_{I} and C_{II} denote the two positively oriented contours shown in Figure 9-5. By Cauchy's theorem for simply connected domains

$$\int_{C_{\mathrm{I}}} f(z)\,dz = 0, \qquad \int_{C_{\mathrm{II}}} f(z)\,dz = 0.$$

Adding, one gets

$$\int_{C_0} f(z)\,dz = \int_{C_1} f(z)\,dz + \int_{C_2} f(z)\,dz.$$

The residue theorem follows from (9.7), as seen by letting $C_1, C_2, \ldots, C_n$ be small circles centered at the singular points α_j. In this case

$$\frac{1}{2\pi i}\int_{C_j} f(z)\,dz = \operatorname{Res} f(\alpha_j)$$

and (9.7) gives Theorem 2.1.

The curves $C_1, C_2, \ldots, C_n$ above are positively oriented with respect to their interior domains but are negatively oriented with respect to the domain D. If all curves C_j are positively oriented with respect to D as shown in Figures 9-5 and 9-6, equation (9.7) takes the form

$$\sum_{j=0}^{n}\int_{C_j} f(z)\,dz = 0.$$

This is equivalent to

$$\int_C f(z)\,dz = 0$$

where $C = C_0 + C_1 + \cdots + C_n$ is the *oriented boundary* of D. In this form the result is a direct generalization of Cauchy's integral theorem to multiply connected domains.

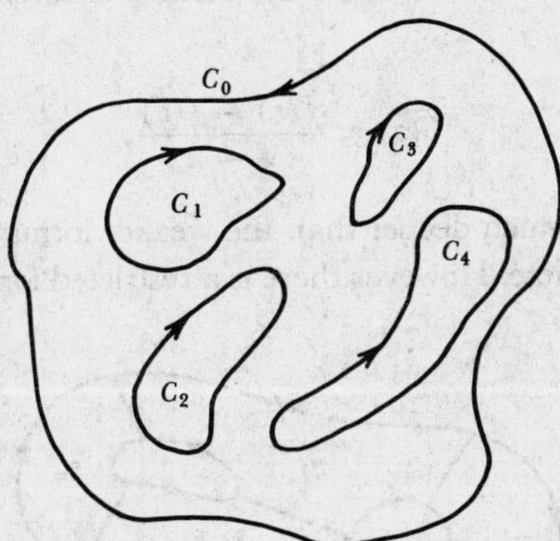

Figure 9-6

Hitherto it has been assumed that $f(z)$ is analytic on C, so that analyticity is required not only in D but in the closed region $\overline{D}$. Actually it suffices to have analyticity in D provided f is continuous. This is the content of the following:

THEOREM 9.2 (*Strong form of Cauchy's theorem*). *Let D be a domain bounded by Jordan contours $C_0, C_1, \ldots, C_n$ as explained above, and let*

$$C = C_0 + C_1 + C_2 + \cdots + C_n$$

be the oriented boundary of D. Then

$$\int_C f(z)\, dz = 0$$

for every function f which is analytic in D and continuous in $\overline{D}$.

Theorem 9.2 simplifies the integration over branch cuts discussed in Section 5 because it shows that extension of the region of analyticity is not necessary. The theorem is also useful when complex analysis is applied to problems of potential theory as in Section 7. Applying Theorem 9.2 to

$$F(z) = f(z) - \sum f_j(z)$$

as in Section 2 gives a strong form of the residue theorem in which analyticity on the boundary is not assumed. A strong form of the Cauchy integral formula is obtained from the residue theorem, or by applying Theorem 9.2 to

$$G(z) = \frac{f(z) - f(a)}{z - a}.$$

Theorem 9.2 is much deeper than the weaker forms discussed hitherto and is not proved here. However there is a restricted form of Theorem 9.2

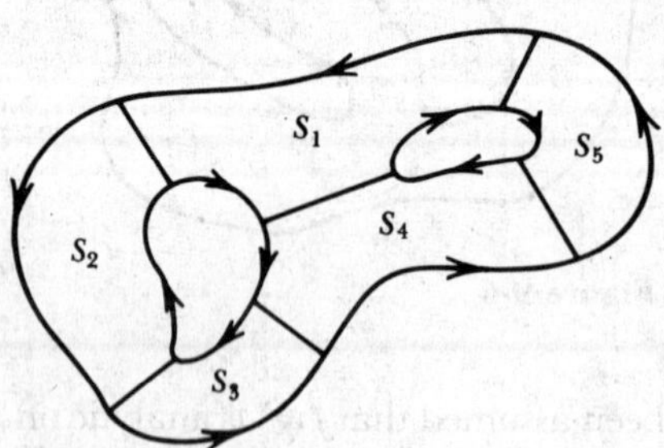

Figure 9-7

which suffices for nearly all applications and is proved with ease. In this restricted form it is supposed that the domain D can be decomposed into star domains S_i as illustrated in Figure 9-7. Thus $\overline{D}$ is the union of the closed regions $\overline{S_i}$,

$$\overline{D} = \overline{S}_1 + \overline{S}_2 + \cdots + \overline{S}_m$$

and the oriented boundary C of D can be written

$$C = T_1 + T_2 + \cdots + T_m$$

where T_j is the oriented boundary of S_j. As the reader will recall from Chapter 3, the latter equation means that

$$\int_C f(z)\, dz = \sum_{j=1}^{m} \int_{T_j} f(z)\, dz \tag{9.8}$$

for every continuous function f. Subject to mild conditions on the contours T_j, it is easily proved that each integral on the right of (9.8) is 0 under the hypothesis of Theorem 9.2, and the restricted form of Theorem 9.2 follows by addition. Details of this development are given in Problems 2 and 3.

Example 9.1. Removable singularities. Let $f(z)$ be analytic in a domain D except possibly on a line segment L lying in D. If $f(z)$ is continuous in D, show that $f(z)$ is analytic on L.

This differs from the theorems of Chapter 3 in that a whole line of singularities is removable, while the singularities in Chapter 3 were isolated. On the other hand, the hypothesis of continuity is more restrictive than the condition

$$\lim_{z \to \alpha} (z - \alpha) f(z) = 0$$

which sufficed for the results of Chapter 3.

For proof, let z_0 be a point interior to the line segment L and let D_0 be a disk centered at z_0. Let D_0 be so small that $\overline{D}_0$ is contained in D and also so small that L bisects D_0 as shown in Figure 9-8. This gives two disjoint domains D_1 and D_2 bounded by semicircular contours C_1 and C_2 as shown in Figure 9-9.

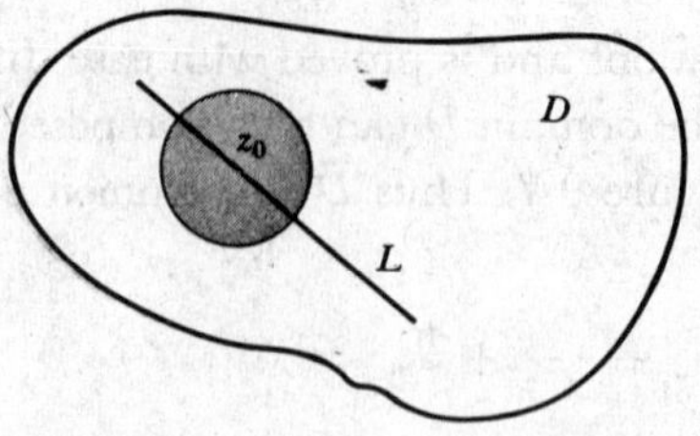

Figure 9-8

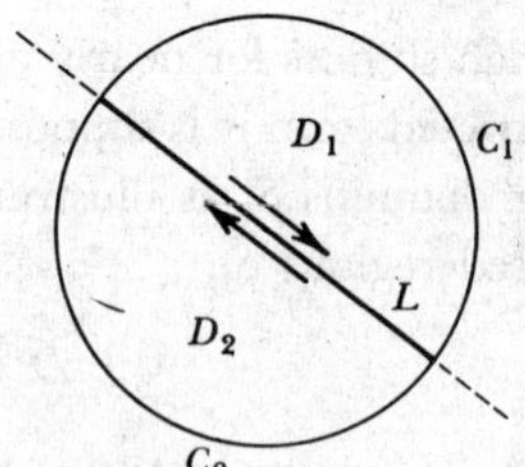

Figure 9-9

If z is in D_1, the Cauchy integral formula and Cauchy integral theorem give, respectively,

$$f(z) = \frac{1}{2\pi i}\int_{C_1}\frac{f(\zeta)}{\zeta - z}\,d\zeta, \qquad 0 = \frac{1}{2\pi i}\int_{C_2}\frac{f(\zeta)}{\zeta - z}\,d\zeta.$$

Hence by addition

$$f(z) = \frac{1}{2\pi i}\int_{C}\frac{f(\zeta)}{\zeta - z}\,d\zeta \tag{9.9}$$

where C is the oriented boundary of the disk D_0. The integrals over L in the two directions cancel because f is continuous. Equation (9.9) is derived in a similar way when z is in D_2, and since both sides of the equation are continuous, it must hold throughout D_0. Since the right side of (9.9) is analytic for z in D_0, the same is true of the left side, and so $f(z)$ is analytic at z_0. Analyticity at the end points of L follows from the results of Chapter 3, or this case can be reduced to the case just considered by extending L.

In Figure 9-9 it is not necessary that L be a straight line. Instead, L could be any simple contour that divides the disk D_0 into two domains bounded by Jordan contours, as shown in Figure 9-10. Such a contour is called a *cross cut* of the disk. The above proof shows that if $f(z)$ is continuous in D_0 and is analytic in D_0 except possibly on the cross cut L, then $f(z)$ is analytic in D_0.

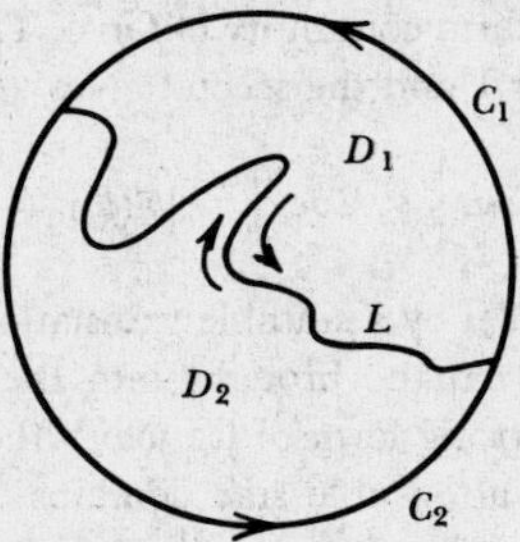

Figure 9-10

Problems

1. Deduce the result of Example 9.1 from Morera's theorem. This method does not readily yield the generalization associated with Figure 9-10 and so the proof based on Cauchy's formula has been preferred here.
2. Let S be a domain which is star with respect to the origin. Suppose the oriented boundary of S is a contour T, which can be described in polar coordinates by $r = h(\theta)$, $0 \leq \theta \leq 2\pi$, where $h'(\theta)$ is piecewise continuous and $h(\theta) > 0$. The interior of T is the set in which $r < h(\theta)$ and coincides with the original domain S. If $f(z)$ is analytic in S and is continuous in $\bar{S}$, prove that

$$\int_T f(z)\,dz = 0.$$

Outline of solution: For $0 < s < 1$ let T_s denote the contour

$$r = sh(\theta), \qquad 0 \leq \theta \leq 2\pi.$$

Then by Cauchy's theorem for the star domain S,

$$\int_T f(z)\,dz = \int_T f(z)\,dz - \frac{1}{s}\int_{T_s} f(z)\,dz$$

the third integral being 0. This is equivalent to

$$\int_T f(z)\,dz = \int_0^{2\pi} \{f[h(\theta)] - f[sh(\theta)]\}\{h'(\theta) + ih(\theta)\}e^{i\theta}\,d\theta$$

Given any $\epsilon > 0$ we can find $\delta > 0$ such that

$$|h(\theta) - sh(\theta)| < \delta \qquad \text{implies} \qquad |f[h(\theta)] - f[sh(\theta)]| < \epsilon.$$

This follows from the uniform continuity of f in S. The first inequality is assured when s is sufficiently near 1 and the second gives

$$\left| \int_T f(z)\, dz \right| \leq 2\pi\epsilon \int_0^{2\pi} \{|h'(\theta)| + |h(\theta)|\}\, d\theta.$$

3. A star domain is *admissible* if by a suitable translation of axes (that is, by a suitable choice of origin) it can be brought into the form of the domain S in Problem 2. Deduce the strong form of Cauchy's theorem when the domain D can be decomposed into admissible star domains S_j as explained in the text.
4. *Laurent's theorem in a ring.* This problem pertains to Chapter 3, Section 9. Let $f(z)$ be analytic for $R_1 < |z - \alpha| < R_2$. Prove that Theorems 9.1 and 9.2 are valid for $R_1 < |z - \alpha| < R_2$. That is, show $f = f_1 + f_2$ where f_1 is analytic for $|z - \alpha| < R_2$ and f_2 for $|z - \alpha| > R_1$. Prove (9.6) where c, the radius of C in (9.5), satisfies $R_1 < c < R_2$.

10. Deformation of contours. Let C_0 and C_1 be contours with the same initial point α and terminal point β. Suppose C_0 and C_1 lie in a domain D, in which a given function $f(z)$ is analytic. It is said that C_0 can be *deformed* into C_1 in D if there is a family of contours C_s with the following properties:

(1) C_s reduces to C_0 when $s = 0$, and to C_1 when $s = 1$
(2) C_s all have the initial point α and terminal point β.
(3) C_s all lie in D, $0 \leq s \leq 1$.
(4) C_s depends continuously on s.

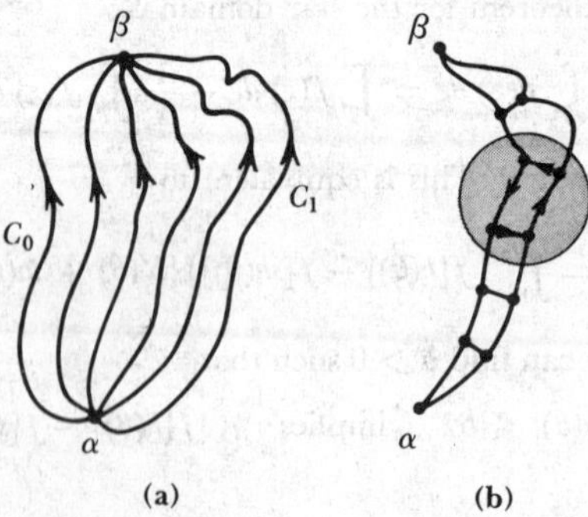

Figure 10-1

These properties are illustrated in Figure 10-1(a). When they hold,

$$\int_{C_0} f(z)\,dz = \int_{C_1} f(z)\,dz.$$

In other words, the path C_0 can be deformed into C_1, without changing the value of the integral, provided the path does not pass over any singularity of f during the deformation. This important fact is called *the principle of deformation of contours.*

The principle of deformation of contours gives added insight into several theorems of this chapter. For instance, in a simply connected domain any closed contour can be deformed into a point and Cauchy's integral theorem follows at once. A version of the residue theorem can be obtained as suggested in Figure 10-2.

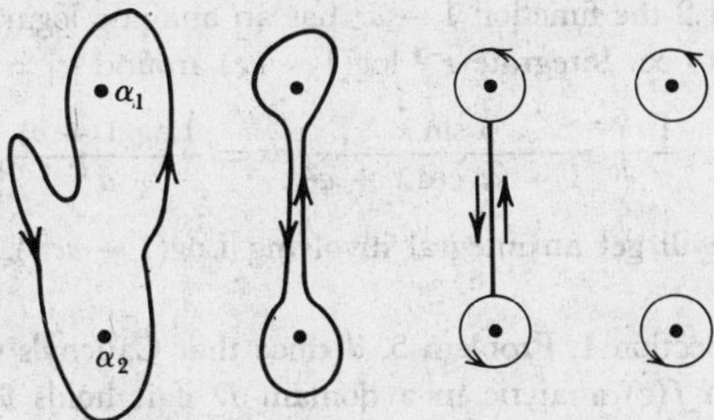

Figure 10-2

The principle of deformation of contours is proved by the construction illustrated in Figure 10-1(b), which shows two contours C_s for different values $s = s_0$ and $s = s_1$. A succession of points on these contours is chosen and these points are joined by straight lines to form the ladder-like configuration shown in the figure. If the successive points are near one another, and the values s_0 and s_1 are also near one another, each mesh of the ladder lies in a disk in which $f(z)$ is analytic, and the integral around a mesh is 0 by Cauchy's theorem for the disk. Addition over all meshes shows that the integrals over the two contours C_s are equal. We can pass from C_0 to C_1 by finitely many steps of this sort, and since each step leaves the integral unchanged, the integral over C_0 must equal the integral over C_1.

For analytic formulation, it is necessary to explain what is meant by a continuous family of curves C_s, and by a succession of points on a curve C_s. The curves form a *continuous family* if C_s can be parameterized in the form

$$z = \zeta(s,t), \qquad 0 \le t \le 1,$$

where $\zeta(s,t)$ is continuous for $0 \le s \le 1$ and $0 \le t \le 1$. A *succession of points* on two curves C_s is obtained by choosing a succession of values $t_0 < t_1 < \cdots < t_n$, the same in both cases. With these definitions, it is not difficult to give the full details of the above argument by using the uniform continuity of $\zeta(s,t)$. The details are not given here.

Additional problems on Chapter 4

1.1. By Example 1.2 the function $1 - az$ has an analytic logarithm in the plane cut from $1/a$ to ∞. Integrate $z^{-1}\log(1 - az)$ around $|z| = 1$ and deduce

$$\frac{1}{\pi}\int_0^\pi \frac{x \sin x}{1 - 2a\cos x + a^2}\,dx = \frac{\operatorname{Log}(1 + a)}{a}, \qquad 0 < a < 1.$$

(At first you will get an integral involving $\operatorname{Log}(1 - ae^{i\theta})$. Integrate this by parts.)

1.2. Referring to Section 1, Problem 5, deduce that Cauchy's theorem holds for every function $f(z)$ analytic in a domain D if it holds for every function $f(z) = 1/(z - a)$ analytic in D.

1.3. If D is a bounded domain, let D_δ denote the larger domain obtained by taking D together with all points whose distance to D is less than δ. Give an example of a domain D such that D is simply connected but D_δ is not simply connected for any small positive δ.

1.4. Show that an entire function $f(z)$ has a finite number of zeros if, and only if, it can be written in the form

$$f(z) = P(z)e^{g(z)}$$

where P is a polynomial and g is entire. (See Example 1.2.)

1.5. *A case of Jensen's formula*. Let $g(z)$ be analytic and nonvanishing for $|z| \le r$. Using Example 1.2 and Cauchy's theorem, show that

$$\operatorname{Log}|g(0)| = \frac{1}{2\pi}\int_0^{2\pi} \operatorname{Log}|g(re^{i\theta})|\,d\theta.$$

2.1. By integrating $z^n(z - a)^{-1}(z - 1/a)^{-1}$ around $|z| = 1$, show that

$$\frac{1}{2\pi}\int_{-\pi}^{\pi}\frac{\cos n\theta}{1-2a\cos\theta+a^2}\,d\theta=\frac{a^n}{1-a^2},\qquad |a|<1,\ n=0,1,2,\ldots.$$

2.2. Obtain the result of Problem 2.1 for $n = 1$ by use of (2.6).

2.3. Obtain the following formulas for $a > 1$ and discuss their extension to complex values of a:

$$\frac{1}{2\pi}\int_0^{2\pi}\frac{\cos\theta\,d\theta}{a+\cos\theta}=1-\frac{a}{\sqrt{a^2-1}},\qquad \frac{1}{2\pi}\int_0^{2\pi}\frac{\sin^2\theta\,d\theta}{a+\cos\theta}=a-\sqrt{a^2-1}.$$

2.4. Show in (2.6) that $2\cos n\theta = z^n + z^{-n}$, and thus get

$$\int_0^{2\pi}\frac{\cos 2\theta}{5-3\cos\theta}\,d\theta=\frac{\pi}{18},\qquad \int_0^{2\pi}\frac{\cos 3\theta}{5-3\cos\theta}\,d\theta=\frac{\pi}{54}.$$

3.1. Determine the range of the complex parameter a for which the upper answer, the lower answer, or neither answer applies:

$$\int_{-\infty}^{\infty}\frac{dx}{(x-a)^2+1}=\begin{cases}0\\ \pi\end{cases},\qquad \int_{-\infty}^{\infty}\frac{8x^2\,dx}{(x^2+a^2)^3}=\begin{cases}\pi a^{-3}\\ \pi(-a)^{-3}\end{cases}.$$

3.2. Show (a) for $n = 4$ and (b) for all integers ≥ 2 that

$$\int_{-\infty}^{\infty}\frac{dx}{(1+x^2)^n}=\frac{1\cdot 3\cdot 5\cdots(2n-3)}{2\cdot 4\cdot 6\cdots(2n-2)}\pi.$$

3.3. Show (a) for $n = 2$ and (b) for all integers ≥ 1 that

$$\int_{-\infty}^{\infty}\frac{dx}{1+x+x^2+\cdots+x^{2n}}=\frac{2\pi}{2n+1}\left(1+\cos\frac{\pi}{2n+1}\right)\csc\frac{2\pi}{2n+1}.$$

(If $\omega = \exp[2\pi i/(2n + 1)]$ use $\omega\omega^{2n} = 1$, $\omega^{1/2}\omega^n = -1$. Part (b) of this problem is not easy.)

3.4. Formulate three general theorems pertaining to the evaluation of integrals of the following types by means of residues:

$$\int_0^{2\pi} f(\cos\theta,\sin\theta)\,d\theta,\qquad \int_{-\infty}^{\infty} f(x)\,dx,\qquad \int_{-\infty}^{\infty} e^{iax}f(x)\,dx.$$

4.1. Let $f(z)$ be analytic for $|z| \leq r$, let $f(0) \neq 0$ and let the zeros of $f(z)$ in $|z| < r$ be $a_1, a_2, \ldots, a_n$, repeated according to their multiplicity. Show that the function

$$g(z)=\frac{r^2-\bar{a}_1 z}{r(z-a_1)}\cdot\frac{r^2-\bar{a}_2 z}{r(z-a_2)}\cdots\frac{r^2-\bar{a}_n z}{r(z-a_n)}f(z)$$

is analytic in $|z| \leq r$ except for removable singularities and does not vanish in $|z| < r$. Show also that $|g(z)| = |f(z)|$ on $|z| = r$. (See Chapter 1, Example 2.1.)

4.2. *Jensen's inequality*. By applying the maximum principle to $g(z)$ above, deduce $|g(0)| \leq M(r)$ where $M(r)$ is the maximum of $|f(z)|$ on $|z| = r$. Thus get

$$\frac{r^n}{|a_1 a_2 \cdots a_n|} \le \frac{M(r)}{|f(0)|}.$$

4.3. *Jensen's formula,* easy case. If $f(z)$ in Problem 4.1 does not vanish on $|z| = r$, apply Problem 1.5 to $g(z)$ and deduce

$$\operatorname{Log} \frac{r^n}{|a_1 a_2 \cdots a_n|} = \frac{1}{2\pi} \int_0^{2\pi} \operatorname{Log} \frac{|f(re^{i\theta})|}{|f(0)|}\, d\theta.$$

4.4. *Jensen's formula,* general case. Let ϕ be any real constant. Deduce from Example 4.2 that

$$\int_0^{2\pi} \operatorname{Log}|re^{i\theta} - re^{i\phi}|\, d\theta = 2\pi \operatorname{Log} r, \qquad r > 0,$$

and hence the formula of Problem 1.5 holds for functions $z - a$ which vanish on $|z| = r$. Thus extend Problem 1.5 to allow $g(z) = 0$ on $|z| = r$. Conclude that the formula of Problem 4.3 remains valid, with a convergent improper integral on the right, if some of the a_j satisfy $|a_j| = r$.

4.5. If b is real, integrate $e^{2biz}/(e^{2\pi z} - 1)$ around an indented rectangle with vertices at $0, R, R + i, i$ and show that

$$\int_0^\infty \frac{\sin 2bx}{e^{2\pi x} - 1}\, dx = \frac{\coth b}{4} - \frac{1}{4b}.$$

5.1. By considering $\log(1 - ix)$ and equating real parts, show that

$$\int_0^\infty \frac{\log(1 + x^2)}{1 + x^2}\, dx = \pi \log 2, \qquad \int_0^\infty \frac{\log(1 + x^2)}{x^2}\, dx = \frac{\pi}{2}.$$

5.2. Do Problem 3.2 by setting $x^2 = t$.

5.3. Using an appropriate semicircle as contour, integrate

$$f(z) = \frac{z \log(1 - iz)}{(1 + 2z^2)^2}$$

by parts and show that

$$\int_0^\infty \frac{x \tan^{-1} x}{(1 + 2x^2)^2}\, dx = \int_0^1 \frac{x \sin^{-1} x}{(1 + x^2)^2}\, dx = \frac{\pi}{8}(\sqrt{2} - 1).$$

(Since $\tan^{-1} x = \sin^{-1} u$ where $u = x/\sqrt{1 + x^2}$, let $x = u/\sqrt{1 - u^2}$ in the first integral to get the second.)

5.4. Formulate three general theorems pertaining to the evaluation of integrals of the following types by means of residues:

$$\int_0^\infty x^a f(x)\, dx, \quad \int_0^\infty f(x) \log x\, dx, \quad \int_0^\infty f(x)(\log x)^2\, dx.$$

6.1. Let $f(z)$ be analytic for $|z| < R$. Let $f(0) = 0, f'(0) \neq 0$. Then there is a $\delta > 0$ and a $\rho > 0$ such that for any complex number α, $|\alpha| < \delta$, the equation

$$f(z) = \alpha$$

has exactly one solution for $|z| < \rho$. In other words, $w = f(z)$ has a unique inverse in the neighborhood of $z = w = 0$ if $f(0) = 0$ and $f'(0) \neq 0$. (Since $f(z)$ is not identically zero, the zero of f at $z = 0$ is isolated. Hence there is $\rho > 0$ such that $f(z) \neq 0$ for $0 < |z| \leq \rho < R$. Let $\min|f(\rho e^{i\theta})| = \delta > 0$. By Rouché's theorem, with $f = f(z)$, $g = -\alpha$, and C the circle $|z| = \rho$, the result follows.)

6.2. Give an alternative proof of the result of Problem 6.1 by noting that

$$\frac{1}{2\pi i}\int_C \frac{f'(z)}{f(z) - \alpha}\,dz$$

is an integer, is analytic in α for $|\alpha| < \delta$, is 1 for $\alpha = 0$, and hence is 1 for all $|\alpha| < \delta$.

6.3. With the hypothesis of Problem 6.1 show that $w = f(z)$ has a unique analytic inverse $z = \phi(w)$ for $|w| < \delta$ where $|\phi(w)| < \rho$. (Let C be $z = \rho e^{i\theta}$ and let

$$\phi(w) = \frac{1}{2\pi i}\int_C \frac{zf'(z)}{f(z) - w}\,dz.$$

With $w = \alpha$, show that $\phi(\alpha)$ must be the solution of $\alpha = f(z)$, the existence of which is proved in Problem 6.1. Show that for $|w| < \delta$, $\phi(w)$ is an analytic function of w.)

7.1. Show that the Hilbert transform of a constant is 0 and hence that the Hilbert formulas can be written

$$u(y) = \frac{1}{\pi}\fint_{-\infty}^{\infty}\frac{v(\omega) - v(y)}{y - \omega}\,d\omega, \qquad v(y) = \frac{1}{\pi}\fint_{-\infty}^{\infty}\frac{u(y) - u(\omega)}{y - \omega}\,d\omega.$$

Under the hypothesis of the text the integrand has only a removable singularity at $\omega = y$ and the bar can often be dropped from the integral.

7.2. Apply the formulas of the preceding problem to e^{-z} and get

$$\cos y = \frac{2}{\pi}\int_0^\infty \frac{\omega\sin\omega - y\sin y}{\omega^2 - y^2}\,d\omega, \qquad \sin y = \frac{2y}{\pi}\int_0^\infty \frac{\cos y - \cos\omega}{\omega^2 - y^2}\,d\omega.$$

The function e^{-z} does not satisfy the condition $M(R) \to 0$, but the integral over C_R can be estimated by Jordan's lemma.

7.3. Let $\sqrt{z}$ be the branch in the plane cut from 0 to $-\infty$ which is positive for $z > 0$, and let c and t be positive constants. Show that

$$\frac{1}{2\pi i}\int_{c-i\infty}^{c+i\infty}\frac{e^{zt}}{\sqrt{z}}\,dz = \frac{1}{\pi}\int_0^\infty \frac{e^{-xt}}{\sqrt{x}}\,dx = \frac{2}{\pi\sqrt{t}}\int_0^\infty e^{-r^2}\,dr.$$

The last integral is known to be $\sqrt{\pi}/2$. (Consider a contour formed by part of $x = c$, by part of $|z| = R$, by $|z| = \epsilon$, and by the upper and lower sides of the cut $-R \le x \le -\epsilon$.)

7.4. Let $G(s)$ be analytic for $\operatorname{Re} s \geq 0$ and satisfy $M(R) \to 0$ so that (7.5) can be used for G. Assuming legitimacy of a change in order of integration, show that the first of the following formulas implies the second for $\operatorname{Re} s > 0$:

$$g(x) = \frac{1}{2\pi} \int_{-\infty}^{\infty} G(i\omega) x^{-i\omega} \, d\omega, \qquad \int_0^1 x^{s-1} g(x) \, dx = G(s).$$

This is a special case of the *Mellin inversion formulas*,

$$G(s) = \int_0^{\infty} x^{s-1} g(x) \, dx, \qquad g(x) = \frac{1}{2\pi i} \int_{c-i\infty}^{c+i\infty} G(s) x^{-s} \, ds,$$

which hold under much less restrictive conditions.

8.1. With a suitable branch of the logarithm in the plane cut from -1 to 1, use Theorem 8.2 or integrate by parts to get

$$\int_C z^2 \log \frac{z+1}{z-1} \, dz = \frac{4\pi i}{3}, \qquad \text{where } C \text{ is } |z| = 2.$$

8.2. By setting $x^4 = t$, show that

$$\int_0^1 x^2 \sqrt[4]{1 - x^4} \, dx = \pi \frac{\sqrt{2}}{16}.$$

9.1. If the branch lines are in the lower half-plane, and if the denominator has the value $\exp(i\pi/4) + \sqrt{3}\exp(i\pi/4)$ at $s = 0$, show that

$$\int_{-\infty}^{\infty} \frac{e^{ixs}}{\sqrt{s+i} + \sqrt{s+3i}} \, ds = \frac{\sqrt{\pi}}{2(-x)^{3/2}} e^{-\pi i/4} (e^{x} - e^{3x})$$

for $x < 0$, and that the integral is 0 for $x > 0$. (Use the relation

$$\frac{1}{\sqrt{s+3i} + \sqrt{s+i}} = \frac{\sqrt{s+3i} - \sqrt{s+i}}{2i}$$

after deforming the contour in the lower half-plane.)

Chapter 5

Conformal mapping

A nonconstant analytic function $w = f(z)$ maps a domain D of the z plane onto another domain $f(D)$ of the w plane. At points where $f'(z) \neq 0$ such a map has the remarkable property that it is conformal. This means that any two smooth curves intersecting in D map into curves which intersect at the same angle in $f(D)$. By means of conformal mapping, problems of fluid flow, electrostatics and other fields can be mapped into simpler problems of the same general sort in $f(D)$. Solution of the problem in $f(D)$ then solves the original problem in D. Conformal mapping also gives geometrical insight into analytical questions. For instance, it is found that $w = f(z)$ can be inverted to give $z = g(w)$ near points where the map is conformal, while the function and its inverse behave much as in the case $w = z^n$ near points where the map is not conformal. This chapter concludes with a statement of the Riemann mapping theorem, which characterizes those domains that can be mapped conformally onto the unit disk.

1. Conformal mapping; bilinear transformations. Let $f(z)$ be analytic in a domain D and hence, of course, single-valued. Then the set of all points $f(z)$ for z in D is called the image of D by the mapping f. The notation $w = f(z)$ designates the mapping of the points of D onto points in the image of D in the w plane by f. In similar fashion, the image of D in the w plane is denoted by $f(D)$.

Examples of mapping have been considered in Chapter 1, where it was shown that a point of $f(D)$ may be the image of more than one point of D. This is the case, for example, if $w = z^2$ and D is the finite z plane. If $f(z_1) = f(z_2)$ implies $z_1 = z_2$, then the mapping of D onto $f(D)$ is said to be *one-to-one*. This means that for each point w_1 of $f(D)$ there is exactly one point z_1 in D such that $w_1 = f(z_1)$. In this case the inverse function $z = f^{-1}(w)$ is defined on $f(D)$. It will be shown later that in the case of a one-to-one mapping, f^{-1} is also an analytic function. The mapping by $w = f(z)$ of a set of points S of the z plane is said to be *onto* a set S^* of the w plane if every point of S^* is the image of a point of S and if S^* contains $f(S)$.

Let $z = \zeta(t)$ for $a \leq t \leq b$ be an *arc* C. Thus $x + iy = \xi(t) + i\eta(t)$ where ξ and η are real differentiable functions of t. By definition

$$\frac{d\zeta}{dt} = \frac{dx}{dt} + i\frac{dy}{dt}$$

and the slope of this vector is dy/dx, which is also the slope of the arc. Therefore, at any point where $\zeta'(t) \neq 0$, the vector $\zeta'(t)$ is tangent to C, and $\arg \zeta'(t)$ measures the angle which the tangent makes with the x axis.

Let $f(z)$ be analytic in a domain D containing the above arc C. The image of C is $w = f[\zeta(t)]$ and is an arc C^*. By the chain rule,

$$\frac{dw}{dt} = f'[\zeta(t)]\frac{d\zeta}{dt}.$$

For t_0 in $[a,b]$ let $z_0 = \zeta(t_0)$ and suppose $f'(z_0) \neq 0$, $\zeta'(t_0) \neq 0$. Then the preceding equation gives $w'(t_0) \neq 0$ and

$$\arg w'(t_0) = \arg f'(z_0) + \arg \zeta'(t_0).$$

Geometrically this means that the angle between the directed tangent to the arc C at t_0 and the directed tangent to the image arc C^*, at the image of z_0, is $\arg f'(z_0)$. In other words, under the mapping, the directed tangent to any arc through z_0 is rotated by an angle $\arg f'(z_0)$ which is independent of the choice of arc $\zeta(t)$ through z_0. In particular, if two arcs intersect at z_0, their angle of intersection (in both magnitude and sign) is preserved by the mapping, as shown in Figure 1-1. A mapping such as this is said to be *conformal*. We have established the following:

THEOREM 1.1. *The mapping $w = f(z)$ is conformal at any point where $f'(z)$ exists and $f'(z) \neq 0$.*

From

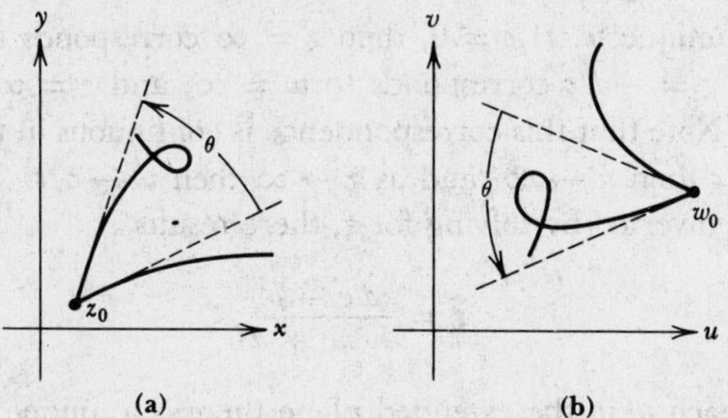

Figure 1-1

$$\lim_{z \to z_0} \frac{|f(z) - f(z_0)|}{|z - z_0|} = |f'(z_0)|$$

it follows that the local scale of the mapping is given by $|f'(z_0)|$ and is independent of direction at z_0.

If $f(z)$ is not given as analytic but only as a function of (x,y) in D with continuous first-order partial derivatives, then if the mapping $w = f(z)$ is known to be conformal, it can be shown that the Cauchy-Riemann equations must be satisfied and hence $f(z)$ must be analytic. Similarly, if the map $w = f(z)$ preserves scale locally, then it can be shown that either $f(z)$ is differentiable or $\overline{f(z)}$ is differentiable at each point of D. Proof of these facts is outlined in Problems 1.3 and 1.4 at the end of this chapter.

A simple but important class of conformal mappings is provided by the *bilinear transformation:*

$$w = \frac{az + b}{cz + d}, \qquad ad - bc \neq 0 \quad (1.1)$$

where a, b, c and d are complex numbers. The transformation is called bilinear because it is equivalent to

$$cwz - az + dw - b = 0$$

and the expression on the left is linear in both z and w. When $c = 0$, the transformation is *linear.*

According to (1.1), to each z in the extended complex plane there corresponds a unique w. If $c = 0$, then $z = \infty$ corresponds to $w = \infty$. If $c \neq 0$, then $z = -d/c$ corresponds to $w = \infty$, and $z = \infty$ corresponds to $w = a/c$. Note that this correspondence is continuous in the sense that as $z \to -d/c$ then $w \to \infty$, and as $z \to \infty$ then $w \to a/c$.

If (1.1) is inverted by solving for z, there results

$$z = \frac{dw - b}{-cw + a},$$

so that for each w in the extended plane there is a unique z. Moreover, as $w \to \infty$ then $z \to -d/c$, and as $w \to a/c$ then $z \to \infty$. Hence (1.1) is a one-to-one mapping of the extended z plane onto the extended w plane. Of all analytic functions, only the bilinear functions give a one-to-one mapping of the extended plane onto itself, as will be proved later.

Since

$$\frac{d}{dz}\left(\frac{az + b}{cz + d}\right) = \frac{ad - bc}{(cz + d)^2},$$

it follows that $f'(z)$ is well-defined and not zero, except at $z = \infty$ and $z = -d/c$ (if $c \neq 0$), where further consideration is necessary. The non-vanishing of f' shows that the bilinear mapping is conformal for all z except possibly at $z = \infty$ and $z = -d/c$, and it will now be shown that even at these points the conformality is preserved.

Indeed, to discuss $z = \infty$ in the case where $c \neq 0$, let $z = 1/\zeta$ and consider $\zeta = 0$. The equation (1.1) becomes

$$w = \frac{b\zeta + a}{d\zeta + c}$$

and the conformality at $\zeta = 0$ is evident at once on differentiating with respect to ζ. To discuss $z = -d/c$, which has $w = \infty$ as its image, let $w = 1/\omega$. Then

$$\omega = \frac{cz + d}{az + b}$$

and here the derivative of the right side, $(bc - ad)/(az + b)^2$, is not zero at $z = -d/c$. Hence the transformation is conformal at $\omega = 0$, the image of $z = -d/c$. In case $c = 0$ (and $d = 1$ with no restriction), the image of

$z = \infty$ is $w = \infty$ and hence the transformations $\zeta = 1/z$ and $\omega = 1/w$ are both made simultaneously in (1.1), giving $\omega = \zeta/(a + b\zeta)$, which is conformal at $\zeta = 0$.

As a special case of (1.1), consider first

$$w = z + b. \tag{1.2}$$

The transformation (1.2) is a translation, since the image of each point z is obtained by adding the fixed vector b to z. A second case of importance is

$$w = az = |a|e^{i\phi}z, \tag{1.3}$$

where $\phi = \arg a$. This transformation is a magnification and rotation. If $w = az$ and $w_0 = az_0$, then

$$|w - w_0| = |az - az_0| = |a||z - z_0|,$$

and so all distances are multiplied by the constant factor $|a|$. Since $\arg w = \phi + \arg z$, the transformation also effects a rotation of the plane about the origin through the angle ϕ.

As the last special case of (1.1) the *inversion*

$$w = 1/z \tag{1.4}$$

is considered. Here $|w| = 1/|z|$ and $\arg w = -\arg z$. Hence the distance from the origin is replaced by its reciprocal, and the argument is replaced by its negative. Under translation, rotation or magnification, triangles are mapped onto similar triangles, and the conformality is obvious. In the case of inversion the conformality is by no means obvious geometrically, but follows from Theorem 1.1.

As explained more fully in the following section, the equation of any line or circle in the complex plane can be put in the form

$$Az\bar{z} + \bar{B}z + B\bar{z} + C = 0 \tag{1.5}$$

where A and C are real and $AC < |B|^2$. For the line, $A = 0$. The verification that (1.5) has the form of the equation of a circle if $A \neq 0$ is trivial. It is also trivial that (1.5) represents a line for $A = 0$, $|B| > 0$.

Lines in the extended plane can be regarded as circles through the point $z = \infty$. Indeed, if z is replaced by $1/\zeta$ in the case $A = 0$, we get

$$C\zeta\bar{\zeta} + B\zeta + \overline{B\zeta} = 0,$$

which for $C \neq 0$ is a circle. Hence it is natural to group circles and lines together.

It is a matter of easy verification that (1.5) retains its form under each of the mappings

$$z = w - b, \qquad z = w/a, \qquad z = 1/w$$

and hence each of these mappings transforms circles and lines into circles and lines. The same is true of the general bilinear transformation, as seen next.

When $c \neq 0$ in (1.1) we can divide the numerator and denominator by c and put (1.1) in the form

$$w = \frac{az + b}{z + d}.$$

The substitution $\zeta_1 = z + d$ is a translation and gives $z = \zeta_1 - d$, so that

$$w = \frac{a(\zeta_1 - d) + b}{\zeta_1} = a + \frac{b - ad}{\zeta_1}.$$

This is an inversion, $\zeta_2 = 1/\zeta_1$, followed by a magnification and rotation, $\zeta_3 = (b - ad)\zeta_2$, followed by another translation, $w = a + \zeta_3$. Hence (1.1) is merely an iteration of (1.2), (1.4), (1.3) and (1.2). When $c = 0$ we can assume without loss of generality that $d = 1$, so that $w = az + b$. This involves only (1.2) and (1.3).

Since each of the transformations (1.2), (1.3) and (1.4) maps lines and circles onto lines and circles, *each bilinear transformation maps lines and circles onto lines and circles.* When $c = 0$ no inversion is needed, the point at ∞ is invariant, and the transformation maps lines onto lines, circles onto circles. An independent proof of the latter property is given in Example 1.1.

The notation $w = Tz$ is often used to describe the transformation T that takes z into w. The inverse transformation taking w into z is denoted by T^{-1} when it exists, and the identity transformation is I. Thus $TT^{-1}z = z$ and $Iz = z$. Two transformations T_1 and T_2 are equal if they are defined for z in the same domain D and if $T_1z = T_2z$ for all z in D. The product of transformations is given by iteration,

$$(T_1T_2)z = T_1(T_2z), \tag{1.6}$$

and is associative; that is,

$$T_1(T_2T_3)z = (T_1T_2)T_3z \tag{1.7}$$

for all z for which either side of this equation makes sense. This is so because both sides of (1.7) mean that z is transformed first by T_3, then by T_2 and finally by T_1. A formal proof can be based on (1.6).

In this notation, the decomposition upon which the foregoing analysis is based can be described as follows: Every bilinear transformation is either linear, or it can be written in the form L_1VL_2 where L_i are linear and V denotes inversion.

Example 1.1. If $a \neq 0$, show that the transformation $w = az + b$ maps the circle $|z - z_0| = \rho$ onto a circle of radius $|a|\rho$, and show that the center of the image circle is the image of the center of the original circle.

If $w = az + b$ and $w_0 = az_0 + b$, then $w - w_0 = a(z - z_0)$. The condition $|z - z_0| = \rho$ is therefore equivalent to

$$|w - w_0| = |a|\rho,$$

which shows that the image is a circle of center w_0 and radius $|a|\rho$. This method is more efficient than that of the text, but does not apply to the transformation $w = 1/z$.

Example 1.2. If z_1, z_2, z_3 and z_4 are four distinct points, their *cross ratio* is

$$X(z_1,z_2,z_3,z_4) = \frac{z_1 - z_2}{z_1 - z_4} \cdot \frac{z_3 - z_4}{z_3 - z_2} \tag{1.8}$$

and is defined by the obvious limit when one of the points z_i is ∞. Show that the cross ratio is invariant under bilinear mappings.

If each z_i is replaced by $z_i + b$, the differences in the numerator and denominator are unchanged, and hence X is unchanged. If z_i is replaced by az_i with $a \neq 0$, then numerator and denominator are both multiplied by a^2 and X is unchanged. Finally, if each z_i is replaced by $1/z_i$ it is found, again, that X is unchanged. This completes the proof.

Example 1.3. Find a bilinear transformation that takes the points $0,1,\infty$ into $-i,1,i$, respectively. If the desired transformation takes z into w, then invariance of the cross ratio gives

$$X(w,-i,1,i) = X(z,0,1,\infty).$$

By (1.8) this is

$$\frac{w+i}{w-i}\cdot\frac{1-i}{1+i} = \frac{z-0}{1-0}$$

or, solving for w,

$$w = \frac{1+iz}{i+z} \tag{1.9}$$

Example 1.4. Find a bilinear mapping that takes the real axis onto the circle $|w| = 1$. The function (1.9) maps the real axis onto a circle or line. This circle or line contains the points $-i,1,i$ and hence it coincides with the unit circle.

Example 1.5. If $w = f(z)$ and $f'(z_0)$ exists, Chapter 2, Section 1 gives

$$\Delta w = f'(z_0)\,\Delta z + \epsilon|\Delta z| \tag{1.10}$$

where $w_0 = f(z_0)$, $\Delta w = w - w_0$, $\Delta z = z - z_0$, and $\epsilon \to 0$ as $\Delta z \to 0$. Discuss from the point of view of linear transformations.

When $f'(z_0) \neq 0$ and $|\Delta z|$ is small, the second term on the right of (1.10) is small compared to the first. If the second term could be neglected, we would get

$$w = f'(z_0)(z - z_0) + w_0.$$

This is a linear transformation $w = az + b$ with $a = f'(z_0)$. Hence it is conformal, represents an expansion in the ratio $|f'(z_0)|$, and involves a rotation through an angle $\arg f'(z_0)$.

Problems

1. Let $z = t$ and $z = e^{i\phi}t$ be two straight lines through the origin meeting at an angle ϕ. Show that their images under $w = z^2$ are straight lines meeting at

an angle 2ϕ. Hence, conformality need not hold at points where $f'(z_0) = 0$.

2. Show that the interior of the circle in Example 1.1 is mapped onto the interior of the image circle, and likewise for the exterior.
3. If the z_i are distinct, show that $(z_1 - z_3)/(z_1 - z_2)$ is invariant under linear mappings.
4. Find the images of 1, i, 0 and $1 + i$ under $w = (z + 2i)/(z + i)$ and check by showing that the cross ratio is unchanged.
5. Find the bilinear mappings taking $(0,1,\infty)$ into:

$$(0,1,2), \qquad (-i,0,i), \qquad (0,1,\infty), \qquad (0,i,\infty), \qquad (-i,\infty,1).$$

6. Verify that $|w| = 1$ for real z in (1.9).
7. If $w = (z - i)/(z + i)$, show that $|w| < 1$ if and only if $\operatorname{Im} z > 0$, and hence this transformation maps the upper half-plane $\operatorname{Im} z > 0$ onto the unit disk $|w| < 1$. (See Chapter 1, Example 2.1.)
8. In Problem 7 let d_1 be the distance from z to the point i and let d_2 be the distance from z to $-i$. Show that $|w| < 1$, $|w| = 1$ and $|w| > 1$ are, respectively, equivalent to $d_1 < d_2$, $d_1 = d_2$ and $d_1 > d_2$, and illustrate by a sketch. From this point of view the result of Problem 7 is obvious.
9. If p and q are unequal complex numbers, the equation of the perpendicular bisector of the segment pq is $|z - p| = |z - q|$. By squaring, reduce this to the form $\bar{B}z + B\bar{z} + C = 0$ where $B = p - q \neq 0$. Also show, conversely, that the latter equation with $B \neq 0$ can be reduced to the former with $p \neq q$.
10. The points p and q in Problem 9 are symmetrically located with respect to the line and are called *inverse points*. If $w = az + b$ with $a \neq 0$, let $p^* = ap + b$ and $q^* = aq + b$. Show that $|z - p| = |z - q|$ implies $|w - p^*| = |w - q^*|$, and conclude that the linear transformation maps lines into lines, inverse points into inverse points.
11. If $T_1z = 1 + z$, $T_2z = 1 + (1/z)$ and $T_3z = (i + z)/(i - z)$, compute the four products needed for the associative law, and verify it in this case.
12. If $Tz = (1 - z)/(1 + z)$, show that

$$T(-z) = \frac{1}{Tz}, \qquad T\left(\frac{1}{z}\right) = -Tz, \qquad T(Tz) = z, \qquad T\frac{a + b}{1 + ab} = (Ta)(Tb).$$

13. A *fixed point* of the bilinear transformation $w = Tz$ in (1.1) is a point z satisfying $z = Tz$. If $c = 0$, we consider that ∞ is a fixed point. Show that there are at most two fixed points unless T reduces to the identity transformation, in which case every point is fixed.
14. By considering $X(w,w_1,w_2,w_3) = X(z,z_1,z_2,z_3)$, show that there is a bilinear mapping taking any three distinct points z_i into any three distinct points w_i. Show also that the mapping is unique; that is, if another bilinear mapping

takes z_i into w_i and z into $\tilde{w}$, then $\tilde{w} = w$. (Invariance of the cross ratio gives

$$X(\tilde{w},w_1,w_2,w_3) = X(z,z_1,z_2,z_3) = X(w,w_1,w_2,w_3)$$

and $\tilde{w} = w$ follows from this.)

15. Deduce the uniqueness in the preceding problem from Problem 13. (If there are two mappings, then one followed by the inverse of the other gives a mapping with three fixed points.)
16. If C is a circle or line in the z plane, and C^* is a circle or line in the w plane, deduce from Problem 14 that there is a bilinear transformation mapping C onto C^*. Is it unique?

2. Bilinear transformations, continued. The equation of the circle with center z_0 and radius ρ is $|z - z_0| = \rho$, or equivalently,

$$(z - z_0)(\bar{z} - \bar{z}_0) = \rho^2, \qquad \rho > 0. \quad (2.1)$$

When $\rho = 0$, the circle reduces to a point and when $\rho < 0$, the locus is empty, that is, has no points. These cases are excluded here.

If the left side of (2.1) is multiplied out, the equation takes the form

$$z\bar{z} + \bar{B}z + B\bar{z} + C = 0 \quad (2.2)$$

where $B = -z_0$ and $C = |z_0|^2 - \rho^2$. The latter implies $C < |B|^2$. Conversely, if (2.2) is given with $C < |B|^2$, we can define z_0 and ρ by

$$z_0 = -B, \qquad \rho^2 = |B|^2 - C, \qquad \rho > 0.$$

Then (2.2) is the same as (2.1) and so represents a circle with center z_0 and radius ρ.

The equation of a straight line is obtained when $B \neq 0$ and $z\bar{z}$ is dropped from (2.2). Both lines and circles are included by writing (2.2) in the form

$$Az\bar{z} + \bar{B}z + B\bar{z} + C = 0, \qquad AC < |B|^2. \quad (2.3)$$

The condition $AC < |B|^2$ comes from the condition $C < |B|^2$ in (2.2), as seen when we divide (2.3) by A. This division reduces (2.3) to the form (2.2) and shows that the center and radius for $A \neq 0$ are

$$z_0 = -\frac{B}{A}, \qquad \rho = \frac{\sqrt{|B|^2 - AC}}{|A|}. \quad (2.4)$$

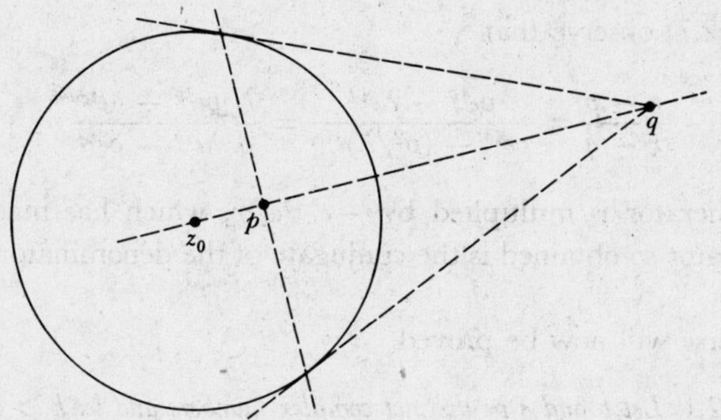

Figure 2-1

If $\lambda > 0$ and ϕ is real, the two points

$$p = z_0 + \lambda e^{i\phi}, \qquad q = z_0 + \frac{\rho^2}{\lambda} e^{i\phi} \tag{2.5}$$

are said to be *inverse points with respect to the circle.* As shown in Figure 2-1, the points p and q lie on the same radial line from z_0. The product of their distances to the center is ρ^2; more specifically,

$$(p - z_0)(\bar{q} - \bar{z}_0) = \rho^2. \tag{2.6}$$

Conversely, if (2.6) holds, we can write $p - z_0 = \lambda e^{i\phi}$ and recover (2.5). Thus (2.6) is necessary and sufficient for p and q to be inverse points in the case $\lambda \neq 0$.

A geometric construction for p and q is suggested by the dotted lines in Figure 2-1. This construction indicates that as q recedes to ∞, the inverse point p tends to the center. The same conclusion is obtained by letting $\lambda \to 0$ in (2.5). If $\lambda = 0$ we therefore set $p = z_0$ and $q = \infty$. Two points are said to be *inverse with respect to a straight line* if the line is the perpendicular bisector of the line segment joining the points.

If $z = z_0 + \rho e^{i\phi}$ is a point on the above circle, it will be proved for $\lambda > 0$ that

$$\left|\frac{z - p}{z - q}\right| = \frac{\lambda}{\rho}. \tag{2.7}$$

To prove (2.7) observe that

$$\frac{z-p}{z-q} = \frac{\rho e^{i\theta} - \lambda e^{i\phi}}{\rho e^{i\theta} - (\rho^2/\lambda)e^{i\phi}} = \frac{\lambda}{\rho}\frac{\rho e^{i\theta} - \lambda e^{i\phi}}{\lambda e^{i\theta} - \rho e^{i\phi}}.$$

If the numerator is multiplied by $-e^{-i\theta}e^{-i\phi}$, which has magnitude 1, the numerator so obtained is the conjugate of the denominator and (2.7) follows.

A converse will now be proved.

THEOREM 2.1. *Let p and q be distinct complex numbers and let $k > 0$. Then the equation*

$$\left|\frac{z-p}{z-q}\right| = k \tag{2.8}$$

represents a circle if $k \neq 1$, with center and radius

$$z_0 = \frac{p - k^2 q}{1 - k^2}, \qquad \rho = \frac{k|p - q|}{|1 - k^2|}. \tag{2.9}$$

The points p and q are inverse points with respect to the circle. If $0 < k < 1$, then p is inside the circle and if $1 < k < \infty$, then p is outside the circle. When $k = 1$, the equation represents a line and p and q are inverse points with respect to the line.

Proof. When $k = 1$ then $|z - p| = |z - q|$, which states that z is equidistant from p and q. This shows that the locus is the perpendicular bisector of the segment pq and gives Theorem 2.1 for $k = 1$.

If $k \neq 1$, the equation $|z - p| = k|z - q|$ becomes, on squaring,

$$(z - p)(\bar{z} - \bar{p}) = k^2(z - q)(\bar{z} - \bar{q}).$$

Transferring all terms to the left gives (2.3) with

$$A = 1 - k^2, \qquad B = k^2 q - p, \qquad C = |p|^2 - k^2|q|^2.$$

By a short calculation, $|B|^2 - AC = k^2|p - q|^2$, and (2.4) therefore gives (2.9). By (2.9),

$$p - z_0 = \frac{k^2(q - p)}{1 - k^2}, \qquad q - z_0 = \frac{q - p}{1 - k^2}$$

and hence $(p - z_0)(\bar{q} - \bar{z}_0) = \rho^2$. Therefore p and q are inverse points. Since $|p - z_0| = k\rho$, clearly p is inside the circle if $k < 1$ and is outside the circle if $k > 1$. This completes the proof.

In the above proof we used the fact that the interior and exterior of the circle are defined, respectively, by

$$|z - z_0| < \rho, \qquad |z - z_0| > \rho.$$

These inequalities lead to corresponding inequalities in (2.2). The interior and exterior of a circle are called the *complementary domains,* because, together with the circle, they complete the extended plane. In the case of a line the complementary domains are two half-planes, one on each side of the line. They are given by inequality in (2.3) with $A = 0$. The point ∞ is interior to one of the complementary domains for a circle, but is a boundary point of both complementary domains in the case of a line.

THEOREM 2.2. *A bilinear transformation maps a circle or line onto a circle or line, and it maps inverse points into inverse points. Furthermore, if the circle or line K is mapped onto the circle or line K^*, then one of the complementary domains of K is mapped onto one of the complementary domains of K^*, and the other complementary domain of K is mapped onto the other complementary domain of K^*.*

Proof. The first part of this theorem was proved in Section 1. The same proof gives the assertion about complementary domains if we use inequality in (2.3) rather than equality. Since the case of a linear transformation can be handled trivially, as in Example 1.1, it is necessary to consider the transformation $w = 1/z$ only.

To prove the statement about inverse points, let p and q be inverse points, both different from ∞. Then the circle or line K can be written as

$$\left|\frac{z - p}{z - q}\right| = k.$$

If $w = az$, $p^* = ap$ and $q^* = aq$, with $a \neq 0$, the factor a drops out of the left side and the resulting equation represents a circle in the w plane with p^* and q^* as inverse points. The transformation $w = z + b$ is handled similarly. If $w = 1/z$, $p^* = 1/p$ and $q^* = 1/q$, substitution gives

$$\left|\frac{w - p^*}{w - q^*}\right| = k\left|\frac{p^*}{q^*}\right|$$

after slight simplification, so that again p^* and q^* are inverse points. This requires $p \neq 0$ and $q \neq 0$. If $p = 0$, the transformed equation is

$$|w - q^*| = \frac{|q^*|}{k}$$

and it is a circle with inverse points ∞ and q^*. The case $q = 0$ is similar.

In the remaining case, $p = \infty$ or $q = \infty$, we can assume $q = \infty$. Then $p = z_0$ and the equation of the circle is $|z - p| = \rho$. It is left for the reader to verify that $w = 1/z$ transforms this circle into another in which $p^* = 1/p$ and $0 = 1/\infty$ are inverse points.

THEOREM 2.3. *A necessary and sufficient condition for a bilinear transformation to map the upper half-plane* Im $z > 0$ *onto the unit disk* $|w| < 1$ *is that it have the form*

$$w = \beta\frac{z - \alpha}{z - \bar{\alpha}} \tag{2.10}$$

where $|\beta| = 1$ *and* Im $\alpha > 0$.

Proof of necessity. Let the bilinear transformation

$$w = \frac{az + b}{cz + d}$$

map $y > 0$ onto $|w| < 1$. It maps $y = 0$ onto some circle or line K^* in the w plane, and it maps $y > 0$ onto one of the complementary domains of K^*. This shows that one of the complementary domains of K^* is $|w| < 1$ and hence K^* is $|w| = 1$. In other words, the real axis is mapped onto the unit circle.

Let α be the point of Im $z > 0$ which maps into $w = 0$. Then $\bar{\alpha}$, the inverse point of α in Im $z = 0$, must map into $w = \infty$, the inverse point of $w = 0$ in $|w| = 1$. Since $w = 0$ and $w = \infty$ correspond to $z = -b/a$ and $z = -d/c$, respectively, it is seen that $a \neq 0$, $c \neq 0$, and $\alpha = -b/a$, $\bar{\alpha} = -d/c$. Hence

$$w = \frac{a}{c} \cdot \frac{z - \alpha}{z - \bar{\alpha}}.$$

Since the image of the point $z = 0$ is on $|w| = 1$, we get $|a/c| = 1$ and this proves the necessity.

Proof of sufficiency. Let (2.10) hold. Since $|x - \alpha| = |x - \bar{\alpha}|$ when x is real, the real axis is mapped onto $|w| = 1$ and hence $y > 0$ is mapped onto one of the complementary domains of $|w| = 1$. Since $z = \alpha$ maps into $w = 0$, this complementary domain contains $w = 0$ and so it must be the domain $|w| < 1$, not the domain $|w| > 1$. This completes the proof.

THEOREM 2.4. *A necessary and sufficient condition for a bilinear transformation to map $|z| < 1$ onto $|w| < 1$ is that it have the form*

$$w = \beta \frac{z - \alpha}{\bar{\alpha} z - 1} \tag{2.11}$$

where $|\alpha| < 1$ and $|\beta| = 1$.

Proof of necessity. Much as in the previous proof, if the mapping takes $|z| < 1$ onto $|w| < 1$ then it takes $|z| = 1$ onto $|w| = 1$. Let α be the point of $|z| < 1$ which maps into $w = 0$. The inverse of α in $|z| = 1$ is $1/\bar{\alpha}$ by (2.6) and must map into $w = \infty$, the inverse of $w = 0$ in $|w| = 1$. Hence $\alpha = -b/a$ and $1/\bar{\alpha} = -d/c$ (unless $\alpha = 0$). Thus the bilinear transformation is

$$w = \frac{a}{c} \cdot \frac{z - \alpha}{z - 1/\bar{\alpha}} = \frac{a\bar{\alpha}}{c} \cdot \frac{z - \alpha}{\bar{\alpha} z - 1}.$$

Since the point $z = 1$ has its image on $|w| = 1$,

$$1 = \left| \frac{a\bar{\alpha}}{c} \cdot \frac{1 - \alpha}{\bar{\alpha} - 1} \right| = \left| \frac{a\bar{\alpha}}{c} \right|,$$

and so the transformation has the form (2.11). If $\alpha = 0$, then $(0,\infty)$ map into $(0,\infty)$ and it is easily verified that $w = \beta z$, $|\beta| = 1$.

Proof of sufficiency. If (2.11) holds, the choice $z = e^{i\theta}$, θ real, gives

$$|w| = \left| \frac{e^{i\theta} - \alpha}{\bar{\alpha} e^{i\theta} - 1} \right| = \left| \frac{\alpha - e^{i\theta}}{\bar{\alpha} - e^{-i\theta}} \right| = 1.$$

Thus $|w| = 1$ for $|z| = 1$ and so $|z| < 1$ is mapped onto one of the complementary domains of $|w| = 1$. Since $z = 0$ maps into $\bar{\alpha}$ and $|\alpha| < 1$, the complementary domain must be $|w| < 1$ and Theorem 2.4 follows.

This proof of sufficiency gives the result of Chapter 1, Example 2.1 with little calculation. Alternatively, Example 2.1 could have been used to prove the sufficiency.

Example 2.1. Find the images of the coordinate lines $x = x_0$ and $y = y_0$ in the right half-plane under the transformation

$$w = \frac{z-1}{z+1}, \qquad z = \frac{1+w}{1-w}.$$

Since we can write $w = (iz - i)/(iz + i)$, Theorem 2.3 with $\beta = 1$ and $\alpha = i$ indicates that this transformation maps $\text{Im}(iz) > 0$ onto $|w| < 1$. Hence it maps $\text{Re}\, z > 0$ onto $|w| < 1$. The same conclusion is evident by inspection of the identities

$$1 - w\bar{w} = \frac{2(z + \bar{z})}{|z+1|^2}, \qquad z + \bar{z} = 2\frac{1 - w\bar{w}}{|1-w|^2},$$

proof of which is left to the reader.

The points $z = 0,1,\infty$ map respectively onto $w = -1,0,1$, and hence the positive real axis maps onto the diameter $-1 < u < 1$ of the circle $|w| = 1$. The lines $x = c$, where c is constant, are orthogonal to the real axis in the z plane, and so they map into lines or circles orthogonal to the segment $-1 < u < 1$ in the w plane. Since the lines $x = c$ pass through ∞ in the z plane, their images pass through the image of ∞ in the w plane. Hence the images are circles through the point 1 and are perpendicular to the real axis, as shown in Figure 2-2. The special case $x = 0$ gives the unit circle $|w| = 1$, as was already known from the fact that $\text{Re}\, z > 0$ maps onto $|w| < 1$. The line $x = 0.2$ maps into the circle labeled 0.2 in the figure, and so on.

The lines $y = c$ pass through $z = \infty$ and hence their images pass through $w = 1$. Since the lines $y = c$ intersect $x = 0$ at right angles, their images intersect $|w| = 1$ at right angles. Hence the images are circles

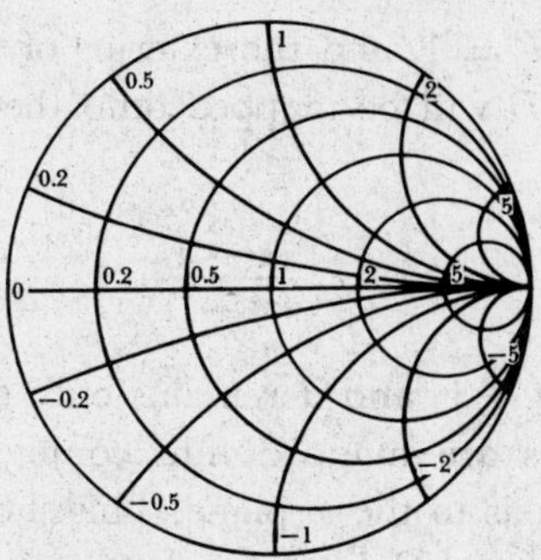

Figure 2-2

tangent to the u axis at the point $w = 1$, as shown in the figure. For instance, the ray $y = -0.2$, $x > 0$ maps into the circular arc labeled -0.2 in Figure 2-2.

Example 2.2. Let D denote the domain bounded by two circles C and C_0, one inside the other. Find a conformal map of D onto an annulus (Figure 2-3).

By rotation and change of scale we can assume that the outer circle is $|z| = 1$ and that the inner circle C has its center at some point $x_0 > 0$. The radius ρ of C satisfies $\rho + x_0 < 1$. We shall map $|z| < 1$ onto $|w| < 1$ in such a way that the image C^* of C is a circle $|w| = \rho^*$ centered at the origin. By Theorem 2.2 it will follow that the interior of $|z| = 1$ is mapped

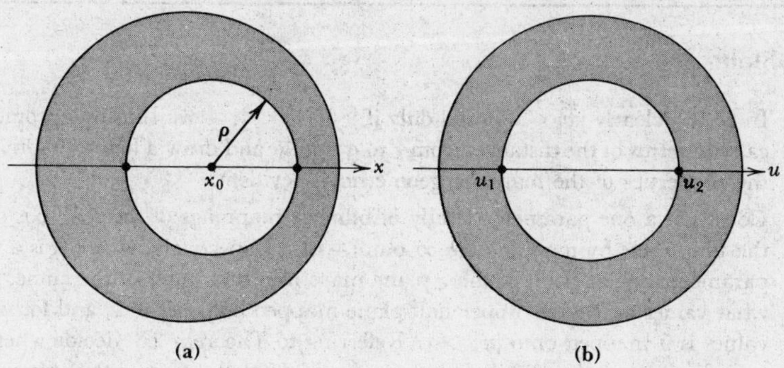

Figure 2-3

onto the interior of $|w| = 1$, and the exterior of C is mapped onto the exterior of C^*. Thus D will be mapped onto the annulus, as required.

By Theorem 2.4

$$w = \frac{\alpha - z}{\alpha z - 1}, \qquad -1 < \alpha < 1, \quad (2.12)$$

maps $|z| < 1$ onto $|w| < 1$, and it remains only to choose α. Since α is real, conjugate points are mapped into conjugate points and C^* is bisected by the real axis in the w plane. This shows that the center w_0 of C^* is real. The center of C does not map into the center of C^*. However, the diameter

$$x_0 - \rho \leq x \leq x_0 + \rho$$

does map into a diameter $u_1 \leq u \leq u_2$ of C^* and hence the center w_0 of C^* satisfies $2w_0 = u_1 + u_2$. That is,

$$2w_0 = \frac{(x_0 - \rho) - \alpha}{\alpha(x_0 - \rho) - 1} + \frac{(x_0 + \rho) - \alpha}{\alpha(x_0 + \rho) - 1}.$$

Setting $w_0 = 0$ leads to the quadratic equation

$$\alpha^2 x_0 - \alpha(1 + x_0^2 - \rho^2) + x_0 = 0,$$

which has distinct real roots if $1 + x_0^2 - \rho^2 > 2x_0$. This holds, because $1 - x_0 > \rho$. The product of the roots is 1, so one of them satisfies $|\alpha| < 1$. Using this root in (2.12) gives the desired transformation.

Problems

1. In (2.10), clearly $|w| < 1$ if and only if $|z - \alpha| < |z - \bar{\alpha}|$. Interpret geometrically in terms of the distances from z to α and $\bar{\alpha}$, and draw a figure that makes the properties of the mapping geometrically evident.
2. Construct a one-parameter family of bilinear mappings of the real axis onto the unit circle by mapping $0,\lambda,\infty$ onto $-i,1,i$, respectively, where λ is a real parameter. What point of the z plane maps into the center of the circle? For what values of λ is the upper half-plane mapped onto $|w| < 1$, and for what values is it mapped onto $|w| > 1$? Referring to Theorem 2.3, decide whether your family includes all bilinear mappings of the real axis onto the unit circle, or only some of them.

3. By applying Theorem 2.4 to z/a and w/b, get all bilinear mappings of $|z| < a$ onto $|w| < b$.
4. Find all bilinear mappings of $|z| < 1$ onto $|w| > 1$.
5. Find all bilinear transformations which map $\text{Im } z > 0$ onto $|w| < 1$ and $z = i$ onto $w = \frac{1}{2}$. (Since inverse points map into inverse points, $z = -i$ maps into $w = 2$. Hence $(z - i)/(z + i) = \beta(w - \frac{1}{2})/(w - 2)$. Show $|w| = 1$ for z real if and only if $|\beta| = 2$.)
6. Find all bilinear transformations that map $|z - 1| < 2$ onto $\text{Im } w > 0$ and $z = 1$ onto $w = i$.
7. Multiplying a, b, c and d by a properly chosen complex number leaves a bilinear transformation unchanged and so, with no restriction, one can assume $ad - bc = 1$. Show that a necessary and sufficient condition for such a transformation to map $\text{Im } z > 0$ onto $\text{Im } w > 0$ is for a, b, c and d to be real.
8. Show that $X(t,0,1,\infty) = t$ and deduce that four distinct points z_i lie on a circle if and only if $X(z_1, z_2, z_3, z_4)$ is real.

Groups of transformations

A set G of one-to-one transformations of a domain D onto itself forms a *group* if G contains with each transformation T its inverse T^{-1}, and with each two transformations T_1, T_2 their product.

9. Show that the set G of all linear transformations $Tz = az + b$ forms a group. Show that the sets in which the coefficients are restricted as follows also form groups:

$$a = 1; \qquad a = 1, \quad b = \text{integer}; \qquad b = 0; \qquad b = 0, \quad |a| = 1.$$

These are called *subgroups* of the original group G.

10. Show that the set of all bilinear mappings of $|z| < 1$ onto itself forms a group.
11. Show that the set of all bilinear mappings forms a group, and that the set for which $ad - bc = 1$ forms a subgroup. (Readers familiar with 2-by-2 matrices can simplify the work by showing that if

$$Tz = \frac{az + b}{cz + d} \quad \text{corresponds to} \quad \begin{pmatrix} a & b \\ c & d \end{pmatrix}$$

then the product $T_1 T_2$ corresponds to the matrix product, and hence the determinant of the product is the product of the determinants.)

12. Referring to Problem 11, show that the set of bilinear mappings where a, b, c and d are integers and $ad - bc = 1$ forms a group. Show that

$$\text{Im } \frac{az + b}{cz + d} = \frac{y}{|cz + d|^2}$$

and deduce that this class of mappings maps the upper half-plane onto itself and also the lower half-plane onto itself. This group is called the *modular group*.

3. Harmonic functions and mappings. It has already been observed that if h is analytic in D, then Re h and Im h are both harmonic functions, that is, they are both solutions of the Laplace equation. This is a consequence of the Cauchy-Riemann equations and of the fact that h and hence Re h and Im h possess higher-order derivatives.

The terminology arising in mathematical physics can be used in the study of mappings and in some cases can provide motivation.[1]

If $\psi(x,y)$ is a solution of the Laplace equation

$$\frac{\partial^2 \psi}{\partial x^2} + \frac{\partial^2 \psi}{\partial y^2} = 0,$$

then the curves $\psi(x,y) = c$, where c is a constant, are the lines of flow of an irrotational, inviscid, incompressible plane flow. These curves are also the lines of force in a plane electrostatic or magnetostatic field; and they are the lines of flow in the steady-state flow of heat in a uniform plane. Hence, if $h(z)$ is analytic in a domain D, the curves Im $h(z) = c$ are the lines of flow or of force as described above. It is a further fact that the orthogonal curves Re $h(z) = c$ are the equipotential lines in the first two cases above and are isothermal lines in the case of heat flow. (The orthogonality of the curves Re $h = c$ and Im $h = c$ for $h'(z) \neq 0$ was deduced from the Cauchy-Riemann equations in Chapter 2, Example 1.1. It also follows from the fact that the map $w = h(z)$ is conformal at points where $h'(z) \neq 0$.)

If $h = \phi + i\psi$, the *complex velocity* in the plane flow described above is $\phi_x + i\phi_y$, where the subscripts denote partial differentiation; thus

$$V = \frac{\partial}{\partial x} \operatorname{Re} h - i \frac{\partial}{\partial x} \operatorname{Im} h.$$

Readers familiar with the concept of gradient will recognize that the

[1] Such terminology is mentioned only to suggest a few applications, and to give a concrete physical model for those who prefer such a model. Readers unfamiliar with some of the terms will find themselves at no disadvantage in following the mathematical development.

components of V are the same as those of grad ϕ, and a detailed discussion of the complex velocity can be found in Chapter 2, Section 6. It suffices here to note that $V = \psi_y - i\psi_x$ by the Cauchy-Riemann equations, and so the slope of V agrees with $-\psi_x/\psi_y$, which is the slope of the curves $\psi(x,y) = c$. Thus V is tangent to the lines of flow. The Cauchy-Riemann equations give $V = \phi_x - i\psi_x$ or

$$V = \overline{h'(z)}. \tag{3.1}$$

The facts described above make functions of a complex variable extremely useful in numerous applications. For example, the function $h(z) = z^2$ leads to the flow curves $2xy = c$ which describe the flow in a quadrant (Figure 3-1). The complex velocity is $2\bar{z}$.

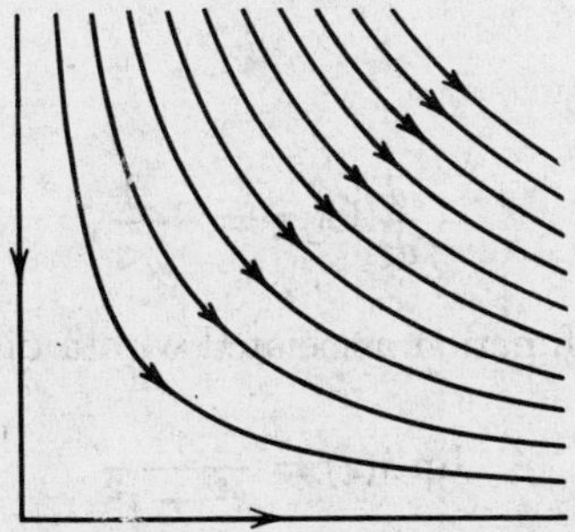

Figure 3-1

As another illustration, consider $h(z) = \log z$. This function is not single-valued, but its derivative, $1/z$, is. The latter determines the complex velocity $V = 1/\bar{z}$ or, in polar coordinates, $V = e^{i\theta}/r$. The flow is outward along radial lines from the origin, and $\log z$ is said to be the analytic function associated with a source at $z = 0$. Similarly, a source at ϵ is associated with $\log(z - \epsilon)$ and a sink at $-\epsilon$ is associated with $-\log(z + \epsilon)$. If such a source and sink are combined, and if their strengths are increased by the factor $1/(2\epsilon)$, the result is

$$\frac{1}{2\epsilon}[\log(z - \epsilon) - \log(z + \epsilon)].$$

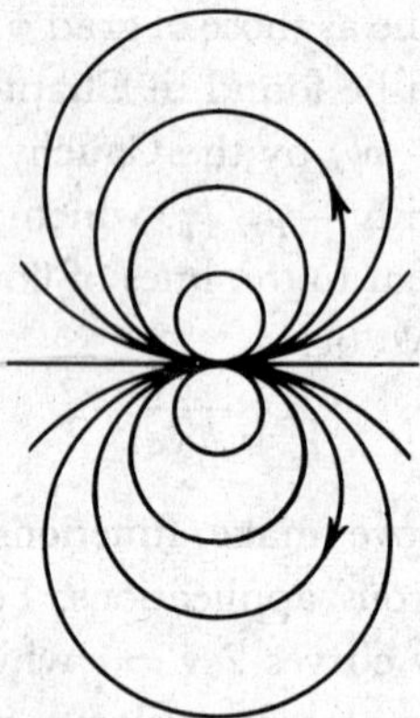

Figure 3-2

As $\epsilon \to 0$ this tends to

$$-\frac{d}{dz}\log z = -\frac{1}{z},$$

which is the analytic function associated with a dipole. Here

$$\operatorname{Im} h(z) = \frac{y}{x^2 + y^2}$$

and the lines of flow are circles tangent to the x axis, as shown in Figure 3-2. The velocity is $1/\bar{z}^2 = e^{2i\theta}/r^2$.

The above theory is very much enriched by the behavior of flows under conformal mapping. Let $w = f(z)$ be a one-to-one mapping from a domain D in the z plane to $f(D)$ in the w plane and let $z = g(w)$ be the inverse function. It is shown in Section 6 that g is analytic. If $h(z)$ is an analytic function associated with a flow in D then, since $h[g(w)]$ is analytic in $f(D)$, the curves $\operatorname{Im} h[g(w)] = c$ and $\operatorname{Re} h[g(w)] = c$, which are the images of $\operatorname{Im} h(z) = c$ and $\operatorname{Re} h(z) = c$, respectively, are again associated with a flow in the w plane. *Hence, under conformal mapping, lines of flow and equipotential lines are mapped into lines of flow and equipotential lines.*

As an example, consider the flow in the z plane associated with $h(z) = \log z$. Here the lines of flow, $\arg z = c$, are the radii extending from $z = 0$

to $z = \infty$. Under the transformation

$$z = \frac{w - 1}{w + 1}$$

the function $h(z)$ becomes

$$\log\left(\frac{w - 1}{w + 1}\right).$$

The point $z = 0$ goes to $w = 1$ and $z = \infty$ goes to $w = -1$. Hence the lines from $z = 0$ to $z = \infty$ map into the arcs of circles joining $w = 1$ to $w = -1$. These circles have the equation $\arg z = c$, or

$$\arg\frac{w - 1}{w + 1} = c.$$

The equipotential curves $\log|z| = c$ are circles and map into the circles

$$\left|\frac{w - 1}{w + 1}\right| = e^c$$

which have $+1$ and -1 as inverse points. The lines of flow and equipotential lines in the w plane are shown in Figure 3-3. This configuration is associated with a number of interesting physical problems.

As another illustration consider the mapping $w = z^{a/\pi}$, where a is

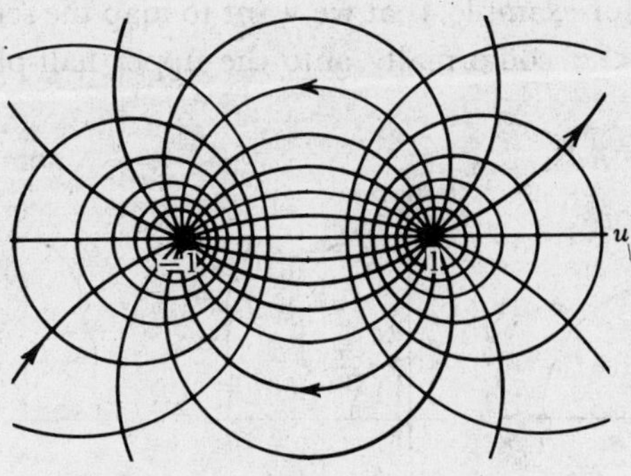

Figure 3-3

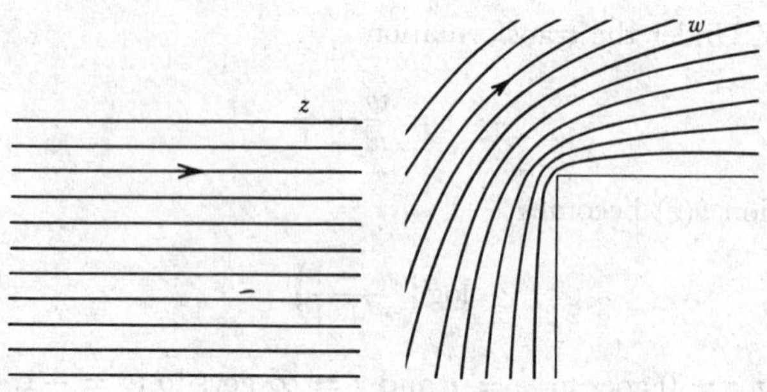

Figure 3-4

constant, $0 < a < 2\pi$. In polar coordinates

$$z = re^{i\theta}, \qquad w = r^* e^{i\theta^*} = r^{a/\pi} e^{i\theta a/\pi},$$

and so the upper half-plane $0 < \theta < \pi$ is mapped one-to-one onto the sector $0 < \theta^* < a$ in the w plane. The function $h(z) = z$ has $y = c$ as its lines of flow and describes uniform flow parallel to the x axis. Under the above mapping the new lines of flow satisfy $\operatorname{Im} w^{\pi/a} = c$ and give the flow around a corner in the w plane as shown in Figure 3-4. For $a \neq \pi$ the mapping $w = z^{a/\pi}$ is not conformal at $z = 0$, since angles with vertex at the origin are changed by the factor a/π. Tangency of curves through the origin is preserved, however.

The above function $w = z^{a/\pi}$ is useful in the solution of mapping problems. Suppose, for example, that we want to map the semicircular region $|z| < 1$, $0 < \theta < \pi$ conformally onto the upper half-plane as shown in

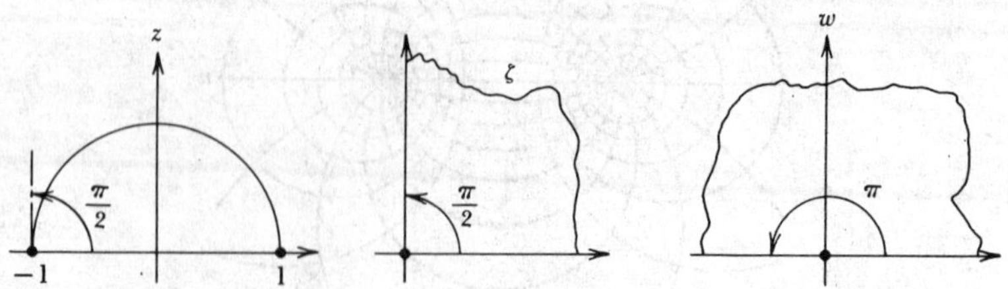

Figure 3-5

Figure 3-5. The bilinear transformation

$$\zeta = \frac{1+z}{1-z}$$

maps -1 onto 0 and 1 onto ∞. Hence the diameter $-1 \leq x \leq 1$ and the semicircle are mapped into circles or lines through 0 and ∞, which is to say, they are mapped into lines. The diameter maps onto $\text{Re}\,\zeta > 0$ and, since angles are preserved, the semicircle must map onto $\text{Im}\,\zeta > 0$. It is easily checked that the semicircular region is mapped onto the first quadrant in the ζ plane. Under $w = \zeta^2$ the angle at the origin is doubled, and so the desired mapping of the semicircular region onto the upper half-plane is

$$w = \left(\frac{1+z}{1-z}\right)^2. \tag{3.2}$$

A more sophisticated application of this technique is given next. Consider the w plane with a cut on the real axis from $w = -1$ to $w = 1$. Let it be required to map the cut w plane onto the exterior of a smooth closed curve C in such a way that the two sides of the slit are mapped onto C. Since C is smooth, it is necessary to map the exterior angles of the cut at $w = +1$ and $w = -1$ onto angles of π. We let these angles of π have their vertices at $z = +1$ and $z = -1$, respectively, and we consider the transformation

$$\frac{z-1}{z+1} = \left(\frac{w-1}{w+1}\right)^{1/2}$$

where the square root is positive for $v = 0$, $u > 1$. This transforms $w = 1$ to $z = 1$ and $w = -1$ to $z = -1$ and also halves the exterior angles at

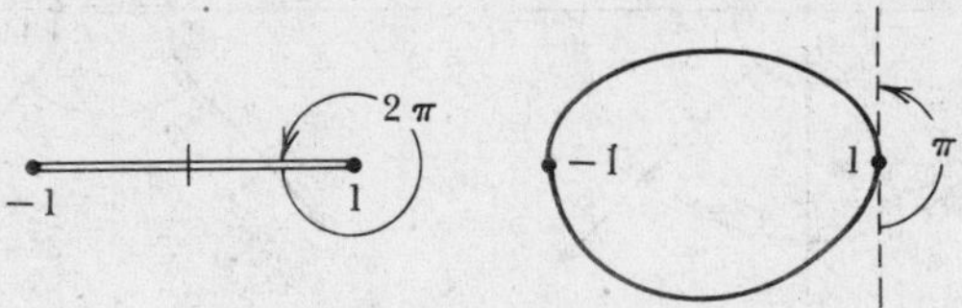

Figure 3-6

$w = 1$ and $w = -1$, as shown in Figure 3-6. Squaring the equation and solving for z gives

$$w = \frac{1}{2}\left(z + \frac{1}{z}\right) \tag{3.3}$$

so that $dw/dz = 0$ only at $z = 1$ and $z = -1$. Setting $z = e^{i\theta}$ gives $w = \cos\theta$ and hence the two sides of the slit, $-1 \leq w \leq 1$, are mapped onto the unit circle $|z| = 1$.

More generally, $z = re^{i\theta}$ gives

$$u = \frac{1}{2}\left(r + \frac{1}{r}\right)\cos\theta, \qquad v = \frac{1}{2}\left(r - \frac{1}{r}\right)\sin\theta \tag{3.4}$$

and hence the circle $r = c > 1$ maps onto an ellipse with major axis $c + 1/c$ and minor axis $c - 1/c$. On the other hand, the radial lines $\theta = c$ map onto quarters of hyperbolas with the equation

$$\frac{u^2}{\cos^2\theta} - \frac{v^2}{\sin^2\theta} = 1.$$

Since the circles $r = c$ and the lines $\theta = c$ are orthogonal in the z plane, the ellipses and hyperbolas are orthogonal in the w plane, as shown in Figure 3-7. The foci are all at $w = \pm 1$. Equation (3.3) provides a one-to-one mapping of the region $|z| > 1$ onto the slit w plane, and the orthogonal net of confocal ellipses and hyperbolas is a *confocal coordinate system* for $|z| > 1$. It is readily ascertained that the inverse mapping is

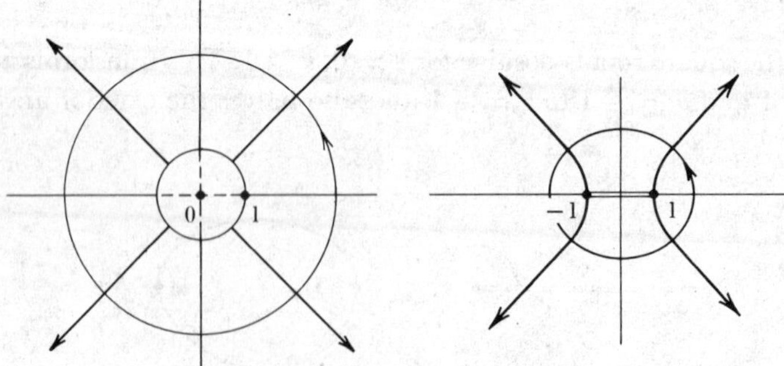

Figure 3-7

$$z = w + (w^2 - 1)^{1/2} \tag{3.5}$$

where the square root is positive for $v = 0$, $u > 1$. A number of flows can be studied by this mapping, as shown in Problems 5–7 at the end of this section.

So far we have taken the view that a flow is given by an analytic function $h = \phi + i\psi$, and hence the mapping of the flow under $z = g(w)$ can be investigated by considering $h[g(w)]$. Sometimes only the harmonic function ϕ or ψ is under discussion and the analytic function $h = \phi + i\psi$ is, apparently, not available. For example, the problem of determining the steady-state temperature in a uniform plate requires construction of a harmonic function ϕ with prescribed boundary conditions, and makes no reference to the conjugate function ψ. Problems of this type can also be studied by conformal mapping.

Under a one-to-one conformal map, $z = g(w)$, $w = f(z)$, a harmonic function ϕ is mapped into another function ϕ^* by the rule that

$$\phi^*(u,v) = \phi(x,y) \quad \text{whenever} \quad x + iy = g(u + iv). \tag{3.6}$$

Then ϕ^* is harmonic in $f(D)$ if ϕ is harmonic in D. This is so because in the neighborhood of any point (x,y) of D we can construct a harmonic conjugate ψ such that

$$h(z) = \phi(x,y) + i\psi(x,y)$$

is analytic. It is easily checked that

$$\phi^*(u,v) = \operatorname{Re} h[g(w)]$$

and since the function on the right is analytic, ϕ^* is harmonic.

The correspondence between ϕ and ϕ^* in (3.6) gives boundary values for ϕ^* from those for ϕ. In particular, if $\phi = c$, where c is a constant, holds on some curve C, then $\phi^* = c$ holds on the image curve C^*. This agrees with the fact noted previously that equipotentials map into equipotentials.

The gist of these remarks is that every harmonic function ϕ is associated locally with an analytic function $h = \phi + i\psi$, and problems involving harmonic functions can be studied by the same conformal mapping techniques as are used in the study of flows. Construction of a harmonic function by conformal mapping is illustrated in the example below.

Example 3.1. Find a function which is harmonic for $|z| < 1$ and assumes values 0 and 1 on the parts of $|z| = 1$ with $x > 0$ and $x < 0$, respectively.

The points $-i,1,i$ belong to the arc associated with the constant boundary value 0 and are mapped respectively onto $0,1,\infty$ by the bilinear transformation

$$w = -i\frac{z+i}{z-i}. \tag{3.7}$$

This transformation maps $|z| < 1$ onto the upper half-plane $\operatorname{Im} w > 0$, and maps ϕ into another harmonic function ϕ^*. Since $\phi = 0$ on the arc $e^{i\theta}$, $-\pi/2 < \theta < \pi/2$, we have $\phi^* = 0$ on the image of this arc, which is the positive real axis $0 < u < \infty$. Similarly, $\phi^* = 1$ on $-\infty < u < 0$, since this is the image of the arc on which $\phi = 1$. See Figure 3-8.

The original problem in $|z| < 1$ has been mapped into a much easier problem in $\operatorname{Im} w > 0$. This second problem requires that ϕ^* change from 0 on the positive real axis to 1 on the negative real axis and has the obvious solution $\phi^* = (1/\pi) \operatorname{Arg} w$. The latter function is harmonic because $\operatorname{Im} \operatorname{Log} w$ is harmonic, and gives

$$\phi(x,y) = \frac{1}{\pi} \operatorname{Arg} \frac{z+i}{z-i}(-i).$$

If the numerator and denominator are multiplied by $\bar{z} + i$, it is found that ϕ satisfies

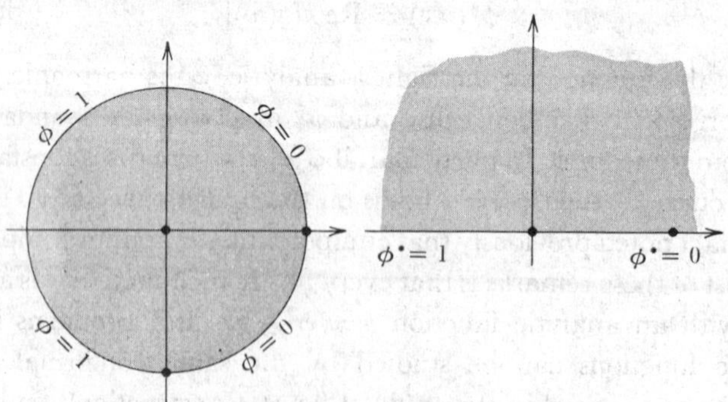

Figure 3-8

$$\tan \pi\phi = \frac{1 - x^2 - y^2}{2x}, \qquad 0 \le \phi \le 1.$$

The formula indicates that the equipotentials $\phi = c$ are circles. The same conclusion is obtained more easily without the formula, as follows: The equipotentials in the z plane are the images of those in the w plane under the mapping (3.7). Since the latter are straight lines through 0 and ∞, the former must be circles through $-i$ and i.

The function ϕ can be interpreted as the steady-state temperature in the disk when half the boundary is kept at the temperature 0 and half at the temperature 1. It can also be interpreted as the electrostatic potential in a split cylinder, and so on.

Problems

1. If λ is a real constant, the function $f(z) = e^{-i\lambda}z$ represents uniform flow in a direction making an angle λ with the x axis. (a) Verify this by finding the lines of flow and the complex velocity. (b) Three flows associated with $\log z$ are discussed in Chapter 2, Example 6.1. Read this discussion.
2. In Example 3.1 show that the value of ϕ at the center of the circle is the average of the values on the circumference. This illustrates a general property of harmonic functions.
3. The conformal transformation $z = e^w$ gives $x = e^u \cos v$, $y = e^u \sin v$. Under this correspondence verify that
$$\phi(x,y) = x^2 - y^2 \quad \text{becomes} \quad \phi^*(u,v) = e^{2u}(\cos^2 v - \sin^2 v).$$
Show that ϕ and ϕ^* are both harmonic, and that the hyperbola C, on which $\phi(x,y) = c$, is mapped onto another curve C^*, on which $\phi^*(u,v) = c$. Show that
$$\phi(x,y) = \operatorname{Re} z^2 \quad \text{and} \quad \phi^*(u,v) = \operatorname{Re}(e^w)^2.$$
4. Let $w = g(z)$ be analytic, and let $\phi^*(u,v) = \phi(x,y)$ as in the text. If ϕ^* is twice differentiable, then $\phi_x = \phi_u{}^* u_x + \phi_v{}^* v_x$ and ϕ_{xx} can be computed from this. Conclude that
$$\phi_{xx} + \phi_{yy} = |g'(z)|^2(\phi^*{}_{uu} + \phi^*{}_{vv}).$$
5. *Flow past a circle.* The function $H(w) = w$ represents uniform flow parallel to the real axis in the w plane and hence it also represents uniform flow in the w plane cut from -1 to 1. By mapping the cut w plane onto the exterior of $|z| = 1$, get the function

$$h(z) = H\left[\frac{1}{2}\left(z + \frac{1}{z}\right)\right] = \frac{1}{2}\left(z + \frac{1}{z}\right)$$

for the corresponding flow in the z plane. The lines of flow $\text{Im } w = c$, where c is a constant, map into lines of flow $\text{Im}(z + 1/z) = c$. Verify directly that $\text{Im } h(z)$ is constant for $|z| = 1$ and that the complex velocity $\overline{h'(z)}$ represents a vector parallel to $|z| = 1$ at points of $|z| = 1$. Since $h(z) \to z/2$, $|z| \to \infty$, the flow is nearly uniform at distant points. As suggested by Figure 3-9, the result describes flow in a deep channel with an obstruction on the bottom. Reflection of Figure 3-9 in the real axis gives the pattern for flow around a circle.

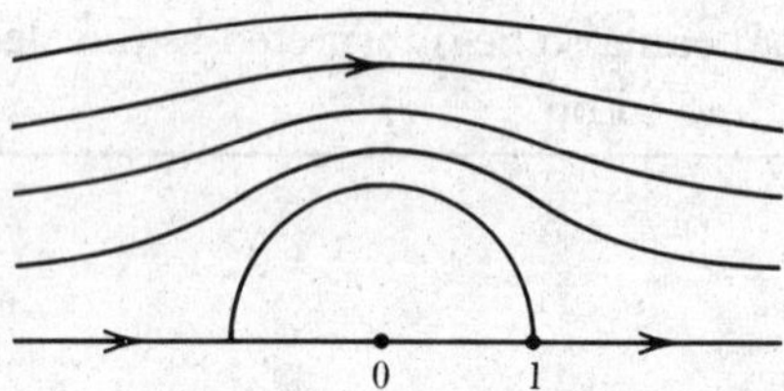

Figure 3-9

6. *Flow past a slit.* Find a flow in the w plane slit from -1 to 1 which tends to $e^{i\lambda}w$ as $|w| \to \infty$, where λ is real. At what points of the top and bottom of the slit is the velocity 0? Outline of solution: Rotate the flow $(½)(z + z^{-1})$ obtained above by writing $e^{i\lambda}z$ in place of z. This gives

$$h(z) = \frac{1}{2}\left(e^{i\lambda}z + \frac{e^{-i\lambda}}{z}\right).$$

Now use the inverse mapping (3.5) to get $H(w)$ from $h(z)$. Since either branch of the square root satisfies

$$(w + \sqrt{w^2 - 1})(w - \sqrt{w^2 - 1}) = 1,$$

it will be found that

$$H(w) = w \cos \lambda + i\sqrt{w^2 - 1} \sin \lambda.$$

7. *Flow past an ellipse.* Use (3.3) to find a function $H(w)$ which gives the flow past an ellipse

$$\frac{u^2}{4} + \frac{v^2}{3} = 1$$

and acts like w at distant points, so that the flow at ∞ is uniform. (Find a flow function in the z plane exterior to the circle which is the image of the desired ellipse.)

4. The Schwarz-Christoffel transformation. An intuitive discussion is now given for determination of an analytic function that maps the interior of a polygon onto the upper half-plane. The application to cases where one or several vertices of the polygon are at ∞ is also considered.

Analytic definition of familiar terms connected with polygons is not difficult, and can be found in Chapter 4, Section 1. For the intuitive discussion given here, however, it is thought that the meaning of terms is conveyed with sufficient clarity by Figure 4-1. The arrows in the figure indicate that the polygon is positively oriented, that is, the edges are traversed in such a way that the interior lies on the left as one faces the direction of traversal.

Consider a positively oriented polygon in the w plane and let the interior angles at successive vertices be $\alpha_1\pi, \alpha_2\pi, \ldots, \alpha_n\pi$. Suppose $w = f(z)$ provides a one-to-one conformal map of the interior of this polygon onto the upper half-plane. Thus $f(z)$ is analytic for $y > 0$ and continuous for

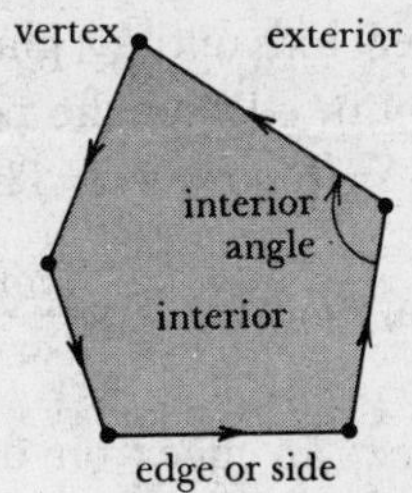

Figure 4-1

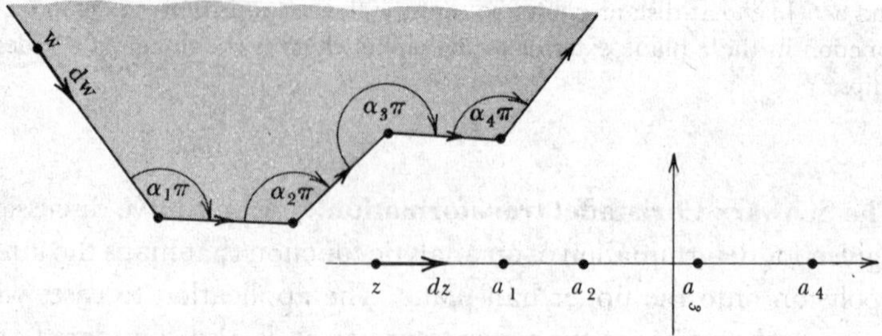

Figure 4-2

$y \geq 0$. We assume also that the inverse function $z = g(w)$ is analytic in the polygon and is continuous in the closed region consisting of the polygon together with its interior. Since the boundary of the polygon is mapped onto the real axis, the successive vertices are mapped onto points $a_1, a_2, \ldots, a_n$ of the real axis, as shown in Figure 4-2.

In terms of the derivatives $f'(z)$, the differential is

$$dw = f'(z)\,dz. \tag{4.1}$$

Assuming that the relation $w = f(z)$, $z = g(w)$ is analytic on the sides of the polygon, and not just at interior points, one can get a clue as to the nature of f by considering the relation of dw to dz on the sides.

Indeed, let w be chosen to be a point on the polygon, not a vertex, and let the image of w under the mapping be z in the z plane. If dz is taken as a positive vector on the real axis of the z plane, it is to be expected that dw would be a vector along an edge of the polygon and pointing in the positive direction as shown in Figure 4-2. This indicates that

$$\arg f'(z) = \arg \frac{dw}{dz}$$

remains constant as w traverses an edge from one vertex to the next. However, when the moving point w passes through a vertex of angle $\alpha_1\pi$, then $\arg dw$ changes by $\pi(1 - \alpha_1)$. (If $\alpha_1 = 1$, there should be no change at all.) Since $\arg dz$ is always 0, in (4.1) one would expect $\arg f'(z)$ to be constant to the left of a_1 and to change at a_1 by the amount $\pi(1 - \alpha_1)$.

We can imitate this behavior by using the function $\arg(z - a_1)$ in the

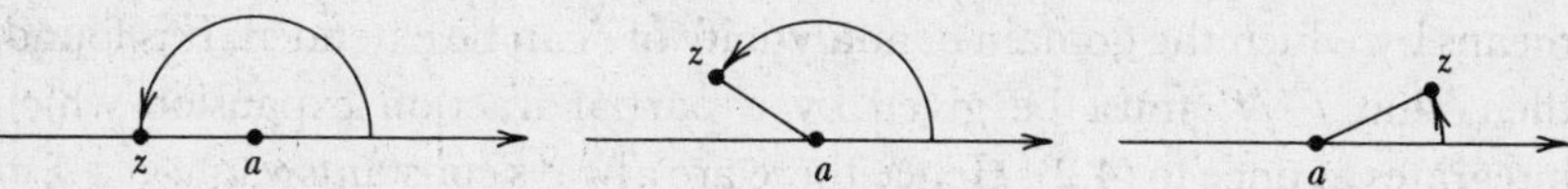

Figure 4-3

neighborhood of a_1 as z moves on the real axis. As illustrated in Figure 4-3, the value of $\arg(z - a_1)$ decreases from π to 0 when z passes from values smaller than a_1 to values larger than a_1, making a small circuit around a_1 in the upper half-plane. Thus $z - a_1$ changes its argument by $-\pi$, and therefore

$$(z - a_1)^{\alpha_1 - 1}$$

changes its argument by $\pi(1 - \alpha_1)$. This function therefore imitates the behavior of $f'(z)$ as found above. If $f'(z)$ has this function as a factor, its argument will change in the desired way at a_1. A similar situation prevails at each vertex and suggests the choice

$$f'(z) = \gamma(z - a_1)^{\alpha_1 - 1}(z - a_2)^{\alpha_2 - 1} \cdots (z - a_n)^{\alpha_n - 1}$$

where γ is constant. Thus it seems plausible that

$$\frac{dw}{dz} = \gamma(z - a_1)^{\alpha_1 - 1}(z - a_2)^{\alpha_2 - 1} \cdots (z - a_n)^{\alpha_n - 1} \tag{4.2}$$

will determine a mapping function for which $\arg(dw/dz)$ has the desired behavior, and hence will map a polygon with the prescribed angles onto a half-plane.

Of course the complex number γ and the real numbers $a_1, a_2, \ldots, a_n$ must then be chosen so that the polygon is the desired one. In other words, even if it does all that can be hoped for, (4.2) merely determines the angles of the polygon and not the lengths of its successive edges.

Nevertheless, (4.2) can, in fact, be justified. The proof of this depends on the Riemann mapping theorem, which is stated in Section 8, and on the Schwarz principle of reflection, which is proved in Section 7. Riemann's theorem guarantees that the required mapping $w = f(z)$ of the polygon onto a half-plane exists, and Schwarz's theorem provides a means by which the domain of analyticity of f can be extended. It is found, then, that f''/f' must be given by a partial fraction expansion which integrates at once to (4.2). Hence there are always constants $\gamma, a_1, a_2, \ldots, a_n$ such that (4.2) achieves the desired mapping.

Although details of the above proof are not given here, it will be shown that (4.2) does give the desired mapping in several cases of interest. In these applications the polygon will not be a simple closed polygon but will have $w = \infty$ as one or several of its vertices. Moreover, in (4.2) a simplification can be made by mapping a vertex into $z = \infty$. In the following applications, this is done.

As the first example, the interior of the strip $\operatorname{Im} w > 0$, $|\operatorname{Re} w| < b$ shown in Figure 4-4 is mapped onto the upper half z plane so that the vertices at $-b$ and b map into $z = -1$ and $z = 1$. Since the angles at $-b$ and b are $\pi/2$, the measures α_1 and α_2 are both ½, and (4.2) gives

$$\frac{dw}{dz} = \gamma(z+1)^{-1/2}(z-1)^{-1/2}.$$

Since dw/dz should be real and positive when w is on the real axis between $-b$ and b, this suggests choosing γ so that

$$\frac{dw}{dz} = \frac{a}{(1-z^2)^{1/2}} \tag{4.3}$$

where $a > 0$ and $(1 - x^2)^{1/2} > 0$ for $|x| < 1$. From (4.3), $w = a \sin^{-1} z$ or $z = \sin w/a$. For the correspondence $w = -b, +b$ with $z = -1, +1$, the choice $a = 2b/\pi$ is indicated. Hence $z = \sin(\pi w/2b)$.

The choice $b = \pi/2$ gives $z = \sin w$, which was studied in Chapter 2, Section 2. There it was found that the half-strip $|u| < \pi/2$, $v > 0$ does

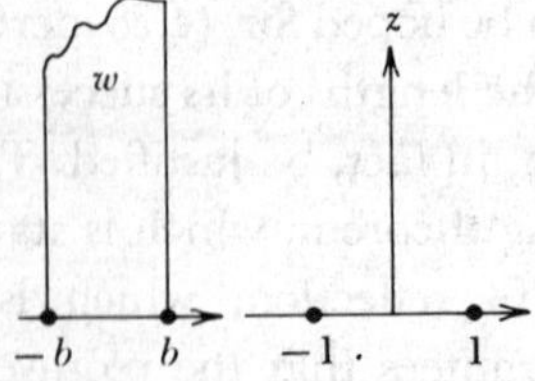

Figure 4-4

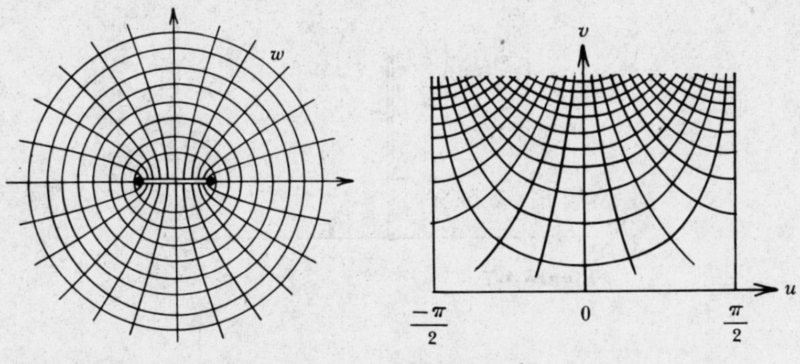

Figure 4-5 **Figure 4-6**

indeed map onto the upper half-plane, and that the lines $u = \text{const}$, $v = \text{const}$ map onto an orthogonal system of confocal coordinates in the z plane, as shown in Figure 4-5.

This transformation can be used to obtain the electrostatic field due to a charged elliptic cylinder, the field due to a charged plate from which a strip has been removed, the circulation of liquid around an elliptic cylinder, and the flow of liquid through a slit in the plane. It is also of interest to consider the images in the w plane of the coordinate lines $x = x_0, y = y_0$ in the z plane. These curves have the respective equations

$$\sin u \cosh v = x_0, \qquad \cos u \sinh v = y_0$$

and are sketched in Figure 4-6. They give the streamlines and equipotentials for flow in a half-strip. Further applications of the mapping $z = \sin w$ are given in Problems 4.1–4.4 at the end of this chapter.

As another example, let the upper half w plane be slit along the imaginary axis for a finite distance and let it be required to map this slit plane onto the upper half of the z plane. When the boundary is traversed in the positive direction, it is seen that the vertices have angles $\pi/2$, 2π and $\pi/2$, respectively, as shown in Figure 4-7. Suppose these vertices are mapped into $z = -1$, 0 and 1, respectively. Then (4.2) gives

$$\frac{dw}{dz} = \gamma(z+1)^{-1/2}z(z-1)^{-1/2} = \frac{\gamma z}{(z^2-1)^{1/2}}.$$

We set $\gamma = a > 0$ and obtain, as one solution,

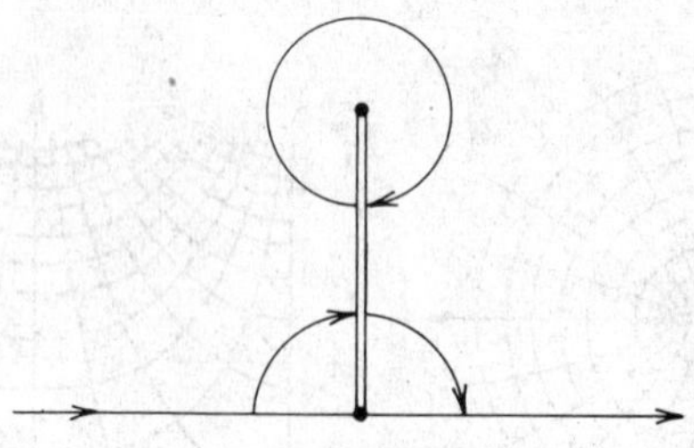

Figure 4-7

$$w = a(z^2 - 1)^{1/2}. \tag{4.4}$$

The square root is chosen to be negative when z is on the part of the real axis where $x < -1$.

Equation (4.4) implies $w^2 = a^2z^2 - a^2$. Hence two values $z_1 \neq z_2$ can give the same w only if $z_1 = -z_2$. Since there is no such pair in Im $z > 0$, the mapping of Im $z > 0$ must be one-to-one. Similarly, there is no pair $w_1 \neq w_2$ in Im $w > 0$ that gives the same z.

For more detailed analysis, write the mapping as a composite of the three mappings

$$\zeta_1 = a^2z^2, \qquad \zeta_2 = \zeta_1 - a^2, \qquad w = \sqrt{\zeta_2}.$$

The first of these fans out the upper half-plane and maps it one-to-one onto the whole ζ_1 plane slit along the positive real axis (Figure 4-8(a)). The second translates the slit ζ_1 plane, as shown in Figure 4-8(b). The third halves angles through the origin and transforms the slit ζ_1 plane, as

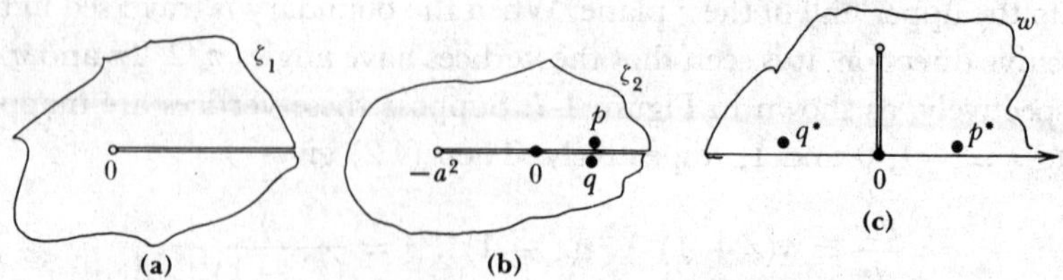

Figure 4-8

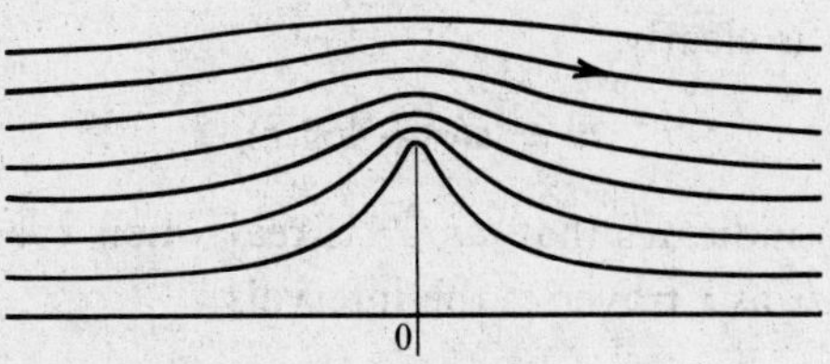

Figure 4-9

shown in Figure 4-8(c). Since each of these mappings is one-to-one, their composition (4.4) is also one-to-one, and (4.4) has the desired properties.

For parallel flow in the z plane associated with $h(z) = z$, the streamlines have as their images in the w plane the curves

$$w = a[(x + iy)^2 - 1]^{1/2}, \qquad -\infty < x < \infty,$$

where y is constant on a given streamline. These describe the flow past a barrier, as shown in Figure 4-9.

As a final example of the Schwarz-Christoffel formula, we map the upper half w plane slit along a half-line parallel to the real axis (Figure 4-10). This domain is regarded as the limit of a domain bounded below by the polygonal line $ABCD$ in the figure as $C \to -\infty$ along the real axis. In the limiting domain A, C and D are considered to be at ∞ and the angles at B and C are considered to be 2π and 0, respectively. If A, B, C and D are mapped respectively onto ∞, -1, 0 and ∞, the formula (4.2) with $\gamma = a > 0$ gives

$$\frac{dw}{dz} = a(z + 1)z^{-1} = a\left(1 + \frac{1}{z}\right) \tag{4.5}$$

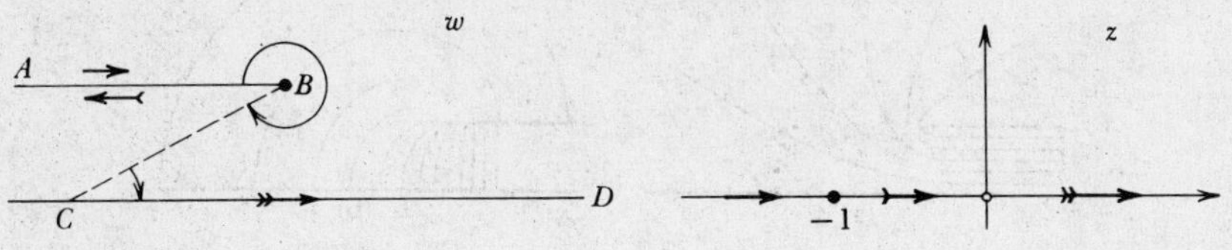

Figure 4-10

and one solution is clearly

$$w = a(z + \log z). \tag{4.6}$$

Equation (4.5) indicates that dw/dz is real when x is real, and hence dw is parallel to dx as x traverses the intervals

$$-\infty < x < -1, \qquad -1 < x < 0, \qquad 0 < x < \infty$$

of the real axis. On these intervals (4.5) gives, respectively,

$$\frac{dw}{dx} > 0, \qquad \frac{dw}{dx} < 0, \qquad \frac{dw}{dx} > 0$$

and the respective values of Im $\log z$ in (4.6) are $i\pi$, $i\pi$, 0. For $x = -1$ we have $w = a(\pi i - 1)$, which is the point B in Figure 4-10.

Using the above information, one can easily verify that the boundary in the w plane corresponds to the real axis in the z plane, as suggested by the arrows in the figure. The boundary correspondence is one-to-one if we agree to distinguish the two sides of the cut that ends at the point B. By Theorem 7.1 it can be shown that the mapping of the interiors is also one-to-one. The lines Im $\log z = c$, or in polar coordinates $\theta = c$, are streamlines from a source at the origin. They are mapped onto the curves

$$w = re^{ic} + \log r + ic, \qquad 0 < r < \infty.$$

These are the lines of flow shown in Figure 4-11(a). The orthogonal curves corresponding to Re $\log z = c$, where c is a constant, of $|z| = c$ are

$$w = ce^{i\theta} + \log c + i\theta, \qquad 0 < \theta < \pi,$$

which may be regarded as the lines of force of a field shown in Figure

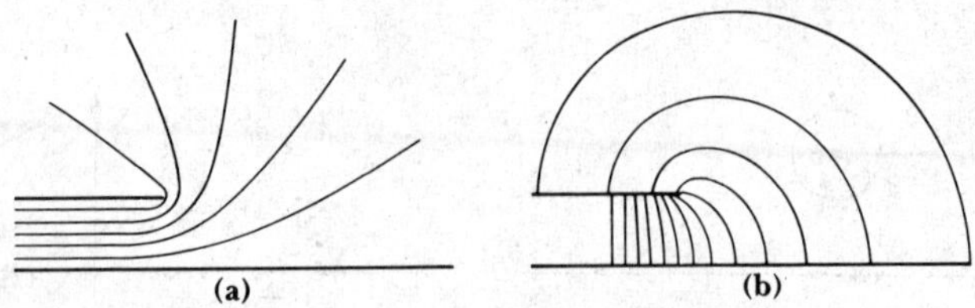

Figure 4-11

4-11(b). If Figure 4-11(a) is reflected in the real axis, the figure, together with its reflection, describes the flow out of a channel. If Figure 4-11(b) is reflected in the real axis, the result describes the fringing at the edges of a parallel plate condenser.

Problems

1. Show that the Schwarz-Christoffel procedure gives $w = \gamma z^\alpha$ for the problem of mapping the interior of an angle α onto a half-plane. What is the effect of the complex constant γ? (Map the vertex of the angle into $z = 0$ and the open end into $z = \infty$.)
2. A strip in the w plane is considered to be a polygon with two angles, each equal to 0. Obtain the function $w = \gamma \log z + \beta$, where γ and β are constant, for mapping this strip onto the upper half-plane in such a way that one "vertex" maps into 0 and the other into ∞. (The case $\gamma = 1$, $\beta = 0$ was considered in Chapter 2.)
3. The mapping of Problem 2 is a composite of $\zeta = \log z$ and $w = \gamma\zeta + \beta$. Show that γ and β can be chosen so as to map any given strip, regardless of its width and position. The constant γ and the integration constant β in the general Schwarz-Christoffel transformation have a similar role; that is, they allow introduction of an arbitrary translation, rotation and magnification.
4. A stream bed with a step has the equation $w = t + i\pi$, $w = (1 - t)i\pi$, $w = t - 1$ for $-\infty < t \leq 0$, $0 \leq t \leq 1$, $1 \leq t < \infty$, respectively. It is required to map the part of the w plane above the stream bed onto $\text{Im } z > 0$ so $w = i$ corresponds to $z = -1$ and $w = 0$ to $z = 1$. Verify that a branch of $f(z) = (z^2 - 1)^{1/2} + \cosh^{-1} z$ satisfies the appropriate differential equation. (It can be shown, in fact, that a branch of $w = f(z)$ provides the desired mapping.)
5. It is desired to map a rectangle into the upper half-plane in such a way that the four vertices correspond respectively to

$$-1/k, \quad -1, \quad 1, \quad 1/k,$$

where $0 < k < 1$. Obtain the differential equation

$$f'(z) = \frac{1}{(1 - z^2)^{1/2}(1 - k^2 z^2)^{1/2}}$$

by suitable choice of γ. Here there is no elementary antiderivative, so that the existence of the solution f is not obvious. However, in Chapter 3 it is shown that a suitable function $f(z)$ is given by

$$f(z) = \int_0^z \frac{d\zeta}{[(1 - \zeta^2)(1 - k^2\zeta^2)]^{1/2}} \tag{*}$$

and that a similar constructive method applies to the general Schwarz-Christoffel transformation. The integral (*) is called an *elliptic integral.*

6. It is desired to map a triangle onto the upper half-plane in such a way that the vertices map into $0, 1, \infty$. Write a differential equation for the mapping function (a) for an equilateral triangle and (b) for a general triangle.
7. Obtain differential equations for mapping functions that map the domains of Figure 4-12 onto the upper half-plane. It is desired that the "vertices" A, B, C and D shall map onto -1, 0, 1 and ∞, respectively.

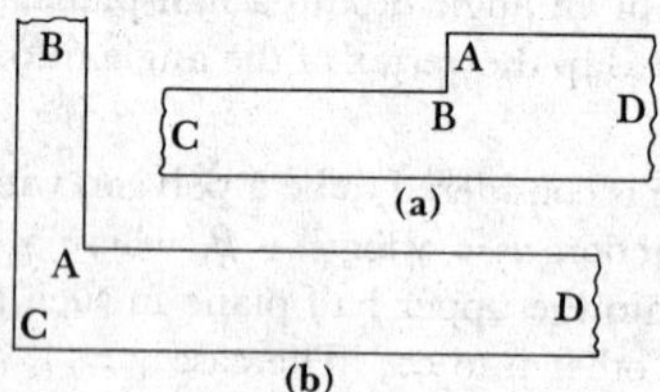

Figure 4-12

5. Hurwitz polynomials and positive functions. A *Hurwitz polynomial* is a nonconstant polynomial that has all its roots in the left half-plane. Thus, if the factored form of the polynomial is

$$f(z) = a_n(z - z_1)(z - z_2) \cdots (z - z_n), \tag{5.1}$$

then $\text{Re } z_k < 0$. Hurwitz polynomials are connected with the problem of stability of mechanical and electrical systems and have numerous practical applications. The same applies to positive functions, considered later.

The study of Hurwitz polynomials leads to a simple operation known as paraconjugation. If $f(z)$ is any rational function, the *paraconjugate* is defined by

$$f^*(z) = \overline{f(-\bar{z})}. \tag{5.2}$$

For instance, when

$$f(z) = a_0 + a_1 z + a_2 z^2 + \cdots + a_n z^n,$$

the paraconjugate is

$$f^*(z) = \bar{a}_0 - \bar{a}_1 z + \bar{a}_2 z^2 + \cdots + (-1)^n \bar{a}_n z^n.$$

When $f(z)$ is given by (5.1), the paraconjugate is

$$f^*(z) = (-1)^n \bar{a}_n (z + \bar{z}_1)(z + \bar{z}_2) \cdots (z + \bar{z}_n), \tag{5.3}$$

and so the zeros of f^* are obtained by reflecting the zeros of f in the imaginary axis. The same conclusion is evident from (5.2), since substitution of $-\bar{z}_k$ on the left gives $\overline{f(z_k)}$ on the right.

Since (5.2) implies $f^*(iy) = \overline{f(iy)}$, it is clear that

$$|f^*(iy)| = |f(iy)|, \qquad -\infty < y < \infty. \tag{5.4}$$

Thus, if $g = f^*/f$, the function $w = g(z)$ maps $\text{Re}\, z = 0$ into $|w| = 1$. We shall presently see that $w = g(z)$ maps $\text{Re}\, z > 0$ into $|w| < 1$ if f is a Hurwitz polynomial, and that the converse is, essentially, true. Further development leads to a class of rational functions that map the right half-plane onto itself and the left half-plane onto itself. A complete characterization of rational functions with this property is given in Theorem 5.2.

In illustration of these remarks, consider the simplest Hurwitz polynomial, $f(z) = 1 + z$. Here $f^*(z) = 1 - z$ and

$$\frac{f^*(z)}{f(z)} = \frac{1 - z}{1 + z}, \qquad \frac{f(z) - f^*(z)}{f(z) + f^*(z)} = z.$$

The transformation

$$w = \frac{1 - z}{1 + z}, \qquad z = \frac{1 - w}{1 + w} \tag{5.5}$$

maps $\text{Re}\, z > 0$ onto $|w| < 1$, as seen by Example 2.1 or by Figure 5-1.

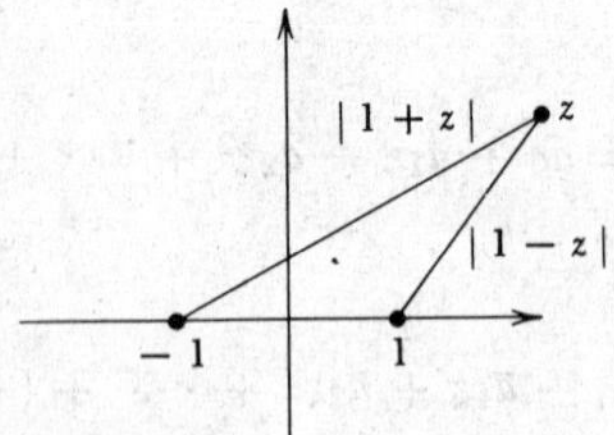

Figure 5-1

On the other hand, the transformation $w = z$ obviously maps the right half-plane onto itself and the left half-plane onto itself.

THEOREM 5.1 *Let f be a nonconstant polynomial such that f and f^* do not vanish simultaneously, and let*

$$g(z) = \frac{f^*(z)}{f(z)}, \qquad h(z) = \frac{f(z) - f^*(z)}{f(z) + f^*(z)}.$$

Then the following statements are equivalent: (1) *f is a Hurwitz polynomial;* (2) *$|g(z)| < 1$ for* $\operatorname{Re} z > 0$; (3) $\operatorname{Re} h(z) > 0$ *for* $\operatorname{Re} z > 0$.

Proof. If $\operatorname{Re} z_k < 0$, then

$$|z - z_k| > |z + \bar{z}_k| \qquad \text{for } \operatorname{Re} z > 0. \quad (5.6)$$

This follows from the fact that z is closer to $-\bar{z}_k$ than to z_k as shown in Figure 5-2. An analytic proof is immediate upon squaring both sides. In any case, (5.6) is valid and gives, in (5.1) and (5.3),

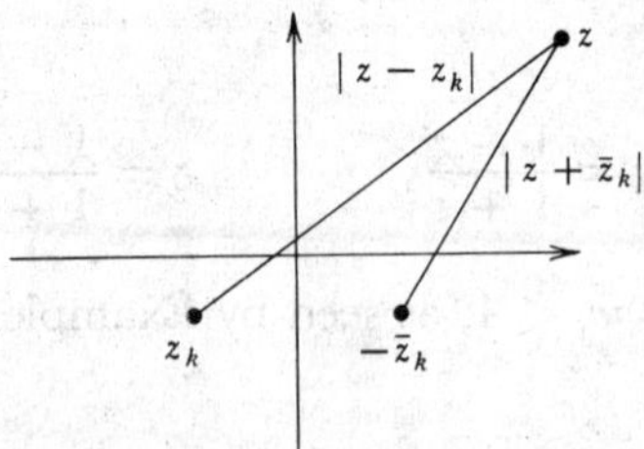

Figure 5-2

$$|f(z)| > |f^*(z)| \qquad \text{for } \operatorname{Re} z > 0. \quad (5.7)$$

This holds for all Hurwitz polynomials $f(z)$. Conversely, if (5.7) holds, then $f(z)$ cannot have any zeros in $\operatorname{Re} z > 0$, and so the only possible zeros in $\operatorname{Re} z \geq 0$ are on the imaginary axis. However, (5.4) shows that any such zero is a common zero of f and f^*, which is excluded in Theorem 5.1. Thus, (5.7) is necessary and sufficient for f to be a Hurwitz polynomial in Theorem 5.1. Since f and f^* do not vanish simultaneously, $|f^*/f| < 1$ if and only if $|f^*| < |f|$. This gives (2). The equivalence with (3) follows from the above discussion of (5.5), which indicates that

$$h = \frac{1-g}{1+g}$$

satisfies $\operatorname{Re} h > 0$ if and only if $|g| < 1$. This completes the proof.

A rational function h is said to be *positive* if $\operatorname{Re} h(z) > 0$ whenever $\operatorname{Re} z > 0$. That is, h is positive if it satisfies condition (3) of Theorem 5.1. Theorem 5.1 reduces the study of Hurwitz polynomials to the study of positive functions, and positive functions are discussed now. We suppose that $h(z)$ is rational and that its numerator and denominator have no common factors.

If α is a zero of $h(z)$, the factorization theorem shows that

$$h(z) = (z - \alpha)^m H(z), \tag{5.8}$$

where $H(\alpha) \neq 0$ and m is an integer. Since $H(z)/H(\alpha) \to 1$ as $z \to \alpha$, we can write

$$H(z) = \rho e^{i(\delta + \theta_0)}$$

where $\theta_0 = \operatorname{Arg} H(\alpha)$ and where $\delta \to 0$, $\rho \to |H(\alpha)|$. If $z - \alpha = re^{i\theta}$, then

$$h(z) = r^m e^{im\theta} \rho e^{i(\delta+\theta_0)} = r^m \rho e^{i(\delta + m\theta + \theta_0)}$$

and hence

$$\operatorname{Re} h(z) = r^m \rho \cos(\delta + m\theta + \theta_0). \tag{5.9}$$

Here $\delta \to 0$ and $\rho \to |H(\alpha)|$ as $r \to 0$.

It is obvious that a positive function cannot have any zeros α with $\operatorname{Re} \alpha > 0$. If $\operatorname{Re} \alpha = 0$ and $m \geq 2$, equation (5.9) for small δ indicates that $\operatorname{Re} h(z)$ changes sign as θ traverses the interval $-\pi/2 < \theta < \pi/2$. Hence the imaginary zeros of a positive function must be simple; that is, they must have $m = 1$. The same consideration shows that $\theta_0 = 0$, so that $H(\alpha) > 0$ at an imaginary zero, α.

For any complex number $\alpha \neq 0$ it is easily checked that $\operatorname{Re} \alpha$ and $\operatorname{Re}(1/\alpha)$ have the same sign. Hence, *$1/h$ is positive if h is positive.* It follows that the denominator of h cannot have any zeros in $\operatorname{Re} z > 0$ and that any imaginary zeros of the denominator must be simple.

The function $h(z)$ in Theorem 5.1 satisfies $h^*(z) = -h(z)$, since $f^{**} = f$, and hence $-h^*$ is positive if h is. For such functions the foregoing analysis can be carried much further and gives the following:

THEOREM 5.2. *Let h be a rational function such that h and $-h^*$ are both positive. Then $h(z)$ can be written in the form*

$$h(z) = a + bz + \frac{b_1}{z - i\omega_1} + \frac{b_2}{z - i\omega_2} + \cdots + \frac{b_n}{z - i\omega_n},$$

where $\operatorname{Re} a = 0$, $b \geq 0$, $b_k \geq 0$, and where the ω_j are distinct real numbers. Conversely, every nonconstant function h of this form satisfies $h^* = -h$ and is positive.

Proof. The converse part of the theorem is trivial, since a sum of positive functions is positive. We therefore assume that h and $-h^*$ are positive and show that h must have the given form.

Since h is positive, there are no zeros in $\operatorname{Re} z > 0$, and since $-h^*$ is positive, there are none in $\operatorname{Re} z < 0$. Thus the zeros are purely imaginary and by the above discussion they must be simple. The partial fraction expansion[1] therefore has the given form, except that the term $a + bz$ may be replaced by a polynomial of higher degree. However if such a polynomial is present, it is easily checked that $\operatorname{Re} h(z)$ changes sign in $\operatorname{Re} z > 0$ for large $|z|$. The same consideration shows that $b \geq 0$, and analysis of the behavior near ω_k shows that $b_k \geq 0$. On the imaginary axis the only contribution to $\operatorname{Re} h$ or $\operatorname{Re} h^*$ comes from a, and by letting $z \to iy$ it is seen that $\operatorname{Re} a = 0$. This completes the proof.

A function satisfying $h^*(z) = -h(z)$ is called *para-odd.* For example, the

[1] Chapter 2, Problem 1.1.

function $h(z)$ in Theorem 5.1 is para-odd, and if h and $-h^*$ are both positive, Theorem 5.2 states that h must be para-odd. A function h with the properties described in Theorem 5.2 is said to be positive para-odd.

The term $a + bz$ in Theorem 5.2 is called the *integral part* of h and is denoted by $[h]$. If h does not coincide with its integral part, Theorem 5.2 indicates that h is positive para-odd if and only if

(1) $[h] = a + bz$ satisfies $\operatorname{Re} a = 0$, $b \geq 0$;
(2) the function $\tilde{h} = h - [h]$ is positive para-odd.

Since $\tilde{h}$ is positive para-odd if and only if $1/\tilde{h}$ is positive para-odd, we can apply the above criteria to the function $h_1 = 1/\tilde{h}$, and so on. The result is a systematic decision procedure and is illustrated in Example 5.1.

Example 5.1. For what values of the real parameter c is

$$f(z) = z^4 + 2z^3 + 3z^2 + 4z + c$$

a Hurwitz polynomial? Here

$$f^*(z) = z^4 - 2z^3 + 3z^2 - 4z + c$$

and, since f is real,

$$h(z) = \frac{f(z) - f^*(z)}{f(z) + f^*(z)} = \frac{\text{odd part of } f}{\text{even part of } f}.$$

Thus,

$$h(z) = \frac{2z^3 + 4z}{z^4 + 3z^2 + c}.$$

We wish to have the polynomial of higher degree in the numerator and so we consider $h_0 = 2/h$,

$$h_0(z) = \frac{z^4 + 3z^2 + c}{z^3 + 2z} = z + \frac{z^2 + c}{z^3 + 2z}. \tag{5.10}$$

The second form is obtained by division. Theorem 5.1 indicates that f is a Hurwitz polynomial if and only if $1/h$ is positive para-odd, and Theorem 5.2 shows that $1/h$ is positive para-odd if and only if the fraction on the right of (5.10) is positive para-odd. Hence we consider

$$h_1(z) = \frac{z^3 + 2z}{z^2 + c} = z + \frac{(2 - c)z}{z^2 + c}.$$

The fraction on the left is positive para-odd if and only if the fraction on the right is positive para-odd, and so we consider

$$h_2(z) = \frac{z^2 + c}{(2 - c)z} = \frac{z}{2 - c} + \frac{c}{(2 - c)z}.$$

The first term shows that $c < 2$ is necessary, and the second shows that $c > 0$ is necessary. The two conditions together are sufficient, by Theorem 5.2, and hence $f(z)$ is a Hurwitz polynomial for $0 < c < 2$.

With $\alpha = 1/(2 - c)$ and $\beta = c/(2 - c)$, the process is summarized as follows:

$$h_2 = \alpha z + \frac{\beta}{z}, \qquad h_1 = z + \frac{1}{h_2}, \qquad h_0 = z + \frac{1}{h_1}, \qquad h = \frac{2}{h_0}.$$

By successive substitution, starting at the right,

$$h(z) = \cfrac{2}{z + \cfrac{1}{z + \cfrac{1}{\alpha z + \cfrac{\beta}{z}}}}.$$

This is the *Stieltjes continued fraction* expansion for h. The same process can be carried out for any rational function, and Theorem 5.2 gives a necessary and sufficient condition for a function to be positive para-odd in terms of the coefficients of its Stieltjes expansion. In particular, *$h(z)$ is a positive real para-odd function if and only if all coefficients in its Stieltjes expansion are positive.* The process involves only rational operations (polynomial division), not extraction of roots, and is well-adapted for programming on a computer.

Problems

1. If $f(z) = (1 + z)(1 - z^2)$, show that $h(z) = z$, which is positive. Hence the condition that f and f^* have no common zeros in Theorem 5.1 is essential.

2. Show that a polynomial f can never be a Hurwitz polynomial if f and f^* have a common zero.
3. (a) The term bz in Theorem 5.2 is called the *principal part of h at* ∞. Show that $b = \lim h(z)/z$ as $z \to \infty$. (b) For real h, $a = 0$ and one can compute $h_1 = h - [h]$ as $h - bz$, avoiding long division. Do Example 5.1 by this method.
4. For what real values c are the following expressions Hurwitz polynomials?

$$3z^3 + 2z^2 + z + c, \quad 4z^4 + z^3 + z^2 + c, \quad z^5 + 5z^4 + 4z^3 + 3z^2 + 2z + c.$$

5. Do part (i) of Problem 4 for complex c. (The case of real c is more important in applications.)
6. Show that $c^* = \bar{c}$ for constants, $z^* = -z$, and at points z where the denominators do not vanish,

$$(f + g)^* = f^* + g^*, \quad (fg)^* = f^*g^*, \quad (f/g)^* = f^*/g^*, \quad (f^*)^* = f.$$

7. Show that h in Theorems 5.1 and 5.2 satisfies $h + h^* = 0$.
8. If $f(z)$ is a real polynomial, its complex roots occur in conjugate pairs. Considering $(z - z_k)$ for $z_k < 0$ and $(z - z_k)(z - \bar{z}_k)$ for $\operatorname{Re} z_k < 0$, show that the coefficients of any real Hurwitz polynomial must be of the same sign. (The polynomial is a product of expressions like those considered here.)
9. Give an example of a cubic polynomial with positive coefficients which is not a Hurwitz polynomial.
10. According to the *Routh-Hurwitz criterion*, the real polynomial

$$a_0z^n + a_1z^{n-1} + \cdots + a_n, \qquad a_0 > 0$$

is a Hurwitz polynomial if and only if

$$a_1 > 0, \quad \begin{vmatrix} a_1 & a_0 \\ a_3 & a_2 \end{vmatrix} > 0, \quad \begin{vmatrix} a_1 & a_0 & 0 \\ a_3 & a_2 & a_1 \\ a_5 & a_4 & a_3 \end{vmatrix} > 0, \ldots .$$

The sequence is continued only to n-by-n determinants, and a_j is replaced by 0 for $j > n$. Verify (a) for $n = 1,2,3$, and (b) for $n = 4$.
11. If h is positive para-odd, show that the degrees of its numerator and denominator differ by 1 at most. (If the degree of the numerator exceeds the degree of the denominator by 2 or more, it is easily checked that $h(z)/z \to \infty$ as $z \to \infty$, while Problem 3 gives $h(z)/z \to b$. The same consideration applies to $1/h$.)
12. Let $h(z)$ be as in Theorem 5.2 with $a = 0$, and let $H(y) = -ih(iy)$. Show that $H'(y) > 0$ for $y \neq \omega_k$. Deduce that $H(y)$ is strictly increasing from $-\infty$ to ∞ in each interval (ω_k, ω_{k+1}) and hence if c is any real constant, the equation $H(y) = c$ has exactly one solution on each of these intervals. The special case

$c = 0$ shows that zeros of the numerator alternate with zeros of the denominator.

13. *Foster's reactance theorem.* An odd positive real function is called a Foster function. Show that every Foster function is positive para-odd. Deduce from Theorem 5.2 that every Foster function has the form

$$f(z) = bz + \frac{c}{z} + \sum_{i=1}^{n} \frac{b_i z}{z^2 + \omega_i^2}$$

where $b \geq 0$, $c \geq 0$, $b_k \geq 0$, and $\omega_i \neq 0$ are real.

14. Let $g(z)$ be a nonconstant rational function such that $|g(z)| = 1$ when $\operatorname{Re} z = 0$. Thus, $z + \bar{z} = 0$ implies $g(z)\overline{g(z)} = 1$. Deduce $g(z)g^*(z) = 1$ for $z = iy$ and hence, since gg^* is rational, this holds everywhere. By factorizing numerator and denominator, show that g admits the representation

$$g(z) = g(\infty)\frac{z + \bar{z}_1}{z - z_1} \cdot \frac{z + \bar{z}_2}{z - z_2} \cdots \frac{z + \bar{z}_n}{z - z_n}.$$

Hence $g = f^*/f$ for a polynomial f.

6. Inverse mappings and univalent functions. General properties of an analytic mapping $w = f(z)$ and its inverse $z = f^{-1}(w)$ are now studied by means of Rouché's theorem and the principle of the argument (see Chapter 4, Section 6).

It has already been shown that where $f'(z) \neq 0$ the mapping $w = f(z)$ is conformal. The following result is valid even at the zeros of $f'(z)$:

THEOREM 6.1 (*Open mapping theorem*). *If f is analytic in a domain D and not constant, then $w = f(z)$ maps open sets of D onto open sets in the w plane. More specifically, if z_0 is in D and if $w_0 = f(z_0)$, then for sufficiently small $\epsilon > 0$ there is a $\delta > 0$ such that the image of $|z - z_0| < \epsilon$ contains the disk $|w - w_0| < \delta$.*

Since $f(z)$ is continuous, all points of D sufficiently close to z_0 are mapped onto points near w_0. The theorem states that some neighborhood of w_0 is actually covered by the image of a neighborhood of z_0.

Proof. Let the zero of $f(z) - w_0$ at $z = z_0$ be of order $n \geq 1$. Let $\rho < \epsilon$ be small enough so that $f(z) - w_0$ does not vanish for $0 < |z - z_0| \leq \rho$. That such a $\rho > 0$ exists follows from the fact that the zeros of a nonconstant analytic function are isolated. Let

$$m = \min|f(z) - w_0| \quad \text{on} \quad |z - z_0| = \rho. \tag{6.1}$$

Then $m > 0$. If α is a complex number such that $|\alpha| < m$, then

$$f(z) - w_0 - \alpha$$

has the same number of zeros in $|z - z_0| < \rho$ as does $f(z) - w_0$. This follows from Rouché's theorem, since

$$|f(z) - w_0| > |\alpha| \quad \text{on} \quad |z - z_0| = \rho.$$

Hence, $f(z) - w_0 - \alpha$ has n zeros in $|z - z_0| < \rho$. Since $n \geq 1, f(z) = w_0 + \alpha$ holds at at least one point in $|z - z_0| < \rho$, and hence the image of $|z - z_0| < \rho$ contains $|w - w_0| < m$. Thus the theorem is proved with $\delta = m$.

Theorem 6.1 has a number of interesting consequences, and more information is contained in its proof than was given in the statement of the theorem. As one application, Theorem 6.1 can be used to obtain the maximum principle for nonconstant analytic functions $f(z)$. Indeed, since the point $w_0 = f(z_0)$ is interior to some disk contained in $f(D)$, there are some points of $f(D)$ that are farther from the origin than w_0 is (see Figure 6-1). Such points satisfy $|w| > |w_0|$ and show that $|f(z_0)|$ is not a maximum. The same applies to every z_0 in D, and so $|f(z)|$ does not attain its maximum in D.

As another application of Theorem 6.1 we shall show that $f(D)$ is a domain. Since Theorem 6.1 already shows that $f(D)$ is an open set, it is only necessary to show that $f(D)$ is connected to prove that it is a domain.

According to Chapter 1, an open set S is connected if each two points of S can be joined by a polygonal line lying in S. Since S is open, we could just as well have said that S is connected if each two points of S can be

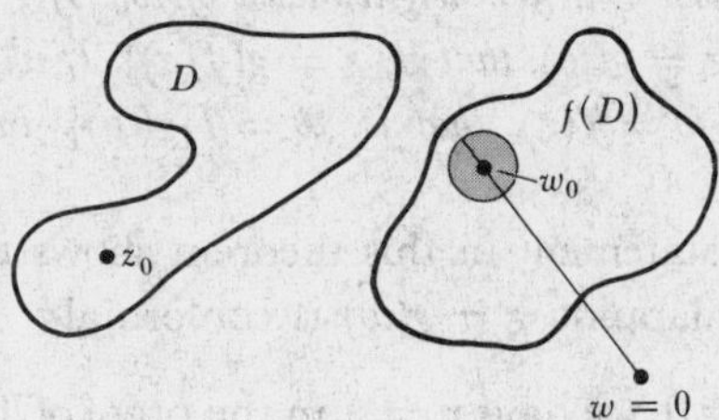

Figure 6-1

joined by a contour lying in S. The equivalence of the two definitions, for open sets, follows from an easy theorem of real analysis, which asserts that any contour can be approximated arbitrarily well by a broken line. The equivalence is taken for granted here.

THEOREM 6.2. *If $f(z)$ is analytic in a domain D, the image $f(D)$ of D under the transformation $w = f(z)$ is a domain in the w plane.*

Proof. Let $w_1 = f(z_1)$ and $w_2 = f(z_2)$ be two points of $f(D)$. Since D is connected, z_1 and z_2 can be joined by a broken line L lying in D. The image of each segment of L is an arc in $f(D)$ (since f is differentiable) and hence the image of L is a contour (see Figure 6-2). This shows that any two points w_1 and w_2 of $f(D)$ can be joined by a contour lying in $f(D)$. Hence $f(D)$ is connected.

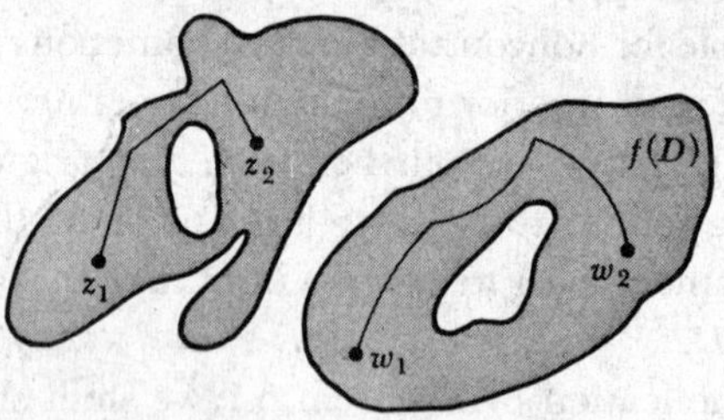

Figure 6-2

As a third application, the proof of Theorem 6.1 shows that an analytic function f possesses an analytic inverse in the neighborhood of any point z where $f'(z)$ does not vanish. We shall establish the following:

THEOREM 6.3. *If f is analytic at z_0, if $f'(z_0) \neq 0$, and if $f(z_0) = w_0$, then f has a unique analytic inverse g in the neighborhood of w_0. If z is sufficiently near z_0 and $w = f(z)$, then $z = g(w)$; that is, $z = g[f(z)]$. If w is sufficiently near w_0 and $z = g(w)$, then $w = f(z)$; that is, $w = f[g(w)]$. In either of these cases $f'(z)g'(w) = 1$.*

The concluding statement in this theorem shows that $g'(w) \neq 0$, and hence the inverse mapping $z = g(w)$ is conformal.

Proof. Since $f'(z_0) \neq 0$, we have $n = 1$ in the proof of Theorem 6.1. Hence each w in the disk K, $|w - w_0| < \delta$, is the image of a unique point z in

$|z - z_0| < \epsilon$ under $w = f(z)$. This uniqueness determines a function $z = g(w)$ defined on K and satisfying $z_0 = g(w_0)$. Moreover, $w = f(z)$ together with $z = g(w)$ on K imply $w = f[g(w)]$ on K.

To prove that g is continuous, let w_1 be a point of K. It is the image of a unique point z_1 of $|z - z_0| < \rho < \epsilon$. By Theorem 6.1, for $\epsilon_1 > 0$ there is a $\delta_1 > 0$ such that the image of $|z - z_1| < \epsilon_1$ contains $|w - w_1| < \delta_1$. Choose δ_1 small enough so that $|w - w_1| < \delta_1$ is in K. For $|w - w_1| < \delta_1$, then, $|g(w) - g(w_1)| < \epsilon_1$. Since ϵ_1 is arbitrary and since there is a $\delta_1 > 0$ corresponding to ϵ_1, it follows that $g(w)$ is continuous on K.

The fact that $g(w)$ is continuous on K is now used to show that $g(w)$ is analytic.[1] Let w_1 be a point of K. Let w_2 be near w_1 and in K. Then w_1 and w_2 are the images of $z_1 = g(w_1)$ and $z_2 = g(w_2)$, respectively. When w_1 is fixed, the continuity of g implies that if $|w_2 - w_1|$ is small, then $|z_2 - z_1|$ must be small. Clearly,

$$\frac{g(w_2) - g(w_1)}{w_2 - w_1} = \frac{z_2 - z_1}{w_2 - w_1} = 1 \div \left(\frac{w_2 - w_1}{z_2 - z_1}\right) = 1 \div \left(\frac{f(z_2) - f(z_1)}{z_2 - z_1}\right).$$

By taking $|w_1 - w_2|$ small enough, one can make $|z_2 - z_1|$ as small as desired. Hence the right side tends to $1/f'(z_1)$ as $|w_2 - w_1| \to 0$. Since $f(z) = w_1$ has only one solution in $|z - z_0| < \rho$, counting multiplicity, and since this solution is given by $z = z_1$, it follows that $f'(z_1) \neq 0$. This proves that $g'(w_1)$ exists and equals $1/f'(z_1)$. Since w_1 is any point of K, the subscripts can be dropped. Hence $g(w)$ is analytic on K and $g'(w) = 1/f'(z)$.

Finally, by the continuity of $f(z)$, every z near z_0 has as its image a point near w_0. Hence for $|z - z_0|$ small enough, $w = f(z)$ is a point of K and thus $z = g(w)$. This implies $z = g[f(z)]$ in a neighborhood of z_0. The equation $1 = g'(w)f'(z)$ follows from the chain rule.

As the reader will recall from Chapter 3, a curve $z = \zeta(t)$ defined for $a \leq t \leq b$ is simple if $t_1 \neq t_2$ implies $\zeta(t_1) \neq \zeta(t_2)$. The condition means that $\zeta(t)$ takes no value more than once for $a \leq t \leq b$. In similar fashion, a function $f(z)$ is said to be *simple* or *univalent* in a domain D if it is analytic in D and assumes no value more than once in D. Thus, for simple functions the condition $z_1 \neq z_2$ implies $f(z_1) \neq f(z_2)$. A simple function pro-

[1] Another proof is given in Chapter 4, Problem 6.3.

vides a one-to-one map of D onto $f(D)$ and hence has a single-valued inverse on $f(D)$.

The existence of an analytic inverse to $f(z)$ fails at points where $f'(z)$ vanishes, and at such points the map is not univalent. The reason is that in the neighborhood of such a point the equation $w = f(z)$, for a given w, has several points in the z plane as solutions and not just one. The situation is essentially the same as that prevailing for $w = z^n$ in the neighborhood of $z = 0$ and $w = 0$, where the integer $n > 1$. This is so because in the neighborhood of z_0 we can write

$$w - w_0 = h(z)(z - z_0)^n, \qquad h(z_0) \neq 0,$$

where $w = f(z)$ and where n is the order of the zero of $f(z) - w_0$ at $z = z_0$ (see Chapter 3, Theorem 7.2). If $h(z)$ could be replaced by $h(z_0)$ we would get

$$w - w_0 = h(z_0)(z - z_0)^n, \qquad h(z_0) \neq 0.$$

Apart from a translation $\tilde{w} = w - w_0$, $\tilde{z} = z - z_0$ in the w and z planes, and apart from the rotation and scale change associated with the constant factor $h(z_0)$, this is just $w = z^n$.

In Chapter 2, Section 4, it was seen that $w = z^n$ provides a one-to-one mapping of the z plane onto an n-sheeted Riemann surface in the w plane. Each sheet of the Riemann surface corresponds to a replica of the w plane cut along the positive real axis,[1] as illustrated for $n = 3$ in Figure 6-3. The planes in Figure 6-3 are considered to be stacked one above the other and the sides of the cuts labeled with the same letter are considered to be joined.

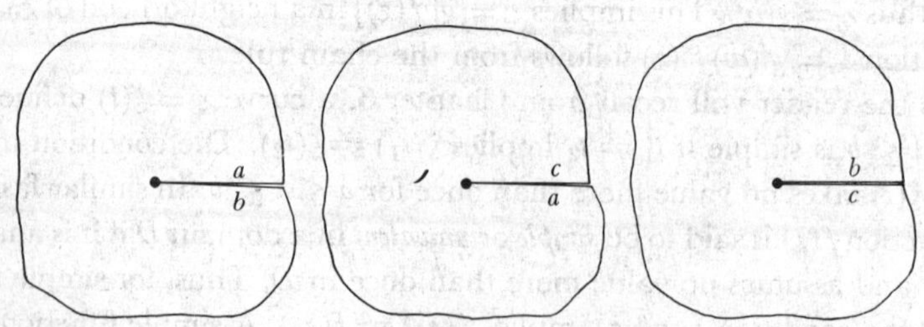

Figure 6-3

[1] In Chapter 2 the cut was along the negative real axis, but the location of the cut is irrelevant.

The first sheet of the w plane corresponds to the sector $0 \leq \theta < 2\pi/n$ of the z plane, the next sheet corresponds to $2\pi/n \leq \theta < 4\pi/n$, and so on. The n-sheeted Riemann surface is thus in one-to-one correspondence with the z plane. If a complex number $w \neq 0$ is given without specification as to the sheet in which it lies, there are n values of z that give this w, and so $w = z^n$ has an n-valued inverse. However, when the Riemann surface is introduced, the correspondence becomes one-to-one, and $w = z^n$ has a single-valued inverse.

It is an important fact that this simple artifice allows the inversion of the general relationship $w = f(z)$ at a point where $f'(z)$ vanishes. Assuming without loss of generality that $z_0 = w_0 = 0$, we shall establish the following:

THEOREM 6.4. *Let $f(z)$ be analytic at $z = 0$ and have a zero of order $n \geq 1$. That is, let $f^{(k)}(0) = 0$ for $k < n$ and $f^{(n)}(0) \neq 0$. Then to each sufficiently small $w \neq 0$ there correspond n distinct points z in the neighborhood of $z = 0$, each of which has w as its image under the mapping $w = f(z)$. Furthermore, the mapping $w = f(z)$ can be decomposed in the form $w = \zeta^n$, $\zeta = g(z)$, where $g(z)$ is univalent near $\zeta = z = 0$.*

If the n-sheeted Riemann surface for w is introduced, then the correspondence between w and ζ becomes one-to-one, and hence the correspondence between w and z becomes one-to-one. The difference between this precise formulation and the approximation $w = h(0)z^n$ discussed above is that here we have a general univalent map $\zeta = g(z)$ near 0, to get $w = \zeta^n$, while the approximate formulation has a linear map $\zeta = \alpha z$. The univalent map introduces only minor distortion of the general picture. See Figure 6-4.

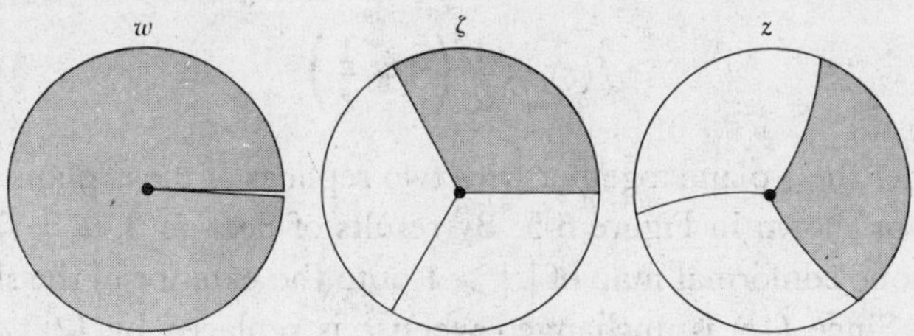

Figure 6-4

Proof. As already observed, $w = z^n h(z)$ near $z = 0$, where h is analytic at $z = 0$ and $h(0) \neq 0$. The fact that $h(0) \neq 0$ makes it possible to construct an analytic nth root $H(z)$ of $h(z)$ near 0, as seen in Problems 5 and 6 below. Thus

$$h(z) = [H(z)]^n,$$

and $H(0) \neq 0$ because $h(0) \neq 0$. The function $g(z) = zH(z)$ satisfies $g(0) = 0$, $g'(0) = H(0) \neq 0$ and so is univalent near 0. Since

$$w = z^n h(z) = [zH(z)]^n = [g(z)]^n,$$

this gives the second assertion in Theorem 6.4. The first assertion follows from properties of $w = \zeta^n$ and from the fact that the mapping $\zeta = g(z)$ is one-to-one near 0.

Example 6.1. If $f(z)$ is simple in a domain D, show that $f'(z) \neq 0$ in D.

Suppose $f'(z_0) = 0$ at a point z_0 in D, and let $w_0 = f(z_0)$. Then the zero of $f(z) - w_0$ has order $n \geq 2$, and Theorem 6.4 indicates that the equation $f(z) = w$ has at least two distinct roots near z_0 for w near w_0. Hence the mapping cannot be simple.

Example 6.2. Prove that if $w = f(z)$ is simple in a domain D then the inverse mapping $z = g(w)$ is simple in $f(D)$.

Since the correspondence between D and $f(D)$ is one-to-one, z is a function of w, $z = g(w)$, and $g(w)$ can assume no value in D more than once. Example 6.1 and Theorem 6.3 together show that $g(w)$ is analytic at each point w_0 of $f(D)$ and hence it is analytic in $f(D)$.

Example 6.3. Describe a Riemann surface for the mapping $w = f(z)$ where

$$f(z) = \frac{1}{2}\left(z + \frac{1}{z}\right).$$

Consider the z plane together with two replicas of the w plane slit from -1 to 1 as shown in Figure 6-5. By results of Section 3, $w = f(z)$ gives a one-to-one conformal map of $|z| > 1$ onto the exterior of the slit in the w plane. Since $f(z)$ is unchanged when z is replaced by $1/z$, $w = f(z)$ also gives a one-to-one conformal map of $|z| < 1$ onto the exterior of the

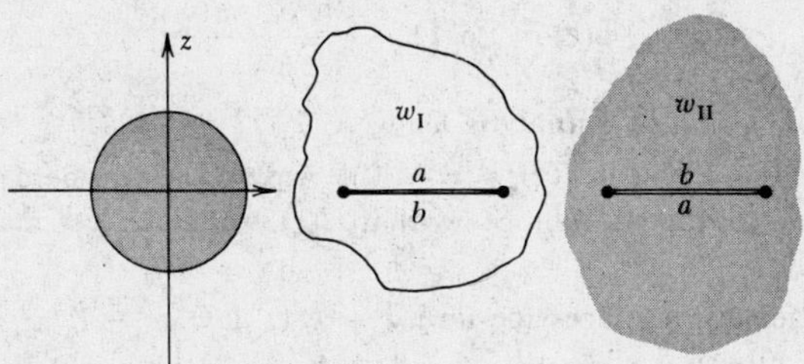

Figure 6-5

slit in the w plane. Let $|z| > 1$ be mapped onto the first slit plane w_I and let $|z| < 1$ be mapped onto the second plane w_{II}. The planes w_I and w_{II} are made into a Riemann surface by joining the sides of slits having the same label a or b in Figure 6-5. (The planes w_I and w_{II} are thought to be one above the other for this joining, but have been drawn side-by-side in Figure 6-5.) Evidently, $f'(z) = 0$ only at 1 and -1, and at these points $f''(z) \neq 0$. Thus Theorem 6.4 applies with $n = 2$, and shows that the Riemann surface has the appropriate structure near the branch points 1 and -1. The points -1 and 1 are excluded from the z plane and also from the Riemann surface. It is not difficult to show, then, that the mapping from the z plane to the surface is one-to-one.

Problems

1. Review Chapter 4, Section 6.
2. By inspection of Figure 6-1 conclude that $|f(z_0)|$ is not a minimum if $f(z_0) \neq 0$; that $\text{Re}\, f(z)$ does not attain a maximum or minimum in D; and that $\text{Im}\, f(z)$ does not attain a maximum or minimum in D.
3. (a) Show that $w = z/(1 - z)^2$ is simple for $|z| < 1$; that is, show that $f(z_1) = f(z_2)$ is possible only if $z_1 = z_2$. (b) Let D be the simply connected domain $1 < |z| < 2$, $|\arg z| < 3\pi/4$. Show that the function z^2 is not simple in D even though its derivative $2z \neq 0$ in D.
4. If $(1 - z)^{1/2} = \exp[(½)\, \text{Log}(1 - z)]$, show that $z/(1 - z)^{1/2}$ is simple for $|z| < 1$.
5. Let $h(z)$ be analytic for small $|z|$ and let $h(0) \neq 0$. Show that $h(z)/h(0)$ has an analytic logarithm near 0, given by

$$L(z) = \operatorname{Log}\left(1 + \frac{h(z) - h(0)}{h(0)}\right).$$

(Note that $\operatorname{Log}(1 + \zeta)$ is analytic for $|\zeta| < 1$.)

6. In the preceding problem let $H(z) = [h(0)]^{1/n} \exp[L(z)/n]$, where $[h(0)]^{1/n}$ is any definite choice of the nth root. Show that $H(z)$ is analytic for small $|z|$ and that $[H(z)]^n = h(z)$.
7. Describe Riemann surfaces for $w = z + 1/z$, $w = z^2 - 1$, $z^2 = (w - 1)/w$.
8. *Joukowski profiles.* Choose small positive numbers a and b, and draw a large circle C centered at $z = a + ib$ and passing through the point $z = -1$. Then sketch the image of C under the transformation $w = (\frac{1}{2})(z + z^{-1})$. These curves were formerly important in aerodynamics. Their interesting shape results from the fact that C passes through the branch point $z = -1$ and from the fact that C lies partly in $|z| < 1$ and partly in $|z| > 1$, both of which are fundamental regions for the transformation. (In polar coordinates $1/z$ is easily found from z, and w is the midpoint of the segment joining $1/z$ to z.)
9. A domain is convex if the line segment joining any two points of the domain necessarily lies in the domain. If $f(z)$ is analytic in a convex domain and satisfies $|f'(z) - 1| < 1$, show that $f(z)$ is simple. (Express $f(z_1) - f(z_2)$ as an integral of $1 + f'(\zeta) - 1$.)

***7. Global mapping theorems.** Most of the results of the preceding section are local, that is, they apply just in a small disk $|z - z_0| < \epsilon$. A global theorem applies to a whole domain D or $f(D)$ which need not be small. For example, Theorem 6.2 pertaining to connectivity of $f(D)$ is a global theorem. Further results of the sort are given now.

From the definition, $f(z)$ is simple in D if the equation $f(z) = w_0$ has at most one solution z in D for each complex number w_0. Hence, Theorem 6.3 indicates that a mapping $w = f(z)$ is simple in the immediate neighborhood of a point z_0 where $f'(z_0) \neq 0$. The question whether a conformal mapping is simple throughout a given domain can sometimes be answered by the following theorem:

THEOREM 7.1. *Let C be a Jordan contour with interior domain D, and let $f(z)$ be analytic in a domain containing C and D. On C let $f(z)$ take no value more than once. Then $f(z)$ is a univalent function in D. The mapping $w = f(z)$ transforms C to a simple closed contour C^* in the w plane. Let D^* be the interior of C^*. Then*

$w = f(z)$ is a one-to-one map of D onto D^*. Also, as z traverses C in the positive direction, $w = f(z)$ traverses C^* in the positive direction.

Proof. The image of C is a closed contour C^*, since $f(z)$ is analytic and single-valued, and C^* is simple because $f(z)$ takes no value more than once on C. Let w_0 be any point not on C^*. Then as in the discussion of (6.2) in Chapter 4, Section 6,

$$\frac{1}{2\pi i}\int_C \frac{f'(z)}{f(z) - w_0}\,dz = \frac{1}{2\pi i}\int_{C^*} \frac{dw}{w - w_0}. \tag{7.1}$$

By Rouché's theorem the left side equals $N - P$, where N is the number of zeros and P is the number of poles of $f(z) - w_0$ within C. Since $f(z)$ is analytic in C, clearly $P = 0$, and the left side equals N.

If w_0 is outside of C^* then $1/(w - w_0)$ is analytic in and on C^*, and hence the integral on the right of (7.1) is 0. This shows that the number of zeros of $f(z) - w_0$ in C is 0, and hence $f(z)$ does not assume the value w_0 in C.

If w_0 is within C^*, the right side of (7.1) is $+1$ or -1 according as C^* is traversed in the positive or negative direction. However the value -1 is excluded, since the left side equals N, which is nonnegative. Thus both sides of (7.1) are $+1$ when w_0 is inside C. This shows that C^* is traversed in the positive direction, and also shows that $f(z) - w_0$ has just one zero within C. Hence the value w_0 is assumed once and only once within C.

If w_0 is on C^*, then w_0 cannot be the image of a point z_0 in D. If it were, then by the fact that open sets map onto open sets, it would follow that some points near z_0 map into the exterior of C^*, which has been shown above to be impossible. This completes the proof.

Another refinement of Theorem 6.3 is given by the following:

THEOREM 7.2. *Let $f(z)$ be analytic and satisfy $|f(z)| \leq M$ for $|z - z_0| < R$, where R and M are constant. Let $f(z_0) = 0$, $|f'(z_0)| = m \neq 0$. Then the function $w = f(z)$ has an inverse $z = g(w)$ which is simple at least in the closed disk*

$$|w - w_0| \leq \frac{1}{4}\frac{(mR)^2}{M}.$$

This theorem is due in the main to Landau, although it is usually stated with a smaller constant than ¼, or with a bound on $|f'(z)|$ which implies the above bound on $|f(z)|$. A somewhat weaker form of Landau's result is proved in Problem 11. A stronger form is proved in Problem 4.6 at the end of Chapter 6.

We now give some uniqueness theorems for simple mappings.

THEOREM 7.3. *A simple function f, such that $w = f(z)$ maps $|z| < 1$ onto $|w| < 1$ and such that $f(0) = 0$, must be of the form $f(z) = \alpha z$ where $|\alpha| = 1$.*

Proof. The function $h(z) = f(z)/z$ has a removable singularity at $z = 0$, which we consider removed. By the maximum principle for h, or more directly by the lemma of Schwarz,[1] $|f(z)| \leq |z|$. Similarly, the inverse function $z = g(w)$ satisfies $|g(w)| \leq |w|$, which is to say, $|z| \leq |f(z)|$. Hence, the function $f(z)/z$ has constant modulus and is constant by Chapter 2, Example 1.2. This gives $f(z) = \alpha z$, and Theorem 7.3 follows.

THEOREM 7.4. *A simple function that maps $|z| < \infty$ onto $|w| < \infty$ must be linear.*

Proof. By the open mapping theorem, the image of $|z| < 1$ under f contains some disk $|w - w_0| < \delta$. If ∞ is an essential singularity for f, then, by the Casorati-Weierstrass theorem, $f(z)$ comes arbitrarily close to w_0 in every neighborhood of ∞. Hence, some values of f in $|z| > 1$ lie in the disk $|w - w_0| < \delta$. But this contradicts the fact that f is simple. Therefore, $z = \infty$ is at worst a pole of f, and f is a polynomial. The polynomial must have degree 1. Otherwise, $f'(z)$ would have at least one zero and, by Example 6.1, the mapping would not be simple.

It follows from Theorem 7.4 that *a simple function which maps $|z| \leq \infty$ onto $|w| \leq \infty$ must be bilinear.* This is proved in Problem 6 below.

A method discussed next is known as the reflection principle because it enlarges the domain of a mapping $w = f(z)$ by reflection in straight lines in the z and w planes. One of the uses of the reflection principle is in extending mappings so that uniqueness theorems, such as Theorem 7.4, can be applied to the extended mapping. The reflection principle also gives insight into periodicity properties of functions defined by mappings, and it justifies some of the analysis of Section 4.

Let $f(z)$ be analytic in a domain D which lies in $\text{Im}\, z > 0$ and which has one or more segments L_k of the real axis as part of its boundary

[1] Chapter 3, Section 7, Problem 8.

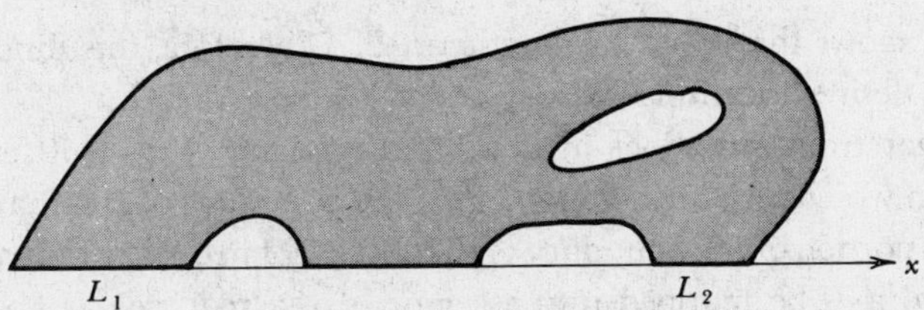

Figure 7-1

(Figure 7-1). Assume further that $f(z)$ is continuous in the region consisting of D together with the segments $L_1, L_2, \ldots, L_n$, and that $f(z)$ is real on each L_k. The case shown in Figure 7-1 has $n = 2$ because the center segment in the figure is not included among the L_k. Such exclusion would be necessary if $f(z)$ is not real or not continuous on the center segment.

Let $\tilde{D}$ be the reflection of D in the real axis. Since $\bar{z}$ is in D whenever z is in $\tilde{D}$, we can define a function $\tilde{f}$ in $\tilde{D}$ by $\tilde{f}(z) = \overline{f(\bar{z})}$, that is, by reflection of $f(\bar{z})$ in the real axis of the w plane. In terms of real and imaginary parts.

$$f(z) = u(x,y) + iv(x,y), \qquad \tilde{f}(z) = u(x,-y) - iv(x,-y).$$

This shows that the real and imaginary parts of $\tilde{f}$ have continuous partial derivatives, and that the Cauchy-Riemann equations for f imply the Cauchy-Riemann equations for $\tilde{f}$. Hence $\tilde{f}$ is analytic in $\tilde{D}$.

We construct $\tilde{f}(z)$ on each L_k by the same method as was used in D. Since $z = \bar{z}$ on L_k, it follows that $\tilde{f}(z) = \overline{f(z)}$ on L_k, and since $f(z)$ is real, $\tilde{f}(z) = f(z)$ on L_k. (This is the reason for assuming f real on L_k.) It is easily checked that the function F defined by

$$F(z) = f(z) \text{ in } D, \quad F(z) = \tilde{f}(z) = f(z) \text{ on } L_k, \quad F(z) = \tilde{f}(z) \text{ in } \tilde{D}$$

is continuous in the region consisting of D, $\tilde{D}$ and the L_k. Moreover, $F(z)$ is analytic provided the segments L_k have the following property: For each point x_0 on L_k there is a $\delta > 0$, depending on x_0, such that the semidisk

$$|z - x_0| < \delta, \qquad \text{Im } z > 0$$

is in D. Segments having this property are called *admissible*. When the L_k are admissible, Chapter 4, Example 9.1 (or an easy application of Morera's theorem) shows that $F(z)$ is analytic on L_k. Therefore, this function provides the desired continuation of f across the segments L_k.

By linear transformations in the z and w planes, the result extends to the case in which any straight portion of the boundary of D is mapped into a straight portion of the boundary of $f(D)$; it is not necessary that a segment of the real axis be mapped into a segment of a real axis as above. This extension is summarized as follows:

THEOREM 7.5 (*Schwarz principle of reflection*). *Let D be a domain which has one or more admissible straight line segments L_k as part of its boundary. Let $w = f(z)$ be analytic in D and continuous in the region consisting of D together with the segments L_k. Suppose each L_k is mapped by f onto a straight line segment L_k^* in the w plane. Then $f(z)$ can be continued analytically across the segments L_k by reflection.*

The phrase "continuation by reflection" means that if $\tilde{z}$ is the mirror image of z in L_k then $\tilde{w} = f(\tilde{z})$ is the mirror image of $w = f(z)$ in L_k^*. As suggested by Figure 7-2, the points z, $\tilde{z}$ need not be close to L_k, and the theorem is not just a local theorem. The reflection takes place in the infinite line of which L_k is a part.

In Figure 7-1, the requirement that $\operatorname{Im} f = 0$ on $\operatorname{Im} z = 0$ means geometrically that the line segments L_k of $y = 0$ map onto segments L_k^* of $v = 0$, and this fact was used in the formulation of Theorem 7.5. Analytic continuation in Figure 7-1 was achieved by assigning to points inverse with respect to the line $y = 0$, points which are inverse with respect to the line $v = 0$. Viewed in this light, the principle of analytic continuation by reflection can be extended to the case where the boundary of D contains segments of circular arcs and the function $f(z)$ maps these arcs

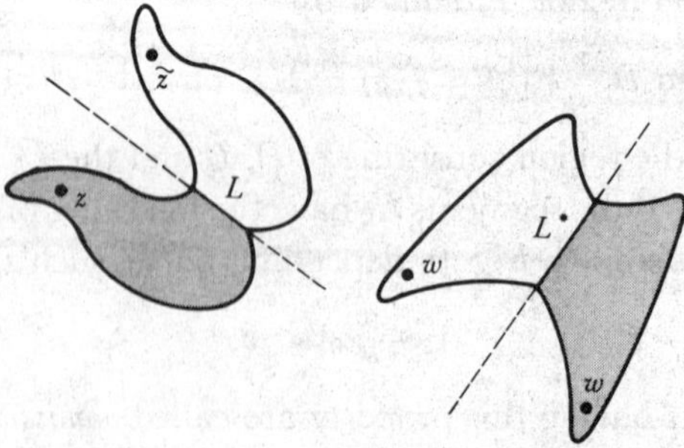

Figure 7-2

onto circular arcs in the w plane. Indeed, a circle in the z plane can be mapped by a linear transformation onto the real axis of a ζ plane and a circle in the w plane can be mapped onto the real axis of an ω plane. Analytic continuation by the reflection principle in the ζ, ω formulation is equivalent to analytic continuation by inverse points in the respective circles in the z, w formulation. This idea can even be extended to the case where the line segment or circular arc is replaced by an analytic arc.

Example 7.1. Let f_1 and f_2 be simple on a domain D, vanish at a point z_0 of D, have positive derivatives at z_0, and map D onto $|w| < 1$. Prove $f_1 = f_2$.

The inverse function f_2^{-1} maps $|w| < 1$ onto D so that $w = 0$ is mapped into $z = z_0$. Hence $f_1[f_2^{-1}(w)]$ maps $|w| < 1$ onto the unit disk and maps $w = 0$ onto the origin. By Theorem 7.3,

$$f_1[f_2^{-1}(w)] = \alpha w$$

where $|\alpha| = 1$. Setting $w = f_2(z)$ gives $f_1(z) = \alpha f_2(z)$, and $\alpha = 1$ because $f_1'(z_0)$ and $f_2'(z_0)$ are both positive.

Example 7.2. A function $f(z)$ is analytic in $\operatorname{Im} z > 0$ and is continuous in the region consisting of $\operatorname{Im} z > 0$ together with a segment $a < x < b$ of the real axis. If $f(x) = x$ for $a < x < b$, show that $f(z) = z$ for $\operatorname{Im} z > 0$.

By reflection, extend the domain of f so that the new function is analytic for $y = 0$, $a < x < b$. The extended function agrees with the analytic function z on this segment, and hence it agrees with z in its whole domain of analyticity. This gives the conclusion.

Example 7.3. A function $w = f(z)$ is continuous in the strip

$$-\pi/2 \leq x \leq \pi/2, \qquad y \geq 0$$

except possibly at the points $-\pi/2$ and $\pi/2$, and provides a simple mapping of the interior of the strip onto the upper half-plane. It also maps the boundary of the strip onto the real axis (Figure 7-3). Extend the function by reflection and show that the extended function has period 2π.

Reflection in the boundary in the z plane and corresponding reflection in the w plane is indicated by the arrows in Figure 7-3. Evidently w re-

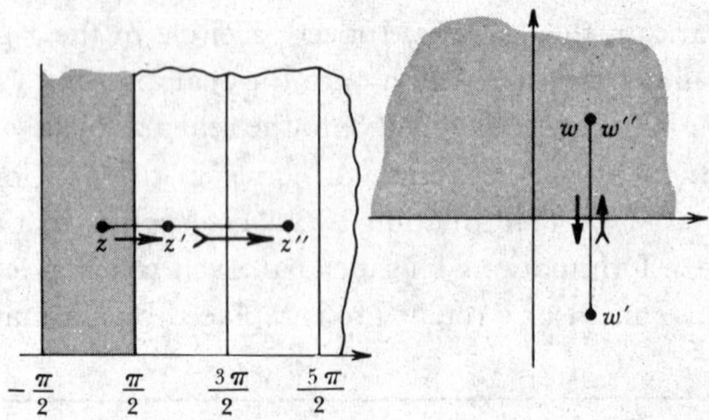

Figure 7-3

turns to its original value after the two reflections shown in the figure, but z becomes $z + 2\pi$. Thus the extended mapping satisfies

$$w = f(z), \qquad w = f(z + 2\pi)$$

and so has period 2π.

This problem pertains to the mapping $w = \sin z$, which can be written as a transformation of Schwarz-Christoffel type,

$$z = \int_0^w \frac{d\zeta}{\sqrt{1 - \zeta^2}}. \tag{7.2}$$

The reflection method applies to integrals of much greater complexity than (7.2) and is a valuable aid in the study of periodic functions.

Problems

1. By considering $w = z + z^2$ in a large disk $|z| < R$, show that the constant ¼ in Landau's theorem is sharp; that is, the theorem fails for some functions $f(z)$ if this constant is replaced by a larger one.
2. Prove that a simple mapping of $|z| < 1$ onto $|w| < 1$ must be bilinear. Solution: Let the given transformation be T and suppose $T\alpha = 0$. Let T_0 denote a bilinear transformation of $|z| < 1$ onto $|w| < 1$ that maps the origin into α, thus $T0 = \alpha$. Then $T_1 = TT_0$ maps $|z| < 1$ onto $|w| < 1$ and maps 0 onto 0. Hence T_1 is linear by Theorem 7.3, and $T = T_1 T_0^{-1}$ is bilinear.
3. Do Problem 2 by use of the functional notation $w = f(z)$, as in Example 7.1, rather than by the notation $w = Tz$.
4. How does the result of Example 7.1 change if $f_k'(z_0) > 0$ is not assumed? Prove your answer by use of the notation $w = Tz$ as in Problem 2.

5. Prove that a simple map of Im $z > 0$ onto Im $w > 0$ which maps $z = i$ onto $w = i$ must have the form

$$w = \frac{z \sin \lambda + \cos \lambda}{\sin \lambda - z \cos \lambda}$$

where λ is real. (Proceed as in Problem 2 or 3 and set $\alpha = e^{2i\lambda}$ in Theorem 7.3.)

6. Show that a simple map of the extended plane onto the extended plane must be bilinear. (If $z = \infty$ and $w = \infty$ correspond, the result follows from Theorem 7.4. If $z = z_0$ maps into $w = \infty$, let $\zeta = 1/(z - z_0)$ and use the previous case to get $w = A\zeta + B$.)

7. Readers familiar with Section 5 will show that the operation of paraconjugation is an instance of the method of reflection.

8. Extend the function considered in Example 7.3 by reflection in the real axes of the z and w planes. By use of two reflections show that the extended function satisfies $f(-z) = -f(z)$ if and only if $f(iy)$ is pure imaginary, $0 < y < \infty$.

9. A certain function $f(z)$ is analytic in Re $z > 0$ and is continuous in the region consisting of Re $z > 0$ together with the segment $0 < y < 1$ of the imaginary axis. If $f(iy) = y^4 - 2y^2$ for $0 < y < 1$, show that $f(1) = 3$.

10. If $(1 - z)^p = \exp[p \operatorname{Log}(1 - z)]$, show that $z/(1 - z)^p$ is simple for $|z| < 1$ and for $0 < p < 2$. The case $p = 2$ was done in Section 6, Problem 2, but this case is much harder. (Let D_ϵ be the set of points in $|z| < 1$ whose distance to the point $z = 1$ exceeds ϵ. If the two arcs forming the boundary of D_ϵ are parametrized by $z = e^{i\theta}$ or $z = 1 - \epsilon e^{-i\phi}$, use the monotony properties of

$$\arg \frac{z}{(1 - z)^p} = \frac{p\pi + (2 - p)\theta}{2}, \quad \arg \frac{z}{(1 - z)^p} = p\phi + \tan^{-1} \frac{\epsilon \sin \phi}{1 - \epsilon \cos \phi}$$

for small ϵ to show that the boundary C_ϵ of D_ϵ is mapped one-to-one onto a simple closed curve C_ϵ^*.)

11. For $|z| \leq R$ let $h(z)$ be analytic and satisfy $h(0) = 0$, $h'(0) = 1$, $|h(z)| \leq M$. Define $A = R^2/(M + R)$. Then if $|w| < A/4$, the equation $h(z) = w$ has one and only one solution z such that $|z| < A/2$. (On a suitable circle $|z| = r$ apply Rouché's theorem with w fixed and

$$f(z) = z - w, \qquad g(z) = h(z) - z.$$

Clearly $|f(z)| \geq r - |w|$ on $|z| = r$. Since $g(z)/z^2$ has a removable singularity at $z = 0$, the maximum principle applied to $g(z)/z^2$ gives $|g(z)| \leq r^2/A$ for $|z| = r$. The condition $|g(z)| < |f(z)|$ holds on $|z| = r$ if $|w| < r - r^2/A$. Choose r so as to maximize the right-hand side.)

8. The Riemann mapping theorem. The finite plane, $|z| < \infty$, is simply connected, but there is no conformal map of $|z| < \infty$ onto the unit disk. This follows from Liouville's theorem, since an analytic function $w = f(z)$ satisfying $|f(z)| < 1$ for all finite z would have to be constant. Of course the same reasoning shows that there is no conformal map of the extended plane $|z| \leq \infty$ onto the unit disk.

It is a remarkable fact that these two domains, $|z| < \infty$ and $|z| \leq \infty$, are the only simply connected domains that cannot be mapped conformally onto the unit disk. This is the content of the *Riemann mapping theorem,* which reads as follows:

THEOREM 8.1 (*Riemann*). *If D is a simply connected domain in the z plane, which is neither the z plane itself nor the extended z plane, then there is a simple function $f(z)$ such that $w = f(z)$ maps D onto the disk $|w| < 1$.*

The hypothesis means that there is at least one finite point α not in D. Since D is simply connected, if there is one point α not in D, there are infinitely many. However, this need not be explicitly stated.

If z_0 is a point of D, and if it is required that $f(z_0) = 0, f'(z_0) > 0$, then the simple function $w = f(z)$ that maps D onto $|w| < 1$ is unique. This is easy to prove and was established in Example 7.1. The existence of $f(z)$ is much harder to prove. Perhaps the simplest proof is one that proceeds as follows:

(1) Consider the set of all simple functions $g(z)$ on D such that $|g(z)| < 1$, $g(z_0) = 0$ and $g'(z_0) > 0$, and show that this set of functions is not empty (which is easy to do).

(2) Show that the above set of functions contains an extremal function $\tilde{g}(z)$ which has the property that $\tilde{g}'(z_0) \geq g'(z_0)$ for any g of the set (which is the major part of the proof and is similar in character to the proof that a bounded equicontinuous set of functions has a convergent subsequence).

(3) Show that if a function g_1 of the set (1) omits a value w_1 in $|w| < 1$, there is a function g_2 of the set such that $g_1'(z_0) < g_2'(z_0)$ (which is easy to do).

Step (3) shows that the extremal function found in step (2) must map D onto $|w| < 1$ and completes the proof. The underlying idea is that *maximizing $g'(z_0)$ forces $w = g(z)$ to cover the whole disk $|w| < 1$.*

Riemann was led to his theorem by physical considerations, but the first complete proof was given by Osgood. Since step (2) is nonconstructive, the proof is not given here. For steps (1) and (3), see Problems 8.1 and 8.2.

Additional problems on Chapter 5

1.1. If a bilinear transformation $w = Tz$ has the distinct finite fixed points z_1 and z_2, it follows from Problem 14 that

$$\frac{w - z_1}{w - z_2} = H\frac{z - z_1}{z - z_2}$$

where H is constant. If two transformations have the same fixed points z_1, z_2 but different constants H_1, H_2 show that their product has the same fixed points and the constant H_1H_2. Extend to n transformations by induction, and thus show how to get the n-times iterated transformation T^nz. This technique is useful in the study of cascaded networks and periodic structures on transmission lines.

1.2. Find the fixed points and thus get T^nz:

$$Tz = \frac{1 - 3z}{z - 3}, \quad Tz = \frac{3 - iz}{z - 2 - i}, \quad Tz = \frac{2iz - z + 2}{2i + 1 - z}.$$

1.3. Let $F(x,y)$ and $G(x,y)$ be real and have continuous first-order partial derivatives on a domain D of the (x,y) plane. Suppose the real transformation $u = F(x,y)$, $v = G(x,y)$ from D to the (u, v) plane is conformal at all points of D. Prove that $F_x = G_y$, $F_y = -G_x$ and hence $F + iG$ is analytic on D. Outline of solution: Let $z = z_0 + te^{i\theta}$ be the straight line through a given point z_0 of D which makes an angle θ with the x axis. If $w = F + iG$, then on this line

$$\frac{dw}{dt} = (F_x + iG_x)\frac{dx}{dt} + (F_y + iG_y)\frac{dy}{dt}. \qquad (*)$$

Substituting

$$\frac{dx}{dt} = \cos\theta = \frac{e^{i\theta} + e^{-i\theta}}{2}, \quad \frac{dy}{dt} = \sin\theta = -i\frac{e^{i\theta} - e^{-i\theta}}{2}$$

into (*) and dividing by $e^{i\theta} = dz/dt$ gives

$$\frac{dw}{dt} \div \frac{dz}{dt} = \frac{1}{2}(F_x + G_y + iG_x - iF_y) + \frac{1}{2}(F_x - G_y + iG_x + iF_y)e^{-2i\theta}.$$

As θ varies, the sum above with $t = 0$ describes a circle with center and radius

$$z_1 = \tfrac{1}{2}(F_x + G_y + iG_x - iF_y), \quad \rho = \tfrac{1}{2}|F_x - G_y + iG_x + iF_y|.$$

If $\arg(dw/dt \div dz/dt)$ is not to change with θ, the radius ρ must be 0, and this gives the Cauchy-Riemann equations.

1.4. If $|dw/dt \div dz/dt|$ is constant at $t = 0$ in the above problem as θ varies, show that either $F + iG$ or $F - iG$ satisfies the Cauchy-Riemann equations at z_0. Show also that if one of the quantities z_1 or ρ is not 0 at z_0 then the other must vanish in a neighborhood of z_0 and hence $f(z)$ or $\overline{f(z)}$ is analytic at z_0.

1.5. This problem requires knowledge of orthogonal matrices. If $f = u + iv$ is analytic, then by the Cauchy-Riemann equations

$$\begin{pmatrix} du \\ dv \end{pmatrix} = \begin{pmatrix} u_x & u_y \\ v_x & v_y \end{pmatrix} \begin{pmatrix} dx \\ dy \end{pmatrix} = \begin{pmatrix} u_x & u_y \\ -u_y & u_x \end{pmatrix} \begin{pmatrix} dx \\ dy \end{pmatrix}.$$

(a) Show that the second matrix is proportional to an orthogonal matrix, and thus get the conformal property of the mapping when $f'(z) \neq 0$. (b) Prove, conversely, that if the first matrix is proportional to an orthogonal matrix, then f or $\bar{f}$ satisfies the Cauchy-Riemann equations at (x,y).

1.6. This problem requires knowledge of Jacobians. (a) If a conformal transformation $w = f(z)$ is written in real form, show that the Jacobian is $|f'(z)|^2$. Conclude that the transformation is locally invertible if $f'(z)$ is continuous and nonzero. (b) Show that the area of $f(D)$ under a conformal map $w = f(z)$ of D is given formally by

$$\iint_{f(D)} du\, dv = \iint_D |f'(z)|^2\, dx\, dy,$$

in agreement with the fact that $|f'(z)|$ describes the local factor of magnification.

2.1. Transformations with $|H| = 1$ in Problem 1.1 are termed *elliptic* and those with H real are *hyperbolic*. Show that the set of all transformations with given fixed points $z_1 \neq z_2$ is a commutative group, in which the sets of elliptic and hyperbolic transformations are subgroups.

2.2. *Steiner's theorem.* Let two circles be given, one inside the other, and let other circles be drawn tangent to them and to one another as shown in Figure 8-1(a). Sometimes the ring closes, that is, the last circle is tangent to the first. Show that this happens for every position of the starting circle if it happens for any position of the starting circle. (By Example 2.2, the given configuration can be mapped conformally onto another in which the two bounding circles are concentric, Figure 8-1(b).)

2.3. Two points p and q are inverse with respect to a circle or line C if and only if every circle or line containing p and q is orthogonal to C. Outline of solution: By $w = (z - p)/(z - q)$, the points p and q transform to $w = 0$ and $w = \infty$. (If q is already at ∞ the transformation $w = z - p$ is used.) Since $w = 0$

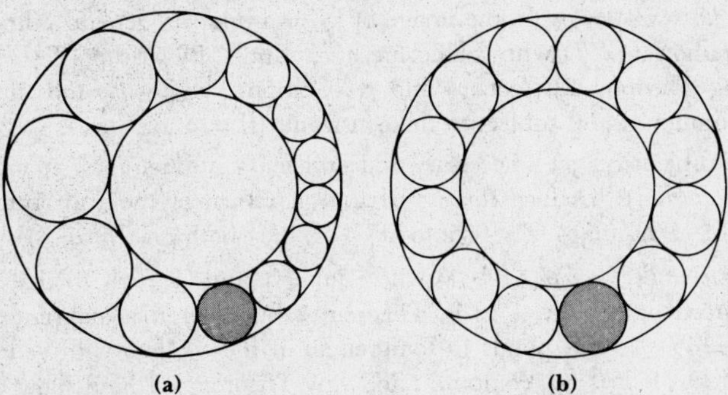

Figure 8-1

and $w = \infty$ are inverse points with respect to the image of C, the image of C must be a circle of the form $|w| = k_1$. Moreover, any circle or line containing both p and q must transform into a radial line $\arg w = k_2$, since p and q transform to 0 and ∞. Clearly, the circles $|w| = k_1$ and the radial lines $\arg w = k_2$ are orthogonal. Hence, their images in the z plane must also be. Thus the families of circles

$$\left|\frac{z-p}{z-q}\right| = k_1, \qquad \arg\left(\frac{z-p}{z-q}\right) = k_2$$

must be mutually orthogonal and the second family must be the circles containing p and q, as shown in Figure 8-2.

2.4. Deduce from Theorem 2.4 that if either of the transformations

$$w = a + \frac{bz}{1 - cz}, \qquad w = c + \frac{bz}{1 - az}$$

maps $|z| < 1$ onto $|w| < 1$, they both do.

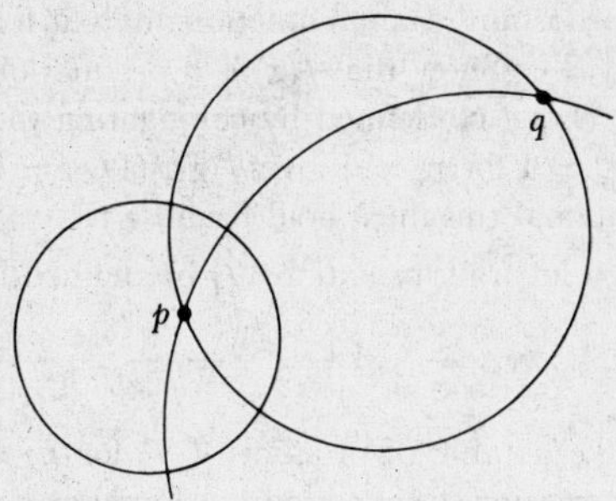

Figure 8-2

2.5. Let $|w - w_0| = \rho_0$ be the image of $|z| = \rho$ under a general bilinear transformation $w = Tz$ with coefficients a, b, c and d. By writing $|T^{-1}w| = \rho$ and using Theorem 2.1, get w_0 and ρ_0. Compute $|w_0| + \rho_0$ and thus get the maximum of $|Tz|$ subject to the constraint $|z| = \rho$.

2.6. By Problem 2.5 get a necessary and sufficient condition for T to map $|z| < 1$ into $|w| < 1$. Deduce for $b \neq 0$ that if either of the transformations of Problem 2.4 maps $|z| < 1$ into $|w| < 1$, they both do.

2.7. Let $w = \beta(z - \alpha)/(\bar{\alpha}z - 1)$ where $|\alpha| < 1$ and $|\beta| = 1$, so that $|z| < 1$ is mapped onto $|w| < 1$ as in Theorem 2.4. Under this mapping the circle $|w| = r < 1$ corresponds to some circle in the z plane with center z_0 and radius s. If $t = |z_0|$, compute t and s by Theorem 2.1, and show that interchanging r and $|\alpha|$ has the effect of interchanging s and t. Show that

$$t^2 = (1 - s/r)(1 - sr)$$

and hence, since s is uniquely determined by r and t, s must be given by the construction of Figure 8-3.

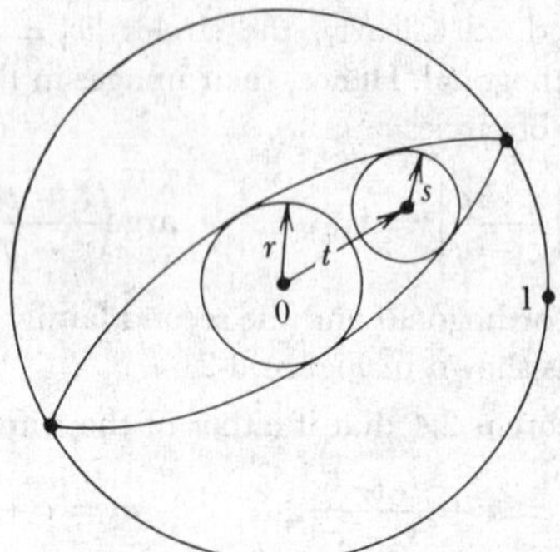

Figure 8-3

2.8. Let $f(z)$ be a nonconstant rational function analytic for $|z| \leq 1$ and satisfying $|f(z)| = 1$ for $|z| = 1$. Show that $f(z)$ is a product of expressions like those in Theorem 2.4. (Let $\bar{f}$ be defined by conjugating the coefficients in f. The equation $f(z)\overline{f(z)} = 1$ for $z\bar{z} = 1$ gives $f(z)\bar{f}(1/z) = 1$ for $|z| = 1$, and since the left side is rational this must hold for all z.)

2.9. For $j = 1,2,\ldots,6$ let transformations T_j be defined by

$$T_jz = \quad z, \quad \frac{1}{z}, \quad 1 - z, \quad \frac{1}{1 - z}, \quad \frac{z}{z - 1}, \quad \frac{z - 1}{z},$$

respectively. Make a table of products T_iT_j for $i,j = 1,2,\ldots,6$ and show that these transformations form a group. A function $J(z)$ is *invariant* if $J(Tz) = J(z)$ for each transformation T of the group. Show that

$$J(z) = f(T_1z) + f(T_2z) + f(T_3z) + f(T_4z) + f(T_5z) + f(T_6z)$$

is invariant whenever f is any function defined in the extended plane, and conversely, every invariant J can be obtained in this way with $f(z) = J(z)/6$. (Replacing z by Tz merely permutes the terms in the sum defining J.)

2.10. Let $z = X(z_1,z_2,z_3,z_4)$ where the z_i are distinct and X is the cross ratio. Show that permutation of the z_i leads to just the six numbers T_jz considered in the preceding problem.

3.1. Let $\phi(x,y) = \text{Re}[(i + z)/(i - z)]$ for $z \neq i$ and let $\phi(0,1) = 0$. Show that ϕ is harmonic in $|z| < 1$, is 0 on $|z| = 1$, and is continuous in $|z| \leq 1$ except at the single point $z = i$. Conclude that the solution of Example 3.1 is not unique.

It is possible to show that uniqueness does hold if the function is required to be bounded in Example 3.1, and the same applies to problems considered below. Uniqueness is trivial when the region is bounded and the boundary values are continuous (Chapter 2, Problem 5.4) but is not proved in other cases here.

3.2. (a) By considering $c_1 + c_2u$, construct a bounded function which is harmonic for $u_0 < u < u_1$ and assumes prescribed real values A and B on $u = u_0$ and $u = u_1$. (b) By considering $c_1 + c_2 \log |w|$, construct a function which is harmonic for $r_0 < |w| < r_1$ and assumes values A and B on $|w| = r_0$ and $|w| = r_1$. (c) By considering $c_1 + c_2 \text{ Arg } w$ construct a bounded function which is harmonic for $\theta_0 < \text{Arg } w < \theta_1$ and assumes values A and B on $\text{Arg } w = \theta_0$ and $\text{Arg } w = \theta_1$.

3.3. Sketch the equipotentials in each of the three cases considered in Problem 3.2. Familiarity with these results is presupposed in the following problems. Problem 3.2 also allows explicit computation of $\phi^*(u,v)$ and hence of $\phi(x,y)$ if explicit formulas are desired; note that the functions of Problem 3.2 have the respective forms

$$\text{Re}(c_1 + c_2w), \quad \text{Re}(c_1 + c_2 \log w), \quad \text{Im}(c_1 + c_2 \log w).$$

3.4. Let D be a region bounded by two circular arcs or by a line segment and an arc as illustrated in Figure 8-4. Assuming that the bounding curves intersect at an angle θ_1 at $z = 0$ and $z = 1$, map the region onto a sector of angle θ_1. If a bounded harmonic function assumes the value A on one bounding circle and B on the other, show that the equipotentials are circles through 0 and 1.

Figure 8-4

3.5. Using Example 2.2, explain how to construct a function that is harmonic in an eccentric annulus and assumes prescribed values A and B on the inner and outer bounding circles, respectively. Show that the equipotentials and lines of flow are arcs of circles, and sketch. This problem pertains to cylindrical condensers, coaxial lines and insulated pipes.

3.6. It is desired to construct a bounded function which is harmonic in the upper half-plane and assumes the value A on a segment $x_0 < x < x_1$ of the real axis and the value B on the rest of the real axis. Describe the equipotentials and flow lines. (Map the upper half-plane onto itself in such a way that x_0 maps into $w = 0$ and x_1 into $w = \infty$.)

4.1. It is desired to find a bounded function harmonic in the half-strip shown in Figure 8-5 which assumes boundary values A, B and C as indicated. By $w = \sin z$, map this problem into another problem in a sector, and by squaring, map onto a third problem in a half-plane. Solve explicitly when:

$$A = 1, B = C = 0; \quad A = B = 0, C = 1; \quad B = 1, A = C = 0.$$

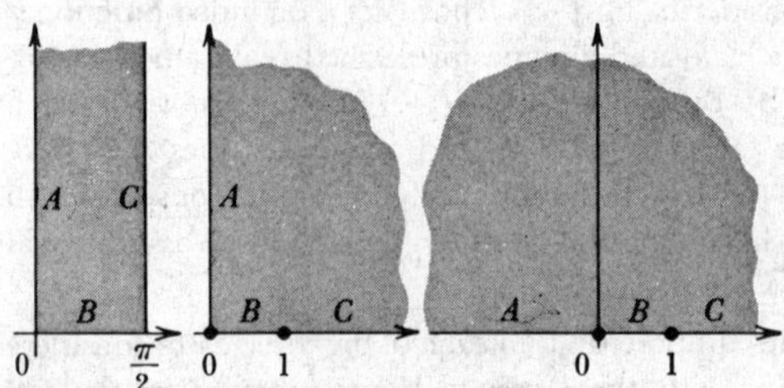

Figure 8-5

4.2. This problem requires familiarity with the concept of directional derivative. If ϕ is differentiable and C is a smooth curve, the derivative of ϕ in a direction normal to C at points of C is called the *normal derivative* and is written $\partial\phi/\partial\nu$ or ϕ_ν. Show that the condition $\phi_\nu = 0$ is preserved under one-to-one conformal maps; that is, if ϕ is mapped onto ϕ^*, as in the text, and C onto C^*, then $\phi_\nu = 0$ on C implies $\phi_\nu{}^* = 0$ on C^*. (The curves on which $\phi_\nu = 0$ are the orthogonal trajectories of the curves on which $\phi = \text{const.}$)

4.3. It is desired to find a bounded function $\phi(x,y)$ which is harmonic in the first quadrant; which assumes the value A on the positive imaginary axis and the value B on the part of the real axis with $x > 1$; and which has 0 normal derivative on the real axis for $0 < x < 1$ (see Figure 8-6(b)). Map this problem into the problem suggested by Figure 8-6(c), solve the latter by Problem 3.2(a), and obtain ϕ explicitly when $y = 0, 0 \leq x \leq 1$. This problem pertains to the steady-state temperature in a uniform plate when part of the boundary is insulated.

4.4. Explain how to reduce the problem suggested by Figure 8-6(a) to Problem 4.3 above.

5.1. This problem requires some acquaintance with linear differential equations

$$y^{(n)} + p_{n-1}y^{(n-1)} + \cdots + p_1 y' + p_0 = 0$$

where the p_j are complex constants and $y' = dy/dt$. The *characteristic polynomial* is

$$P(z) = z^n + p_{n-1}z^{n-1} + \cdots + p_1 z + p_0$$

and the equation is said to be stable if every solution y satisfies $y(t) \to 0$ as $t \to \infty$. (a) By trying $y = e^{st}$, show that the equation is not stable if P is not a Hurwitz polynomial. (b) Show that the equation is stable if P is a Hurwitz polynomial such that no two of the roots have equal real parts. The latter condition is introduced here to simplify the proof and is, in fact, superfluous. (If $P(s) = 0$, the substitution $y = e^{st}Y$ leads to an equation of lower order for Y'. Use induction.)

6.1. According to Chapter 2, Section 2, the strips

$$-\infty < x < \infty, \qquad (2n - 1)\pi < y < (2n + 1)\pi$$

are fundamental regions for $w = e^z$, and each of these strips is mapped onto the w plane cut along the negative real axis. Depict the Riemann surface by making replicas of the cut w plane, side-by-side, and label the edges of slits that are to be joined with corresponding letters.

6.2. According to Chapter 2, Section 2, the strip

$$-\tfrac{1}{2}\pi < u < \tfrac{1}{2}\pi, \qquad -\infty < v < \infty \qquad (*)$$

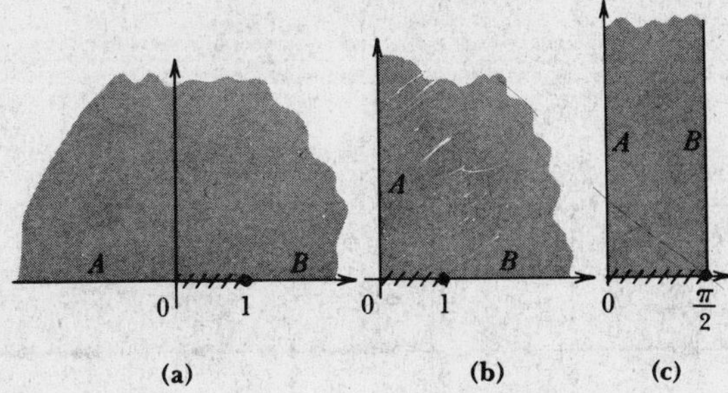

Figure 8-6

is a fundamental region for the mapping $z = \sin w$, and each of these strips is mapped onto the z plane cut along the real axis from $-\infty$ to -1 and from 1 to ∞. Depict the Riemann surface as suggested in Problem 6.1. (It may be helpful to consider the mapping $z = \sin w$ to be a composite of $\zeta = e^{iw}$, $2iz = \zeta - \zeta^{-1}$.)

6.3. Construct a single-valued branch of $\tan^{-1}z$ in the z plane when the latter is cut by deleting the part of the imaginary axis with $|y| > 1$. (Show that

$$\tan^{-1}z = \frac{i}{2}\log\zeta \quad \text{where} \quad \zeta = \frac{i+z}{i-z}.$$

By properties of bilinear mappings, the ζ plane cut along the negative real axis is in one-to-one correspondence with the cut z plane.)

6.4. Show that a fundamental region for $z = \tan w$ is given by (*) and that this strip maps onto the cut z plane of Problem 6.3. By introducing replicas of the cut z plane, one for each strip, describe the Riemann surface for $z = \tan w$.

7.1. Let $f(z)$ be an entire function which is real on a segment of the real axis and purely imaginary on a segment of the imaginary axis. Deduce from the reflection principle that f is odd.

7.2. Let $f(z)$ be analytic for $|z| \leq 1$ and satisfy $|f(z)| = 1$ for $|z| = 1$, Prove f is rational. (Let $\bar{f}(\zeta) = \overline{f(\bar{\zeta})}$ for $|\zeta| \leq 1$. Using the reflection principle for circular boundaries, deduce that $f(z) = 1/\bar{f}(1/\bar{z})$ for $|z| \geq 1$. Hence the only singularities α in $|z| > 1$ are poles corresponding to zeros $1/\bar{\alpha}$ of f.)

7.3. By a technique similar to that in Chapter 4, Problem 4.1, extend Problem 7.2 to meromorphic functions.

7.4. A function $f(z)$ is analytic in the rectangle $0 < x < a$, $0 < y < b$ and is continuous in the closed rectangle except perhaps at the corners. It maps the interior onto the first quadrant and the boundary onto the axes, as in Figure 8-7. Extend the mapping by reflection, and show that the extended

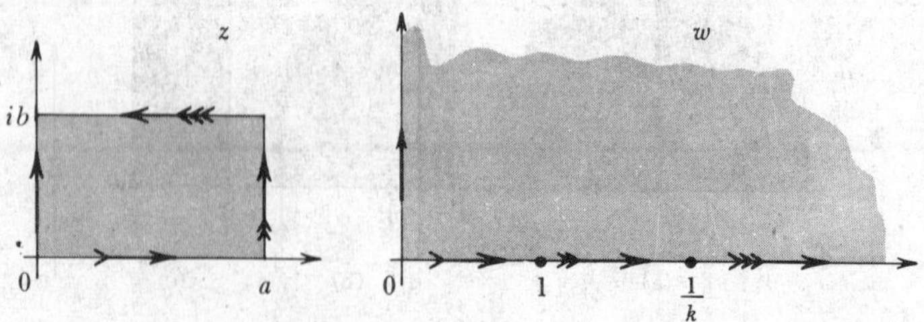

Figure 8-7

function has the two periods $\omega_1 = 4a$, $\omega_2 = 2ib$. (Four successive reflections in vertical lines in the z plane correspond to successive reflection in the real, imaginary, real, imaginary axes of the w plane.)

7.5. A function with the properties described in Problem 7.4 is given implicitly by an elliptic integral

$$z = \int_0^w \frac{d\zeta}{[(1-\zeta^2)(1-k^2\zeta^2)]^{1/2}}, \qquad 0 < k < 1.$$

(Compare Section 4, Problem 5.) With a suitable choice of branch for the integrand, show that this function actually does map the boundary of the rectangle as described above, and that

$$a = \int_0^1 \frac{dx}{[(1-x^2)(1-k^2x^2)]^{1/2}}, \quad b = \int_0^\infty \frac{dy}{[(1+y^2)(1+k^2y^2)]^{1/2}}.$$

Also express b as an integral from $1/k$ to ∞.

7.6. Show that the mapping $w = z + \log z$ of the slit plane in Figure 4-10 on the upper half-plane is univalent. (Let D_ϵ be the set of points of the upper half-plane in $|z| < 1/\epsilon$ whose distance from the segment $-1 < x < 0$ of the real axis exceeds ϵ, and proceed as in Section 7, Problem 10. An estimate of the variation of $\arg[f(z) - w_0]$ to within an accuracy of 2π suffices, since the variation is a multiple of 2π.)

8.1. Show that the set of functions $\{g\}$ considered in connection with the Riemann mapping theorem is not empty. (If α is not in D, Chapter 4, Example 1.2 gives a single-valued analytic logarithm in D,

$$L(z) = \log \frac{z-\alpha}{z_0 - \alpha}, \qquad L(z_0) = 0.$$

Taking exponentials shows $L(z_1) = L(z_2)$ only if $z_1 = z_2$, so L is simple. Similarly, if $L(z_n) + 2\pi i \to 0$ for some sequence $\{z_n\}$, then taking exponentials gives $z_n \to z_0$. But continuity of L shows that $L(z_n) \to L(z_0) = 0$ if $z_n \to z_0$. This contradiction shows that $L(z) + 2\pi i$ is bounded away from 0 in D and hence the function

$$h(z) = \frac{1}{L(z) + 2\pi i}$$

is bounded in D. Consideration of $Ah(z) + B$ yields the desired function g.)

8.2. If $g_1(z)$ is simple on the domain D of Problem 8.1, $g_1(z_0) = 0$, $g_1'(z_0) > 0$, $|g_1(z)| < 1$, and g_1 omits a value w_1, $|w_1| < 1$, show that there exists a function $g_2(z)$ with all the properties of $g_1(z)$ except the last and $g_2'(z_0) > g_1'(z_0)$. In other words, establish (3) for proof of Riemann's theorem. (Let

$$h_1(z) = \frac{g_1(z) - w_1}{1 - g_1(z)\overline{w}_1}, \qquad h_2(z) = \sqrt{h_1(z)}, \qquad h_3(z) = \frac{h_2(z) - h_2(z_0)}{1 - h_2(z)\overline{h_2(z_0)}}.$$

Show h_1 is simple in D, $|h_1| < 1$, and h_1 omits the value 0. Show h_2 is simple in D and $|h_2| < 1$, and likewise for h_3. Also, $h_3(z_0) = 0$,

$$h_3'(z_0) = \frac{g_1'(z_0)}{2\sqrt{-w_1}}(1 + |w_1|) = \frac{g_1'(z_0)}{\alpha}\frac{1 + t^2}{2t}$$

where $|\alpha| = 1$ and $t^2 = |w_1|$. Since $1 + t^2 > 2t$, $0 < t < 1$, the function $g_2 = \alpha h_3$ has the required properties.)

Note that in taking $\sqrt{h_1(z)}$ above, since $|h_1| < 1$, the image of D is dilated yet remains in the unit disk. The construction of g_2 from g_1 is of some practical interest because it gives a concrete way of getting a sequence of functions tending to the extremal function.

Chapter 6

Uniform convergence

If each term of a finite sum is analytic in a domain D, the same is true of the sum, and the derivative can be found by differentiating each term. The question arises as to conditions under which an infinite series of analytic functions is analytic, and whether term by term differentiation is justified for infinite series. Similar questions arise for integrals containing a parameter and for infinite products. Analysis of these questions depends on a type of convergence, called uniform convergence, which pertains not just to one point, but to a whole region. If z is restricted to a sufficiently small region, many calculations can be justified with ease, by uniform convergence. The results are then extended to larger regions by analytic continuation. Uniform convergence is applied here to the solution of differential equations, to the expansion of integrals as series, to series and product representation of analytic functions, and to the solution of the Dirichlet problem for a disk. The chapter concludes with a discussion of the monodromy theorem, which gives conditions under which analytic continuation leads to a single-valued function.

1. Convergence of sequences. For $n = 1,2,3,\ldots$ let $s_n(z)$ be a complex-valued function of z defined in some region G of the complex plane. Using these functions we can form an infinite sequence

$$s_1(z), \quad s_2(z), \quad s_3(z), \ldots, \quad s_n(z), \ldots.$$

The sequence is also denoted by $\{s_n(z)\}$, $n = 1,2,3,\ldots$, or just by $\{s_n(z)\}$ if the range of values of n is understood. The sequence is infinite because it has infinitely many terms. (In general, a sequence is a function $s(n,z)$ of the two variables n and z, where n ranges over some set of integers and z ranges over some set in the complex plane. The notation $s_n(z)$ is used because n and z have different roles.)

It is said that the above sequence converges to $s(z)$, and we write

$$\lim_{n\to\infty} s_n(z) = s(z) \quad \text{or} \quad s_n(z) \to s(z),$$

if the following is true: For each $\epsilon > 0$ and for each z there is an N, depending on ϵ and z, such that

$$n \geq N \quad \text{implies} \quad |s_n(z) - s(z)| \leq \epsilon. \tag{1.1}$$

The convergence is *uniform* if it is possible to choose N independently of z; that is, if one and the same N will do for all z in G. For convergence $N = N(\epsilon,z)$, but for uniform convergence N depends only on ϵ, thus $N = N(\epsilon)$. The same definitions are used when G is not a region but is any set of points in the complex plane. For instance, G could be the set of points occupied by a contour C. In this case, it is said that the sequence converges, or converges uniformly, on C.

A sequence of nonconstant functions can converge without converging uniformly. In illustration of this fact let $s_n(z) = 1/(nz)$ in the punctured disk $0 < |z| < 1$. The sequence $\{s_n(z)\}$ is

$$\frac{1}{z}, \frac{1}{2z}, \frac{1}{3z}, \ldots, \frac{1}{nz}, \ldots$$

and $s_n(z) \to 0$ as $n \to \infty$. With $s(z) = 0$ we have

$$|s_n(z) - s(z)| = \frac{1}{|z|n}.$$

Hence (1.1) holds if and only if $n \geq 1/(|z|\epsilon)$. Thus N can be any number satisfying $N \geq 1/(|z|\epsilon)$ and no smaller N will do. In this case the value of N necessarily depends on z; if ϵ and N are fixed, we could falsify the condition $N \geq 1/(|z|\epsilon)$ by choosing z so that $|z| < 1/(\epsilon N)$. The sequence converges for $0 < |z| < 1$, but not uniformly.

On the other hand, if the region is $\epsilon_0 < |z| < 1$, with $\epsilon_0 > 0$ fixed, the convergence would be uniform. This is so because we could choose $N = 1/(\epsilon_0\epsilon)$, which does not depend on z.

The above example of nonuniform convergence depended on the fact that z could be changed. The concept of uniform convergence is very essentially associated with a set of points or region in the complex plane.

If z is not allowed to vary, the distinction between convergence and uniform convergence disappears.

Several proofs in Chapters 3 and 4 were based on the idea of uniform convergence. (This was done, in part, so that the reader would be somewhat familiar with the concept when reaching this chapter.) In each of these cases, the analysis was based directly on the definition. It is a major objective here to obtain general theorems from which earlier results follow as special cases. The systematic development not only is more efficient, but leads to results of greater scope and power.

THEOREM 1.1. *Let $s_n(z)$ be continuous and let $s_n(z) \to s(z)$ uniformly in a region G. Then $s(z)$ is continuous and, if C is any contour lying in G,*

$$\lim_{n\to\infty} \int_C s_n(z)\, dz = \int_C s(z)\, dz.$$

Proof. Let z_0 and z be in G. Then

$$s(z) - s(z_0) = s_n(z) - s_n(z_0) + s_n(z_0) - s(z_0) + s(z) - s_n(z). \quad (1.2)$$

Given $\epsilon > 0$, we can choose N independent of z so that $n \geq N$ implies

$$|s_n(z_0) - s(z_0)| \leq \epsilon, \qquad |s(z) - s_n(z)| \leq \epsilon.$$

We chose $n = N$, which clearly satisfies $n \geq N$. Since $s_n(z)$ is continuous, we can then choose $\delta > 0$ so that

$$|s_n(z) - s_n(z_0)| < \epsilon \quad \text{for} \quad |z - z_0| < \delta.$$

Then (1.2) gives

$$|s(z) - s(z_0)| < 3\epsilon \quad \text{for} \quad |z - z_0| < \delta$$

and $s(z)$ is continuous at z_0. Since z_0 was any point of G, $s(z)$ is continuous in G.

The fact that $s(z)$ is continuous in G shows that $s(z)$ is integrable. We shall establish the second assertion of Theorem 1.1 in the following strengthened form: Given $\epsilon > 0$, there is an $N = N(\epsilon, L)$ independent of C such that $n > N$ implies

$$\left| \int_C s_n(z)\, dz - \int_C s(z)\, dz \right| < \epsilon \quad (1.3)$$

for all contours C in G having length $\leq L$. It is said, briefly, that the convergence is uniform with respect to C for contours of the specified class. To prove this, choose N so that

$$|s_n(z) - s(z)| < \epsilon/L$$

for $n > N$ and for all z in G. Then the left side of (1.3) does not exceed $(\epsilon/L)L$, and the result follows.

THEOREM 1.2. *Let $s_n(z)$ be analytic in a disk $|z - z_0| < R$ and let $s_n(z) \to s(z)$ uniformly in each disk $|z - z_0| \leq R - \delta$, where $\delta > 0$ is constant. Then $s(z)$ is analytic for $|z - z_0| < R$, and $s_n'(z) \to s'(z)$ uniformly in each disk $|z - z_0| \leq R - \delta$.*

This is a remarkable result. It is not at all the case that a uniformly convergent sequence of real differentiable functions must have a convergent sequence of derivatives, as shown by the example

$$s_n(x) = \frac{\sin nx}{n}.$$

Here $s_n'(x) = \cos nx$ and the limit does not exist for any $x \neq 2k\pi$. Nevertheless, $\{s_n(x)\}$ converges uniformly for $-\infty < x < \infty$.

Proof. Let C be any triangular contour lying in $|z - z_0| < R$ and choose $\delta > 0$ so small that C is in the disk $|z - z_0| \leq R - \delta$. Then

$$\int_C s(z)\,dz = \lim_{n\to\infty} \int_C s_n(z)\,dz = 0$$

where the first equality follows from Theorem 1.1 and the second from Cauchy's theorem. By Morera's theorem, $s(z)$ is analytic in $|z - z_0| < R$.

Next, let C be any circle $|z - z_0| = R - \delta$ where $\delta > 0$. If z is in C, Cauchy's formula gives

$$s'(z) - s_n'(z) = \frac{1}{2\pi i} \int_C [s(\zeta) - s_n(\zeta)] \frac{d\zeta}{(\zeta - z)^2}. \tag{1.4}$$

For any $\epsilon > 0$ we can choose N so that, in (1.4),

$$n \geq N \quad \text{implies} \quad |s(\zeta) - s_n(\zeta)| \leq \frac{\epsilon\delta^2}{R}.$$

If z is not only within C but satisfies the stronger condition

$$|z - z_0| \leq R - 2\delta, \tag{1.5}$$

then $|\zeta - z|^2 \geq \delta^2$ in (1.4), and hence

$$|s'(z) - s_n'(z)| \leq \frac{1}{2\pi} 2\pi R \frac{\epsilon\delta^2}{R} \frac{1}{\delta^2} = \epsilon.$$

This establishes uniform convergence of $s_n'(z)$ to $s'(z)$ in the disk (1.5). Since δ is arbitrary, Theorem 1.2 follows.

THEOREM 1.3. *Let $s_n(z)$ be analytic in a domain D and let $s_n(z) \to s(z)$ uniformly in D. Then for z in D*

$$s_n'(z) \to s'(z), \quad s_n''(z) \to s''(z), \quad s_n'''(z) \to s'''(z),$$

and so on. (Note that D is arbitrary.)

Proof. Let z_0 be any point of D and let $|z - z_0| < R$ be a disk contained in D. Then $s_n(z)$ satisfies the hypothesis of Theorem 1.2 in this disk. The conclusion of Theorem 1.2 is that $s_n'(z)$ satisfies the hypothesis of Theorem 1.2 in this disk. Applying Theorem 1.2 again, we see that $s_n''(z)$ satisfies the hypothesis of Theorem 1.2 in the disk, and so on. Theorem 1.2 thus gives the conclusion of Theorem 1.3 at the point $z = z_0$. Since z_0 was an arbitrary point of D, this completes the proof.

The foregoing results assume that the limit $s(z)$ is already given. We now establish a convergence criterion that can be used when no function $s(z)$ such that $s_n(z) \to s(z)$ is at hand.

A sequence of complex constants s_n forms a Cauchy sequence if the following is true: For every $\epsilon > 0$ there is an N depending on ϵ such that

$$n \geq N \text{ and } m \geq N \quad \text{implies} \quad |s_n - s_m| \leq \epsilon.$$

The same definition is used for a sequence of functions $s_n(z)$ except that N can depend on z as well as on ϵ. If N can be chosen independently of z, for z in a region G, the sequence $\{s_n(z)\}$ is said to be a uniform Cauchy sequence on G.

For example, suppose $s_n(z) \to s(z)$ on G and let $\epsilon > 0$ be given. In this case we can pick $N = N(\epsilon,z)$ so that

$$|s_n(z) - s(z)| \leq \tfrac{1}{2}\epsilon, \quad |s(z) - s_m(z)| \leq \tfrac{1}{2}\epsilon$$

for $n \geq N$ and $m \geq N$, respectively. The above implies

$$|s_n(z) - s_m(z)| \leq \epsilon$$

and so $\{s_n(z)\}$ is a Cauchy sequence. Furthermore, if $s_n(z) \to s(z)$ uniformly on G, then we can pick $N = N(\epsilon)$ independently of z, and the sequence is a uniform Cauchy sequence.

A converse will now be proved. The converse depends on a result of real analysis which asserts that every real Cauchy sequence has a limit. In some developments of real analysis this statement is accepted as an axiom, and in others it is proved as a theorem. In any case, it is assumed here.

THEOREM 1.4. *If $\{s_n(z)\}$ is a Cauchy sequence in a region G, then there exists a function $s(z)$ such that $s_n(z) \to s(z)$ on G. Furthermore, if $\{s_n(z)\}$ is a uniform Cauchy sequence on G, then $s_n(z) \to s(z)$ uniformly on G.*

Proof. Let $s_n(z) = u_n(z) + i\,v_n(z)$. If z is a point of G, then

$$|u_n(z) - u_m(z)| \leq |s_n(z) - s_m(z)|$$

and so $\{u_n(z)\}$ is a real Cauchy sequence. Hence $\lim u_n(z) = u(z)$ exists. A similar result holds for $v_n(z)$ and existence of $\lim s_n(z) = s(z)$ follows with $s(z) = u(z) + i\,v(z)$. To establish uniformity of the convergence, let $\epsilon > 0$ be given and choose $N = N(\epsilon)$ so that

$$|s_n(z) - s_m(z)| \leq \epsilon$$

for $n \geq N$, $m \geq N$, and for all z in G. Letting $m \to \infty$ gives

$$|s_n(z) - s(z)| \leq \epsilon$$

for $n \geq N$, which is the condition for uniform convergence.

Example 1.1. If R is fixed and $0 < R < 1$, show that

$$\lim_{n\to\infty} z^n = 0 \qquad \text{uniformly for } 0 \leq |z| \leq R.$$

Since $1/R > 1$, we have $1/R = 1 + \delta$ where $\delta > 0$. Then

$$|z| \leq \frac{1}{1+\delta}, \qquad |z|^n \leq \frac{1}{1+n\delta}$$

where the first equation follows from $|z| \leq R$ and the second follows from the binomial theorem (Chapter 1, Example 1.6). The desired conclusion is obvious from this. If, however, we want to prove the conclusion, we reason as follows: Given $\epsilon > 0$ let $N = 1/(\epsilon\delta)$, which is independent of z. Then $n \geq N$ implies $n\delta \geq 1/\epsilon$, and by the above

$$|z|^n \leq \frac{1}{1+(1/\epsilon)} = \frac{\epsilon}{1+\epsilon} < \epsilon.$$

Example 1.2. A sequence $\{s_n(z)\}$ is defined by

$$s_n(z) = 1 + z + z^2 + \cdots + z^n.$$

Show that $s_n(z) \to (1-z)^{-1}$ for $|z| < 1$ and that the convergence is uniform for $|z| \leq R$, if $R < 1$ is constant.

By Chapter 1, Example 1.5,

$$s_n(z) = \frac{1}{1-z} - \frac{z}{1-z} z^n.$$

Hence, both statements follow from Example 1.1 above. As explained in Chapter 3, Section 6, the conclusion is written

$$\sum_{n=0}^{\infty} z^n = \frac{1}{1-z}, \qquad |z| < 1, \quad (1.6)$$

and it is said that (1.6) holds uniformly for $|z| \leq R$.

The series (1.6) is called a geometric series, as in real analysis. We have shown that familiar results of real analysis for geometric series continue to hold for complex values of z. Many other real analysis results for sequences and series extend with equal ease from the real to the complex case, and are not emphasized here.

Problems

1. (a) Sketch the graph of $s(z) = \lim z^n$ for $0 \leq z \leq 1$ and deduce from Theorem 1.1 that the convergence is not uniform on this interval. (b) Show also that the convergence is not uniform on $0 \leq z < 1$.

2. Let $s_n(z) \to s(z)$ and $t_n(z) \to t(z)$ uniformly on a region G and let $f(z)$ be bounded on G; that is, $|f(z)| <$ constant. Show that

$$s_n(z) \to s(z), \quad s_n(z) + t_n(z) \to s(z) + t(z), \quad s_n(z)f(z) \to s(z)f(z)$$

uniformly on G. Show also that the hypothesis that f is bounded is essential. (Consider $f(z) = 1/z$, $s_n(z) = 1/n$.)

3. Suppose $s_n(z) \to s(z)$ and $t_n(z) \to t(z)$ uniformly in a region G, and suppose $s(z)$ and $t(z)$ are bounded in G. Prove that

$$s_n(z)t_n(z) \to s(z)t(z)$$

uniformly in G. (Consider $s_n(t_n - t) + t(s_n - s)$.)

4. For $p > 0$ a sequence $\{s_n\}$ is defined by

$$s_n = 1 + \frac{1}{2^p} + \frac{1}{3^p} + \cdots + \frac{1}{n^p}.$$

Referring to Figure 1-1, show for $n > m \geq 2$ that

$$\int_m^{n+1} \frac{dx}{x^p} < s_n - s_m < \int_{m-1}^{n} \frac{dx}{x^p}.$$

Conclude that $\{s_n\}$ converges if and only if $p > 1$.

A review of series

Here $\{a_j\}$ and $\{b_j\}$ are sequences of complex constants, $j \geq m$.

5. As in real analysis, if $n \geq m$,

$$\sum_{j=m}^{n} a_j = a_m + a_{m+1} + \cdots + a_n.$$

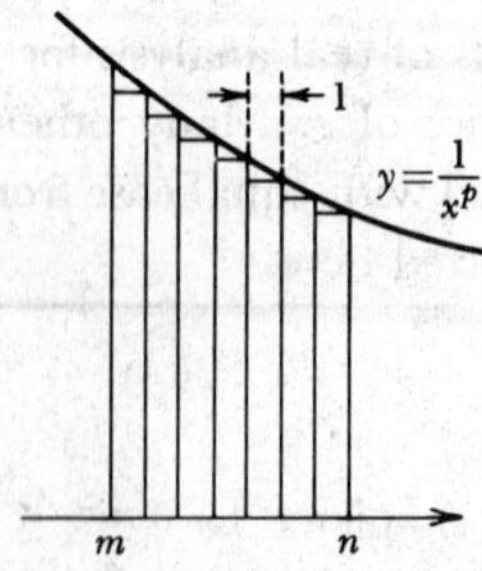

Figure 1-1

Show that this operation is linear, in the sense that

$$\sum_{j=m}^{n} (\alpha a_j + \beta b_j) = \alpha \sum_{j=m}^{n} a_j + \beta \sum_{j=m}^{n} b_j \qquad (*)$$

for complex constants α and β.

6. As in real analysis, the value for $n = \infty$ in the preceding problem is obtained by taking limits as $n \to \infty$. Thus, if the limit exists,

$$\sum_{j=m}^{\infty} a_j = \lim_{n\to\infty} \sum_{j=m}^{n} a_j.$$

The infinite series on the left is then said to be convergent. If the two series on the right are convergent when $n = \infty$, extend the result (*) to the case $n = \infty$.

7. By choosing $\beta = 0$ in (*), or $\alpha = 1, \beta = 1$, or $\alpha = 1, \beta = -1$, deduce that a convergent series can be multiplied termwise by any constant, two convergent series can be added term by term, and two convergent series can be subtracted term by term.

8. Let $a_i(z)$ be terms of an infinite series defined in a region G. If finitely many of the $a_i(z)$ are replaced by other functions $b_i(z)$, also defined in G, prove that the convergence is not affected. (Of course the value of the sum may be affected.)

2. Convergence of series. An infinite sequence of function is equivalent to the sequence of partial sums of an infinite series. If

$$\sum_{j=1}^{\infty} a_j(z) = a_1(z) + a_2(z) + \cdots + a_n(z) + \cdots$$

is an infinite series, the nth partial sum is, by definition,

$$s_n(z) = a_1(z) + a_2(z) + \cdots + a_n(z). \qquad (2.1)$$

The sequence of partial sums $\{s_n(z)\}$ is therefore

$$a_1(z),\ a_1(z) + a_2(z),\ a_1(z) + a_2(z) + a_3(z), \ldots$$

and is determined if the sequence of terms $\{a_n(z)\}$ is known. Conversely, given the sequence of partial sums, we can find the sequence of terms by

$$a_1(z) = s_1(z), \qquad a_j(z) = s_j(z) - s_{j-1}(z) \qquad (2.2)$$

for $j \geq 2$. Because of this correspondence, no logical distinction is made between the series and the sequence of partial sums which it determines.

There is a substantial formal theory of series which does not require that the series have any number attached to it as the sum. However, the series commonly used in analysis do have a sum, which is calculated by considering the limit of the sequence $\{s_n(z)\}$. If the partial sums satisfy $s_n(z) \to s(z)$ on a region G, it is said that the series converges to the sum $s(z)$ on G and we write

$$\sum_{j=1}^{\infty} a_j(z) = s(z). \tag{2.3}$$

When the limit fails to exist at a point z, the series diverges at that point, and no value $s(z)$ is assigned to it. If $s_n(z) \to s(z)$ uniformly in G, then it is said that the series converges uniformly in G. Of course, similar conventions apply when G is any point set, not necessarily a region, and when the sum starts at some other number than $j = 1$.

If the summation range is $(-\infty,\infty)$ instead of $(0,\infty)$ or $(1,\infty)$, the definitions for series are analogous to those used for integrals in Chapter 4, Section 4. Thus,

$$\sum_{n=-\infty}^{\infty} a_n = a_0 + \lim_{N_1\to\infty} \sum_{n=1}^{N_1} a_n + \lim_{N_2\to\infty} \sum_{n=-N_2}^{-1} a_n$$

provided the two limits on the right exist. The series is then said to be convergent. The Cauchy principal value is

$$P\sum_{n=-\infty}^{\infty} a_n = \lim_{n\to\infty} \sum_{n=-N}^{N} a_n$$

and can exist when the former limits do not. However, if a_n is an even function of n, convergence and Cauchy convergence are equivalent.

Taking the limit as $j \to \infty$ in (2.2) gives

$$\lim_{j\to\infty} a_j(z) = \lim_{j\to\infty} s_j(z) - \lim_{j\to\infty} s_{j-1}(z) = s(z) - s(z) = 0$$

if the series converges to $s(z)$ at z, and hence the general term of a convergent series must tend to 0.

From Theorem 1.3 it follows that if each term $a_j(z)$ of a series is an analytic function in a domain D, and if the series (2.3) is uniformly convergent in D, then the sum $s(z)$ is analytic in D, and the series can be differentiated term by term to yield

$$\sum_{j=0}^{\infty} a_j'(z) = s'(z), \qquad \sum_{j=0}^{\infty} a_j''(z) = s''(z), \tag{2.4}$$

and so on, in D. Also, if C is any contour in D, the series can be integrated term by term to give

$$\int_C s(z)\, dz = \sum_{n=1}^{\infty} \int_C a_j(z)\, dz. \tag{2.5}$$

This follows from Theorem 1.1. The result for integration is less deep than that for differentiation, and applies if the terms are only continuous rather than analytic.

In order to use these results it is necessary to know when a series converges uniformly. A useful sufficient condition is given by the Weierstrass M test, which reads as follows:

THEOREM 2.1 (*Weierstrass*). *Suppose* $|a_j(z)| \le M_j$ *in a region* G, *where the* M_j *are constant and*[1]

$$\sum_{j=1}^{\infty} M_j < \infty. \tag{2.6}$$

Then the series (2.3) *converges uniformly in* G.

Proof. Clearly, for $n > m$,

$$|s_n(z) - s_m(z)| \le \sum_{m+1}^{n} |a_j(z)| \le \sum_{m+1}^{n} M_j \le \sum_{m+1}^{\infty} M_j.$$

Because of (2.6), the right side can be made as small as desired by taking m large enough. Hence $\{s_n(z)\}$ is a uniform Cauchy sequence on G, and Theorem 2.1 follows from Theorem 1.4.

In using the Weierstrass test, it is important to be familiar with some common convergent series. By Example 1.2 and by Section 1, Problem 4,

$$\sum_{n=0}^{\infty} r^n < \infty, \quad 0 \le r < 1; \qquad \sum_{n=1}^{\infty} \frac{1}{n^p} < \infty, \quad p > 1.$$

[1] Equation (2.6) means that the series converges. We use this notation only for series with nonnegative terms and integrals with nonnegative integrands.

Knowledge of these two series suffices for the needs of this book.

Example 2.1. One of the most challenging and important functions of mathematics is the Riemann zeta function, which is defined by

$$\zeta(z) = \sum_{n=1}^{\infty} \frac{1}{n^z}, \qquad \text{Re}\, z > 1.$$

Show that $\zeta(z)$ is analytic for $\text{Re}\, z > 1$.

Each term $n^{-z} = \exp(-z \,\text{Log}\, n)$ is an entire function. Also,

$$|n^{-z}| = |e^{-(x+iy)\,\text{Log}\, n}| = |e^{-x \,\text{Log}\, n} e^{-iy \,\text{Log}\, n}| = n^{-x}.$$

If $x \geq p > 1$, then $|n^{-z}| \leq n^{-p}$, and the Weierstrass test shows that the series converges uniformly. Therefore it represents an analytic function in $x \geq p$ for every $p > 1$, and the result follows.

Historically the zeta function is written with $s = \sigma + it$ in place of z and is designated by

$$\zeta(s) = \sum_{n=1}^{\infty} \frac{1}{n^s}, \qquad \text{Re}\, s > 1.$$

We have shown that $\zeta(s)$ is analytic for $\text{Re}\, s > 1$. Actually $\zeta(s)$ can be continued analytically for $\text{Re}\, s \leq 1$, and it is found that the function so obtained has, as sole finite singularity, a simple pole at $s = 1$.

THEOREM 2.2 (*Ratio test*). *In a region G let $a_j(z)$ be defined for $j \geq 1$, let $|a_N(z)|$ be bounded for some fixed $N \geq 1$, and let*

$$\left|\frac{a_{n+1}(z)}{a_n(z)}\right| \leq R < 1, \qquad n \geq N,$$

where R is constant. Then the series (2.3) converges uniformly in G.

Proof. Since the first few terms of an infinite series do not affect the convergence, it suffices to prove the theorem for $N = 1$. Clearly

$$a_n(z) = a_1(z) \frac{a_2(z)}{a_1(z)} \frac{a_3(z)}{a_2(z)} \cdots \frac{a_n(z)}{a_{n-1}(z)}.$$

Taking absolute values gives $|a_n(z)| \le MR^{n-1}$, where M is a bound for $|a_1(z)|$ in G, and the result follows from the M test.

The following special case of Theorem 2.2 is not a test for uniform convergence, but is often useful. At a given point $z = z_0$, suppose

$$\lim_{n\to\infty} \left| \frac{a_{n+1}(z)}{a_n(z)} \right| = R_0. \tag{2.7}$$

Then the series (2.3) converges at z_0 if $R_0 < 1$ and diverges if $R_0 > 1$. Indeed, if $R_0 < 1$, we can pick R so that $R_0 < R < 1$. It is easily checked that the criterion of Theorem 2.2 holds at z_0 for sufficiently large N, and so the series converges. On the other hand, if $R_0 > 1$, the general term does not tend to 0, and the series diverges.

Example 2.2. Show that the function

$$f(z) = \frac{1}{z} + \sum_{n=1}^{\infty} \frac{2z}{z^2 - n^2} \tag{2.8}$$

is meromorphic in the z plane and has, as its only finite singularities, simple poles of residue 1 at the integers.

We shall prove that the function has these properties in $|z| < R$ for every R. Let $R > 0$ be given and choose an integer $N > 2R$. Then $f(z) = J_1(z) + J_2(z)$ where

$$J_1(z) = \frac{1}{z} + \sum_{n=1}^{N} \frac{2z}{z^2 - n^2}, \qquad J_2(z) = \sum_{n=N+1}^{\infty} \frac{2z}{z^2 - n^2}.$$

Clearly, $J_1(z)$ is analytic in $|z| < R$ except for simple poles with residue 1 at the integers. For $|z| < R$, the terms of the series $J_2(z)$ satisfy

$$\left| \frac{2z}{z^2 - n^2} \right| \le \frac{2R}{n^2 - R^2} = \frac{1}{n^2} \frac{2R}{1 - (R/n)^2}.$$

The factor $1 - (R/n)^2$ in the denominator is least when $n = 2R$, and its value then is ¾. Therefore

$$\left| \frac{2z}{z^2 - n^2} \right| \le \frac{8R}{3n^2}, \qquad (n \ge 2R)$$

and the series for $J_2(z)$ converges uniformly in $|z| < R$ by the M test. This shows that $J_2(z)$ is analytic in $|z| < R$ and completes the proof.

The function $\pi \cot \pi z$ has the same properties as the function $f(z)$ considered here and it is now seen that $f(z) = \pi \cot \pi z$. The series is called the Mittag-Leffler expansion of $\pi \cot \pi z$

Example 2.3. If α is not an integer, show that

$$\sum_{-\infty}^{\infty} \frac{1}{\alpha^2 - n^2} = \pi \frac{\cot \pi\alpha}{\alpha}. \tag{2.9}$$

As suggested by Chapter 4, Section 6, consider

$$I = \frac{1}{2\pi i} \int_C \frac{\pi \cot \pi z}{\alpha^2 - z^2}\, dz$$

where C is a simple closed contour. The integrand can be written

$$\frac{\pi \cos \pi z}{\sin \pi z} \frac{1}{\alpha^2 - z^2}$$

and by Chapter 4, Equation (2.4), the residue at $z = n$, an integer, is

$$\frac{\pi \cos \pi z}{\pi \cos \pi z} \frac{1}{\alpha^2 - z^2}\bigg|_{z=n} = \frac{1}{\alpha^2 - n^2}.$$

The corresponding contribution to the integral is the sum of terms of the series (2.9) for which n is within the contour C.

For C we use the oriented boundary of the rectangular domain

$$-(N + ½) < x < N + ½, \qquad -N < y < N$$

where $N \geq 2$ is an integer (see Figure 2-1). If $N > |\alpha|$, the residue theorem gives

$$I = \sum_{n=-N}^{N} \frac{1}{\alpha^2 - n^2} - \pi \frac{\cot \pi\alpha}{\alpha}$$

where the last term arises from the residues at $-\alpha$ and α. In Problem 6 it is shown that

$$|\cot \pi z| \leq \coth \pi, \qquad z \text{ on } C. \tag{2.10}$$

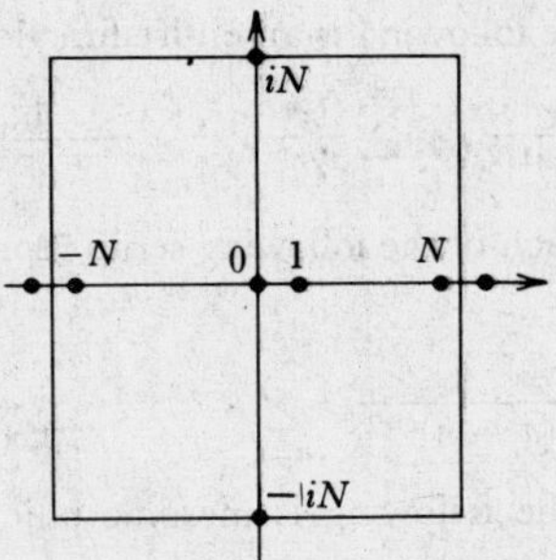

Figure 2-1

Since the length of C is $4N + 1$, and since $|z| \geq N$ on C, it follows that

$$|I| \leq \frac{4N+1}{2\pi} \frac{\pi \coth \pi}{N^2 - |\alpha|^2}, \qquad N \geq 2, \quad N > |\alpha|.$$

This tends to 0 as $N \to \infty$ and gives the desired result as a Cauchy principal value. The series converges, since n^2 is even.

If we set $\alpha = z$ in (2.9) and multiply by z, the resulting series agrees with (2.8). Hence, $f(z) = \pi \cot \pi z$.

Problems

1. If a is any real constant and n^a has its usual real-variable meaning, show that
$$\sum_{n=1}^{\infty} \frac{z^{2n}}{4^n n^a}$$
converges for $|z| < 2$ and diverges for $|z| > 2$. Solution: If $z \neq 0$, the ratio of the $n + 1$th to the nth term satisfies
$$\left|\frac{a_{n+1}(z)}{a_n(z)}\right| = \left|\frac{z^{2n+2}}{4^{n+1}(n+1)^a} \frac{4^n n^a}{z^{2n}}\right| = \frac{|z|^2}{4} \left|\frac{n}{n+1}\right|^a.$$
As $n \to \infty$ the limit is $|z|^2/4$, and the conclusion follows from (2.7).
2. For each of the following find a number R such that the series converges for $|z| < R$ and diverges for $|z| > R$:
$$\sum_{n=0}^{\infty} 2^n z^n, \qquad \sum_{n=0}^{\infty} n^2 z^n, \qquad \sum_{n=1}^{\infty} \frac{2^n z^{2n}}{n^2+n}, \qquad \sum_{n=0}^{\infty} \frac{3^n z^n}{4^n+5^n}, \qquad \sum_{n=1}^{\infty} \frac{z^n n!}{n^n}.$$

3. Show that each of the following is an entire function of z:

$$\sum_{n=1}^{\infty} \frac{z^n}{(n!)^{1/2}}, \quad \sum_{n=1}^{\infty} \frac{z^n}{2^{n^2}}, \quad \sum_{n=1}^{\infty} \frac{1}{2^n}\frac{1}{n^z}, \quad \sum_{n=1}^{\infty} \frac{\sin nz}{n!}.$$

4. If $a > 0$, show that each of the following series represents an analytic function in $\operatorname{Re} z > 0$:

$$\sum_{n=1}^{\infty} \frac{1}{(a+n)^{z+1}}, \quad \sum_{n=1}^{\infty} e^{-n^2 az}, \quad \sum_{n=1}^{\infty} \frac{e^{-anz}}{(a+n)^z}.$$

5. Show that each of the following is meromorphic in $|z| < \infty$ and find the residues at the poles:

$$\sum_{n=0}^{\infty} \frac{(-1)^n}{n!(n+z)}, \quad \sum_{n=1}^{\infty} \frac{1}{z^2+n^2}, \quad \sum_{n=1}^{\infty} \frac{1}{(z+n)^2}, \quad \sum_{n=0}^{\infty} \frac{\sin nz}{n!(z^2+n^2)}.$$

Summation of series

6. In Figure 2-1, on the vertical sides, $\pi x = \pi m + \pi/2$, where m is an integer, and hence

$$|\cot \pi(x+iy)| = |\tan \pi iy| = |\tanh \pi y|.$$

Since $|e^{\pi i(x+iy)}| = e^{-\pi y}$ and $|e^{\pi i(x-iy)}| = e^{\pi y}$, show also, for all $y \neq 0$, that

$$|\cot \pi(x+iy)| \leq |\coth \pi y|.$$

Show that $|\tanh \pi y| \leq 1$ and that $|\coth \pi y|$ decreases as $|y|$ increases, hence attains its maximum for $|y| \geq 1$ at $y = 1$. Thus get (2.10).

7. Except at the poles, show that

$$\sum_{n=-\infty}^{\infty} \frac{\alpha}{n^2+\alpha^2} = \frac{\pi}{\tanh \pi\alpha}, \quad \sum_{n=-\infty}^{\infty} \frac{1}{(n+\alpha)^2} = \left(\frac{\pi}{\sin \pi\alpha}\right)^2.$$

8. Rewrite the first series in Problem 7 as $1/\alpha$ plus a sum from 1 to ∞ by combining terms for n and $-n$. Thus evaluate the second series in Problem 5 and, setting $z = 0$, get

$$\frac{1}{1} + \frac{1}{4} + \frac{1}{9} + \frac{1}{16} + \frac{1}{25} + \cdots = \frac{\pi^2}{6}.$$

9. Show that $|\pi \csc \pi z|$ is bounded on the contour of Figure 2-1 and discuss the relation of

$$\int_C f(z)\pi \csc \pi z \, dz \quad \text{and} \quad P\sum_{n=-\infty}^{\infty} (-1)^n f(n).$$

10. Referring to Problem 9, show for $\alpha \neq$ integer and $-\pi < \beta < \pi$ that

$$\sum_{n=-\infty}^{\infty} \frac{(-1)^n}{(\alpha+n)^2} = \pi^2 \frac{\cot \pi\alpha}{\sin \pi\alpha}, \quad \sum_{n=-\infty}^{\infty} \frac{(-1)^n n \sin n\beta}{\alpha^2 - n^2} = \frac{\pi \sin \alpha\beta}{2 \sin \alpha\pi}.$$

3. Power series. If α and a_n are complex constants, the series

$$f(z) = \sum_{n=0}^{\infty} a_n(z - \alpha)^n \tag{3.1}$$

is a power series expanded about the point $z = \alpha$. By using $z - \alpha$ as a new variable, it is always possible to replace such a series by a series of the form

$$\sum_{n=0}^{\infty} a_n z^n \tag{3.2}$$

in which the expansion is about the origin. This is done in the proofs below.

THEOREM 3.1. *If the series (3.1) converges at some value $z_0 \neq \alpha$, then it converges for all z in the disk $|z - \alpha| < |z_0 - \alpha|$ and it converges uniformly in $|z - \alpha| \leq r$ for each $r < |z_0 - \alpha|$.*

Proof. Let $\alpha = 0$. Convergence of (3.2) at z_0 shows that the general term tends to 0 and hence $|a_n z_0^n| < 1$ for all large n; say, for $n > N$. Then for $n > N$

$$|a_n z^n| = |a_n z_0^n| \left|\frac{z}{z_0}\right|^n \leq \left|\frac{z}{z_0}\right|^n. \tag{3.3}$$

If $|z/z_0| \leq R < 1$, we can use $M_n = R^n$ in the Weierstrass M test, and both results follow from this.

THEOREM 3.2. *Let $R > 0$ and let $|b_n| \leq |a_n|$ for all large n. Then if the first of the following series converges for all z satisfying $|z - \alpha| < R$, the same is true of the second:*

$$\sum_{n=0}^{\infty} a_n(z - \alpha)^n, \qquad \sum_{n=0}^{\infty} b_n(z - \alpha)^n. \tag{3.4}$$

Proof. Let $\alpha = 0$ and let the first series converge at $z_0 \neq 0$. Then (3.3) leads to an estimate of the same sort for $|b_n z^n|$, under the hypothesis $|b_n| \leq |a_n|$, and so the second series converges for $|z| < |z_0|$. Since z_0 can be any number satisfying $|z_0| < R$, this completes the proof.

THEOREM 3.3. *A power series (3.1) must be of one of the following three types:*

(1) *The series converges for $z = \alpha$ only.*

(2) *The series converges for all $|z| < \infty$.*

(3) *There exists an $R > 0$ such that the series converges for $|z - \alpha| < R$ and diverges for $|z - \alpha| > R$.*

The number R is called the radius of convergence. The cases (1) and (2) are sometimes designated by $R = 0$ and $R = \infty$, respectively. If $|a_n| = |b_n|$, Theorem 3.2 shows that the two series (3.4) have the same radius of convergence. Hence, R depends only on the sequence of absolute values $\{|a_n|\}$ and not on $\{\arg a_n\}$.

Proof. Without loss of generality let $\alpha = 0$ so that the series is given by (3.2). If cases (1) and (2) do not occur, then the series converges at some value $z_0 \neq 0$ and diverges at some other value z_1. Let S be the set of all numbers ρ such that the series converges for $|z| < \rho$. Clearly S contains no numbers larger than $|z_1|$, and so S is bounded above. On the other hand $|z_0|$ is in S, and so S is not empty. (This follows because convergence at z_0 implies convergence for $|z| < |z_0|$.) By real analysis[1] the set S has a least upper bound R. It is easily checked that the series diverges for $|z| > R$ and converges for $|z| < R$. Indeed, if the series converges at some number z_2, with $|z_2| > R$, then it would have to converge for $|z| < |z_2|$, by Theorem 3.1, and R would not be an upper bound for the set S. On the other hand, if $|z_3| < R$, then by the definition of least upper bound we can find a number ρ in S such that $|z_3| < \rho < R$. This shows that the series converges at z_3 and completes the proof.

If a power series is differentiated term by term, the differentiated series has the same radius of convergence as the original series. To see this, let (3.2) converge at $z \neq 0$. Then (3.3) gives

$$|na_n z^{n-1}| \leq nR^{n-1}$$

where $R = |z/z_0| < 1$. The series with $M_n = nR^{n-1}$ converges by the ratio test, and so the radius of convergence is not diminished by the differentiation. Theorem 3.2 shows that the radius is not increased either.

The following theorem asserts that the differentiated series not only converges, but gives the derivative of the original function:

[1] Chapter 3, Section 7.

THEOREM 3.4. *If the power series (3.1) has the radius of convergence $R > 0$, then it represents an analytic function in $|z - \alpha| < R$ and $f'(z), f''(z), \ldots$ can be calculated by differentiating term by term.*

Proof. Let $\alpha = 0$. If $|z_1| < R$, the series converges at z_1 and so it converges uniformly in $|z| < |z_0|$ for each $|z_0| < |z_1|$. The terms $a_j(z) = a_j z^j$ are, of course, analytic for all z, and so the term by term differentiation is justified by (2.4). Since $|z_0|$ is as close as desired to R, this completes the proof.

From Theorem 3.4 it follows that every power series with nonzero radius of convergence is a Taylor series, namely, the Taylor series of its sum. Indeed, differentiating (3.1) n times gives

$$f^{(n)}(z) = n!a_n + \cdots$$

where the terms not written involve $z - \alpha$ or higher powers. If $z = \alpha$, these terms drop out and hence

$$a_n = \frac{f^{(n)}(\alpha)}{n!}. \tag{3.5}$$

Equation (3.5) shows that a power series is uniquely determined by its sum in any disk $|z - \alpha| < \epsilon$. That is, if $f(z)$ in (3.1) satisfies

$$f(z) = \sum_{n=0}^{\infty} b_n(z - \alpha)^n$$

for $|z - \alpha| < \epsilon$, then $b_n = f^{(n)}(\alpha)/n!$ and so $b_n = a_n$. Another consequence of (3.5) is that a power series converges "out to the nearest singularity" in a sense made precise by the following theorem:

THEOREM 3.5. *Let $f(z)$ be represented by a power series (3.1) with radius of convergence $R > 0$. Then if $R_0 > R$ it is impossible for $f(z)$ to be analytic in the disk $|z - \alpha| < R_0$.*

Proof. If $f(z)$ is analytic for $|z - \alpha| < R_0$, then Chapter 3, Section 6 shows that the Taylor series for f converges for $|z - \alpha| < R_0$. However, the Taylor series coincides with the series (3.1), as we have just seen, and so (3.1) would also have to converge for $|z - \alpha| < R_0$. This contradicts the hypothesis $R < R_0$.

Theorem 3.5 indicates that, given an analytic function f, it is possible to get the radius of convergence of its Taylor series about α by determining the largest circle with center α inside of which f is analytic. For example, the series

$$\frac{1}{1+x^2} = 1 - x^2 + x^4 - x^6 + \cdots$$

diverges for $|x| \geq 1$, because the general term does not tend to 0. If $1/(1+x^2)$ is regarded as a function of the real variable x, there seems to be no reason why the series should diverge, for $1/(1+x^2)$ has derivatives of all orders at every real value of x. But when we regard x as a complex variable, the divergence is explained by the fact that the denominator vanishes at $\pm i$. Clearly, if the circle of convergence cannot contain the points $\pm i$, then the radius of convergence cannot exceed 1. In fact $R = 1$ by the ratio test, and this is the distance from $\alpha = 0$ to the points $\pm i$.

In Problem 3.7 another proof of Theorem 3.4 will be given, which is independent of the theory of integration. This makes possible an alternative approach to complex analysis, initiated in part by Lagrange, but due in the main to Weierstrass. In the Weierstrass approach the power series is regarded as fundamental, and the main use of the Cauchy theory is to show that analytic functions do, in fact, have power series expansions.

For example, if $f(z)$ is analytic for $|z - \alpha| < R_0$ and if C is the circle $|\zeta - \alpha| = R < R_0$, Cauchy's integral formula is

$$f(z) = \frac{1}{2\pi i}\int_C \frac{f(\zeta)}{\zeta - z}\,d\zeta, \qquad |z - \alpha| < R.$$

By expanding $(\zeta - z)^{-1}$ in geometric series it was seen in Chapter 3, Section 6 that (3.1) holds with

$$a_n = \frac{1}{2\pi i}\int_C \frac{f(\zeta)}{(\zeta - \alpha)^{n+1}}\,d\zeta.$$

From Theorem 3.4 it follows without further discussion that $f'(z), f''(z)$, and so on, are analytic, and the fact that the above formula must agree with (3.5) gives also

$$f^{(n)}(\alpha) = \frac{n!}{2\pi i}\int_C \frac{f(\zeta)}{(\zeta - \alpha)^{n+1}}\,d\zeta.$$

Further examples of the power-series approach are given in Section 4.

Example 3.1. If $f(t)$ is piecewise continuous for $0 \leq t \leq a$, show that

$$F(z) = \int_0^a e^{-zt} f(t)\, dt \tag{3.6}$$

is an entire function of z and find its power series expansion.

For any fixed value of z the series

$$e^{-zt} = 1 - zt + \frac{(-zt)^2}{2!} + \cdots + \frac{(-zt)^n}{n!} + \cdots$$

converges uniformly in $0 \leq t \leq a$. Hence, it can be multiplied by the bounded function $f(t)$ and integrated term by term. The result is

$$F(z) = \int_0^a f(t)\, dt - z \int_0^a tf(t)\, dt + \cdots + \frac{(-z)^n}{n!} \int_0^a t^n f(t)\, dt + \cdots$$

which is the required series. If $|f(t)| \leq M$, a constant, then

$$\left| \int_0^a t^n f(t)\, dt \right| \leq \int_0^a t^n M\, dt = \frac{a^{n+1}}{n+1} M.$$

The choice $M_n = Ma(aR)^n/(n+1)!$ in the Weierstrass test shows that the above power series converges for all z satisfying $|z| \leq R$. Since R is arbitrary, $F(z)$ is entire.

Example 3.2. Show that the function

$$g(z) = \int_0^1 e^{-t} t^{z-1}\, dt, \qquad \operatorname{Re} z > 1, \tag{3.7}$$

can be continued analytically so as to be meromorphic in the whole plane. More precisely, show that there is a meromorphic function $G(z)$ which agrees with $g(z)$ for $\operatorname{Re} z > 1$. The function $G(z)$ is said to give an analytic continuation of $g(z)$.

Substituting the power series for e^{-t} and integrating, we get

$$g(z) = \sum_{n=0}^{\infty} \int_0^1 \frac{(-t)^n}{n!} t^{z-1}\, dt = \sum_{n=0}^{\infty} \frac{(-1)^n}{n!(n+z)} \tag{3.8}$$

for Re $z > 1$. The term by term integration is justified because the series for e^{-t} converges uniformly and because

$$|t^{z-1}| = e^{(x-1)\,\mathrm{Log}\,t} \leq 1.$$

It is easily shown that the function on the right of (3.8) is meromorphic, hence provides the desired analytic continuation $G(z)$.

The integral (3.7) diverges for $z = x < 0$. Thus, not only does (3.8) give G explicitly, but it really extends the domain in which g was known to be analytic in a way that could not be determined from (3.7) at first glance. The continuation is unique because, if G_0 is another such continuation, the fact that $G(z) = G_0(z)$ for Re $z > 1$ makes $G(z) = G_0(z)$ on the whole domain of analyticity.

Example 3.3. The Bessel equation of order 0 is

$$z^2Y'' + zY' + z^2Y = 0$$

where the primes denote differentiation with respect to z. Obtain a power series solution of this equation satisfying $Y(0) = 1$, $Y'(0) = 0$.

We define $a_n = 0$ for $n < 0$. If

$$Y(z) = \sum_{n=0}^{\infty} a_n z^n, \qquad Y'(z) = \sum_{n=0}^{\infty} n a_n z^{n-1}, \qquad Y''(z) = \sum_{n=0}^{\infty} n(n-1)a_n z^{n-2}$$

for $|z| < \epsilon$, then substitution into the differential equation gives

$$\sum_{n=0}^{\infty} [n(n-1)a_n + na_n + a_{n-2}]z^n = 0, \qquad |z| < \epsilon.$$

By uniqueness of power series expansions, the coefficient of z^n must be 0. Hence a_n satisfies the recursion relation $n^2 a_n = -a_{n-2}$, or

$$a_n = -\frac{a_{n-2}}{n^2}, \qquad n = 2,3,4,\ldots. \qquad (3.9)$$

The condition $Y'(0) = 0$ gives $a_1 = 0$ and (3.9) then gives, in succession, $a_3 = 0$, $a_5 = 0$, and so on. Similarly, $Y(0) = 1$ gives $a_0 = 1$ and hence

$$a_0 = 1, \quad a_2 = -\frac{1}{2^2}, \quad a_4 = \frac{1}{4^2 2^2}, \quad a_6 = -\frac{1}{6^2 4^2 2^2}$$

and so on. The solution is $Y = J_0(z)$ where

$$J_0(z) = 1 + \sum_{n=1}^{\infty} \frac{(-1)^n z^{2n}}{(2\cdot 4\cdot 6 \cdots 2n)^2}.$$

By the ratio test this converges for $|z| < \infty$ and represents an entire function. The term by term differentiation is therefore justified by Theorem 3.4, and $Y = J_0(z)$ actually satisfies the differential equation.

The function $J_0(z)$ is called the Bessel function of order 0.

Problems

1. Show that an even function $f(z)$ can have only even powers in its Taylor expansion about $z = 0$, and similarly for odd functions. (Use $f(z) - f(-z) = 0$ and uniqueness.)
2. The Bessel function of order m is defined by

$$J_m(z) = \sum_{n=0}^{\infty} \frac{(-1)^n (z/2)^{2n+m}}{n!(m+n)!}, \qquad m = 0,1,2,3,\ldots.$$

By the ratio test, show that $J_m(z)$ is an entire function. Also show that $J_0(z)$ agrees with the function obtained in Example 3.3.

3. Let $f(t)$ be continuous and let $t^z = e^{z \operatorname{Log} t}$. Show that the first two of the following integrals represent entire functions, and that the third represents a function analytic at least in $|z| < 1$:

$$\int_{-\pi}^{\pi} f(t) \cos zt\, dt, \quad \int_1^2 t^z f(t)\, dt, \quad \int_0^1 \frac{f(t)}{1 - zt}\, dt.$$

4. (a) If $f(z)$ has a power series expansion (3.2) with $a_n \geq 0$, prove that $|f(w)| \leq f(|w|)$ for $|w| < R$. (b) Show that $|e^w - 1| \leq e^{|w|} - 1$.
5. For $|w| \leq ½$, show that $|\operatorname{Log}(1 - w)| \leq 2|w|$ and $|\operatorname{Log}(1 - w) + w| \leq |w|^2$. (By Taylor's series,

$$|\operatorname{Log}(1-w)| = \left| -w - \frac{w^2}{2} - \frac{w^3}{3} - \cdots \right| \leq |w|\left(1 + \frac{1}{2} + \frac{1}{4} + \cdots\right).$$

The second case is similar.)

6. If $P(n) \not\equiv 0$ is a polynomial in n, show that $P(n+1)/P(n) \to 1$ as $n \to \infty$. Conclude that the power series with coefficients $a_n = P(n)$ has radius of convergence $R = 1$.

Power series definition of elementary functions

7. *The exponential function.* Let $E(z) = 1 + z + z^2/2! + z^3/3! + \cdots$, and let no other properties of $E(z)$ be given. Show that $E(z)$ is an entire function satisfying $E'(z) = E(z)$, $E(0) = 1$. Show that $f(z) = E(-z)E(z)$ satisfies $f'(z) = 0$, hence $f(z) = f(0) = 1$. Thus, $E(-z)E(z) = 1$, and $E(z)$ never vanishes. If a is any complex constant, $f_a(z) = E(-z)E(z + a)$ satisfies $f_a'(z) = 0$, hence $f_a(z) = f_a(0) = E(a)$. Multiply by $E(z)$ and get

$$E(z + a) = E(z)E(a). \qquad (*)$$

8. *The graph of $E(x)$.* From $E'(x) = E(x)$ get $E'(x) > 0$ for $x > 0$ and hence $E(x)$ is strictly increasing. From $E(-x)E(x) = 1$ deduce that $E(x)$ is strictly increasing for all x, and hence $E(x) > 1$ for $x > 0$, $E(x) < 1$ for $x < 0$.

9. *Sine and cosine.* Let $2C(z) = E(iz) + E(-iz)$, $2iS(z) = E(iz) - E(-iz)$. Get the Taylor series for $C(z)$ and $S(z)$ from that for $E(w)$. Also show, by the chain rule, that

$$C'(z) = -S(z), \qquad S'(z) = C(z).$$

Show that $g(z) = [C(z)]^2 + [S(z)]^2$ satisfies $g'(z) = 0$, hence $g(z) = g(0) = 1$. Thus

$$[C(z)]^2 + [S(z)]^2 = 1. \qquad (**)$$

From $E(iz) = C(z) + iS(z)$ get also $|E(ix)| = 1$, $-\infty < x < \infty$.

10. *Periods.* A period of $E(z)$ is a number $T \neq 0$ such that $E(z + T) = E(z)$ for all z. By (*) with $a = T$, show that T is a period if and only if $E(T) = 1$. If $T = \sigma + i\tau$, show that $|E(T)| = |E(\sigma)||E(i\tau)| = E(\sigma)$ and from Problem 8 deduce $\sigma = 0$. Thus $T = i\tau$, and every period is purely imaginary. From $E(i\tau) = C(\tau) + iS(\tau)$, show that $T = i\tau$ is a period if and only if $\tau \neq 0$, $C(\tau) = 1$, $S(\tau) = 0$.

11. *Existence of a period.* Using the Taylor series, show that

$$C(1) > 1 - \tfrac{1}{2} - \tfrac{1}{4} - \tfrac{1}{8} \cdots = 0,$$
$$C(2) < 1 - 2 + \tfrac{2}{3} + \tfrac{1}{4} + (\tfrac{1}{4})^2 + (\tfrac{1}{4})^3 + \cdots = 0.$$

Hence, by real analysis, the real continuous function $C(x)$ has a zero x_0 between 1 and 2. Deduce $S(x_0) = \pm 1$ from (**) and hence

$$E(4ix_0) = [E(ix_0)]^4 = (\pm i)^4 = 1.$$

Thus $i\tau$ is a period, where $\tau = 4x_0$. The smallest $\tau > 0$ for which $i\tau$ is a period is denoted by 2π. (Actually $2\pi = 4x_0$). From $E(z + 2\pi i) = E(z)$ deduce $C(z + 2\pi) = C(z)$, $S(z + 2\pi) = S(z)$.

12. The unit circle is defined by $z = E(it)$, $0 \leq t \leq 2\pi$. From the fact that 2π is a period, and is the smallest period, deduce that the unit circle is a closed curve,

and is simple. Also show that $|z| = 1$ on the unit circle, and show that the arc of the circle from $t = 0$ to θ is

$$s = \int_0^\theta |iE(it)|\, dt = \int_0^\theta dt = \theta, \qquad 0 \le \theta \le 2\pi.$$

13. If e^z is the function considered in Chapter 2, verify that $h(z) = E(-z)e^z$ satisfies $h'(z) = 0$, hence $h(z) = h(0) = 1$, and thus $e^z = E(z)$. In the problems following Chapter 2, Section 2, the main properties of $\cos z$ and $\sin z$ were derived from (*).

4. Formulas of Parseval, Schwarz and Poisson.

The following theorem leads to simpler proofs, and sharper forms, of several results that were deduced from the Cauchy integral formula in Chapter 3.

THEOREM 4.1. *Let $R > 0$, and for $|z| < R$ let*

$$f(z) = \sum_{n=0}^{\infty} a_n z^n. \tag{4.1}$$

Then if $0 \le r < R$,

$$\frac{1}{2\pi}\int_0^{2\pi} |f(re^{i\theta})|^2\, d\theta = \sum_{n=0}^{\infty} |a_n|^2 r^{2n}. \tag{4.2}$$

Proof. Let $s_n(z)$ denote the nth partial sum of (4.1), so that

$|s_n(re^{i\theta})|^2 = s_n(re^{i\theta})\overline{s_n(re^{i\theta})} =$

$$= (a_0 + a_1 re^{i\theta} + \cdots + a_n r^n e^{in\theta})(\bar{a}_0 + \bar{a}_1 re^{-i\theta} + \cdots + \bar{a}_n re^{-in\theta}).$$

If this is multiplied out and integrated from 0 to 2π, most of the terms integrate to 0, because

$$\int_0^{2\pi} e^{ij\theta} e^{-ik\theta}\, d\theta = \begin{cases} 0 & (j \ne k) \\ 2\pi & (j = k) \end{cases} \tag{4.3}$$

The surviving terms are of the form $a_k r^k \bar{a}_k r^k$, and give

$$\int_0^{2\pi} |s_n(re^{i\theta})|^2\, d\theta = 2\pi \sum_{k=0}^{n} |a_k|^2 r^{2k}. \tag{4.4}$$

Since $s_n(z) \to f(z)$ uniformly on $|z| = r < R$, and since $f(z)$ is bounded on $|z| = r$, it is easily shown that $|s_n(z)|^2 \to |f(z)|^2$ uniformly on $|z| = r$. Hence letting $n \to \infty$ in (4.4) gives (4.2).

Equation (4.2) is sometimes called the Parseval equality because of its resemblance to an equation for Fourier series which was stated by Parseval in 1805. When $|f(z)| \leq M(r)$ for $|z| = r$, the left side of (4.2) does not exceed $[M(r)]^2$ and the Parseval equality gives

$$\sum_{n=0}^{\infty} |a_n|^2 r^{2n} \leq [M(r)]^2. \tag{4.5}$$

To illustrate the use of (4.5), suppose $f(z)$ is entire and satisfies $|f(z)| \leq M$, where M is constant. Then the left side of (4.5) can have no term $|a_n|r^{2n}$ with $n \geq 1$, and $f(z)$ reduces to a constant a_0. This is Liouville's theorem. The same reasoning shows $|a_n|r^n \leq M(r)$, with strict inequality unless $f(z)$ reduces to the single term $a_n z^n$. Since $a_n = f^{(n)}(0)/n!$, this gives Cauchy's inequality

$$|f^{(n)}(0)| \leq n!\frac{M(r)}{r^n}, \qquad 0 < r < R.$$

As a third application, if $|f(0)| = M(r)$ for some r, $0 < r < R$, then

$$|a_0|^2 = [M(r)]^2$$

and hence the left side of (4.5) cannot have any terms except $|a_0|^2$. Thus $a_n = 0$ for $n \geq 1$ and $f(z)$ is constant. This leads to a simple proof of the maximum principle.

A technique similar to that in the proof of Theorem 4.1 gives a result of great importance known as the Schwarz formula. Let $f(z)$ be analytic for $|z| < R$, where $R > 0$, and let the power series (4.1) converge uniformly for $|z| = R$. If $F(\zeta) = \operatorname{Re} f(\zeta)$ for $|\zeta| = R$, then

$$2F(\zeta) = f(\zeta) + \overline{f(\zeta)}$$

and hence

$$2F(\zeta) = \sum_{m=0}^{\infty} a_m \zeta^m + \sum_{m=0}^{\infty} \bar{a}_m \bar{\zeta}^m.$$

To determine a_n, multiply by $\zeta^{-n}\zeta^{-1}$ and integrate over $|\zeta| = R$. By uniform convergence we can integrate term by term. Thus,

$$\int_C \frac{2F(\zeta)}{\zeta^n}\frac{d\zeta}{\zeta} = \sum_{m=0}^{\infty} a_m \int_C \frac{\zeta^m}{\zeta^n}\frac{d\zeta}{\zeta} + \sum_{m=0}^{\infty} \bar{a}_m \int_C \frac{\overline{\zeta^m}}{\zeta^n}\frac{d\zeta}{\zeta}$$

where C is $|\zeta| = R$. Setting $\zeta = Re^{i\theta}$ and $\bar{\zeta} = Re^{-i\theta}$ reduces the integrals to those of the form (4.3), and it is found that the only surviving terms are those with $m = n$. The result after division by $2\pi i$ is

$$\frac{1}{\pi i}\int_C F(\zeta)\frac{1}{\zeta^n}\frac{d\zeta}{\zeta} = \begin{cases} a_0 + \bar{a}_0 & (n = 0) \\ a_n & (n \geq 1) \end{cases} \tag{4.6}$$

Substitution of these results into (4.1) leads to

$$f(z) = \frac{1}{\pi i}\sum_{n=0}^{\infty} \int_C F(\zeta)\left(\frac{z}{\zeta}\right)^n \frac{d\zeta}{\zeta} - \bar{a}_0, \tag{4.7}$$

where the term $\bar{a}_0$ must be subtracted because the first term of the series (4.7) gives $a_0 + \bar{a}_0$ instead of a_0. For $|z| < R$, clearly $|z/\zeta| < 1$ and

$$\sum_{n=0}^{\infty}\left(\frac{z}{\zeta}\right)^n = \frac{1}{1 - (z/\zeta)} = \frac{\zeta}{\zeta - z}.$$

This series converges uniformly for $|z| \leq \rho < R$ and so the order of summation and integration in (4.7) can be changed. This gives

$$f(z) = \frac{1}{\pi i}\int_C \frac{F(\zeta)}{\zeta - z}\,d\zeta - \bar{a}_0 \tag{4.8}$$

which is one form of the Schwarz formula. By (4.6)

$$\operatorname{Re} a_0 = \frac{1}{2\pi i}\int_C F(\zeta)\frac{d\zeta}{\zeta}. \tag{4.9}$$

Since $-\bar{a}_0 = -\operatorname{Re} a_0 + i \operatorname{Im} a_0$, substitution into (4.8) gives

$$f(z) = \frac{1}{2\pi i}\int_C F(\zeta)\frac{\zeta + z}{\zeta - z}\frac{d\zeta}{\zeta} + i \operatorname{Im} a_0 \tag{4.10}$$

which is another form of the Schwarz formula. The equation determines $f(z)$, apart from an imaginary constant, by knowledge of $F(\zeta) = \operatorname{Re} f(\zeta)$ on the boundary. This is a more subtle result than the Cauchy integral formula, since the latter involves both real and imaginary parts of $f(\zeta)$.

With $f(re^{i\theta}) = u(r,\theta) + i\,v(r,\theta)$ and $F(Re^{i\theta}) = u(\theta)$, (4.10) can be written

$$u(r,\theta) + i\,v(r,\theta) = \frac{1}{2\pi}\int_0^{2\pi} \frac{Re^{i\phi} + re^{i\theta}}{Re^{i\phi} - re^{i\theta}} u(\phi)\, d\phi + ic$$

where c is a real constant. If the numerator and denominator of the integrand are multiplied by $Re^{-i\phi} - re^{-i\theta}$, it is easy to equate real parts. The result is

$$u(r,\theta) = \frac{1}{2\pi}\int_0^{2\pi} \frac{R^2 - r^2}{R^2 - 2Rr\cos(\theta - \phi) + r^2} u(\phi)\, d\phi \tag{4.11}$$

and is known as the Poisson formula for a circle. The Poisson formula has been obtained here under the assumption that $f(z)$ is analytic for $|z| < R$ and has a uniformly convergent power series on $|z| = R$. For example, $f(z) = 1$ satisfies these conditions and gives

$$1 = \frac{1}{2\pi}\int_0^{2\pi} \frac{R^2 - r^2}{R^2 - 2Rr\cos(\theta - \phi) + r^2}\, d\phi. \tag{4.12}$$

We now change our point of view. Suppose $F(\zeta)$ is any continuous function on $|\zeta| = R$, and let $f(z)$ be defined by the integral (4.10). Then, as in Chapter 3, Section 5, it follows that $f(z)$ is analytic. (An independent proof is given in Section 5 of this chapter.) Since $f(z)$ is analytic, its real part $u(r,\theta)$ is harmonic. In Section 5 it will be seen that $u(r,\theta) \to u(\theta)$ as $r \to R-$, and that if $u(R,\theta) = u(\theta)$, the resulting function is continuous in the closed disk $|z| \leq R$. Hence, the Poisson formula solves the Dirichlet problem for a disk. Of course, the same is true of the Schwarz formula. Because of the difficulty of evaluating real integrals, the Schwarz formula is often more useful, and it has the distinct advantage that it gives, not only u, but the analytic function $u + iv$.

THEOREM 4.2. *If $H(x,y)$ is harmonic in a domain D, then $H(x,y)$ coincides locally with the real part of an analytic function $f(z)$, and hence H has partial derivatives of all orders at every point of D.*

Proof. Let (x_0,y_0) or $z_0 = x_0 + iy_0$ be a point of D. Since z_0 is interior, there is some disk $|z - z_0| \leq R$ contained in D. A translation allows us to assume $z_0 = 0$, so that the disk is $|z| \leq R$. The given function H is continuous in D and therefore continuous on $|z| = R$. By the Poisson formula, we can construct a function u which is continuous in $|z| \leq R$, harmonic in $|z| < R$, and agrees with H on $|z| = R$. By uniqueness,[1] $H = u$ for $|z| \leq R$ and hence H coincides with the real part of an analytic function in $|z| < R$. Existence of the derivatives now follows from Chapter 3, Theorem 5.3.

Example 4.1. For $0 \leq r < \infty$, show that

$$\sum_{n=0}^{\infty} \frac{(r/2)^{2n}}{(n!)^2} = \frac{1}{2\pi}\int_0^{2\pi} e^{r\cos\theta}\, d\theta. \tag{4.13}$$

If $z = x + iy = re^{i\theta}$, then $|e^{z/2}|^2 = e^x = e^{r\cos\theta}$. The result follows by substitution of the Taylor series for $e^{z/2}$ into the Parseval equality.

The series (4.13) is $J_0(ir)$, where J_0 is the Bessel function introduced in Example 3.3. In Problem 1 it will be seen that the integral in (4.13) represents an entire function, and the same is true of the series by the ratio test. Since these entire functions agree for real $z = r \geq 0$, they must agree for all z. Hence,

$$J_0(iz) = \frac{1}{2\pi}\int_0^{2\pi} e^{z\cos\theta}\, d\theta, \qquad |z| < \infty.$$

Example 4.2. Let α and β be points on $|z| = R$, as shown in Figure 4-1. Find a function $H(x,y)$ which is harmonic in $|z| < R$, assumes the value 1 on the arc $\alpha\beta$, and assumes the value 0 on the arc $\beta\alpha$. The function $H(x,y)$ is called the harmonic measure of arc $\alpha\beta$ with respect to the disk $|z| < R$.

Equation (4.9) gives

$$\operatorname{Re} a_0 = \frac{1}{2\pi i}\int_C \frac{F(\zeta)}{\zeta}\, d\zeta = \frac{1}{2\pi i}\int_\alpha^\beta \frac{d\zeta}{\zeta} = \frac{1}{2\pi i}\log\zeta\Big|_\alpha^\beta$$

where the path of integration is along the circle in the positive sense from α to β. The result is independent of the branch of log chosen. Similarly, by (4.8),

[1] Chapter 2, Problem 5.4.

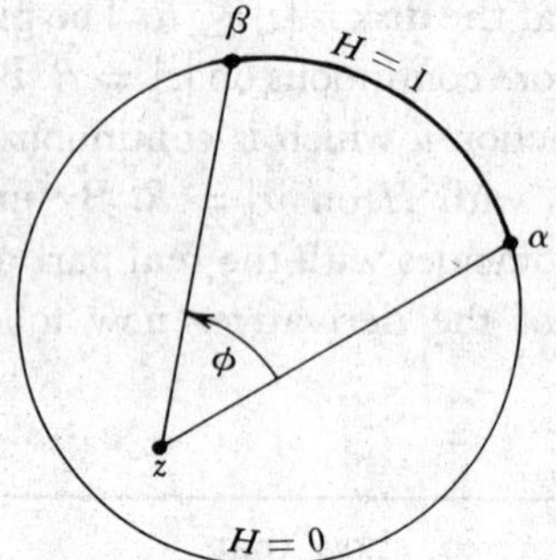

Figure 4-1

$$f(z) = \frac{1}{\pi i} \int_\alpha^\beta \frac{d\zeta}{\zeta - z} - \bar{a}_0 = \frac{1}{\pi i} \log(\zeta - z)\Big|_\alpha^\beta - \bar{a}_0.$$

Substituting the above value of Re a_0 and taking real parts gives

$$H(x,y) = \frac{1}{\pi} \arg(\zeta - z)\Big|_\alpha^\beta - \frac{1}{2\pi} \arg \zeta\Big|_\alpha^\beta$$

where $z = x + iy$ and where arg z denotes the branch associated with the chosen branch log z. Geometrically, this means that

$$H(x,y) = \frac{1}{\pi}\left(\phi - \frac{1}{2}\phi_0\right)$$

where ϕ is the angle subtended by arc $\alpha\beta$ at z and ϕ_0 is the angle subtended by $\alpha\beta$ at 0. The boundary conditions are readily verified by plane geometry.

Problems

1. (a) By the Taylor series for e^w with $w = r \cos \theta$, show that the integral in Example 4.1 satisfies

$$\frac{1}{2\pi} \int_0^{2\pi} e^{r \cos \theta} \, d\theta = \sum_{n=0}^{\infty} \frac{r^n}{n!} \frac{1}{2\pi} \int_0^{2\pi} (\cos \theta)^n \, d\theta$$

and hence represents an entire function. (b) What integral formula do you get from the fact that this power series must agree with that in Example 4.1?

2. By letting $z = re^{i\theta}$ in the geometric series for $1/(1 - z)$ and equating real parts, deduce that

$$\frac{1 - r^2}{1 - 2r\cos\theta + r^2} = 1 + r\cos\theta + r^2\cos 2\theta + \cdots, \qquad 0 \le r < 1.$$

Does this hold for complex $r = w$, $|w| < 1$?

3. By use of $2\cos k\theta = e^{ik\theta} + e^{-ik\theta}$ deduce for $m,n = 1,2,3,\ldots$ that

$$\int_0^{2\pi} \cos n\theta \cos m\theta \, d\theta = \begin{cases} \pi & (m = n) \\ 0 & (m \neq n). \end{cases}$$

Using this together with the series expansion of Problem 2, get

$$\int_0^{2\pi} \frac{1 - r^2}{1 - 2r\cos\theta + r^2} \cos n\theta \, d\theta = \pi r^n, \qquad n = 1,2,3,\ldots.$$

4. Formulate and solve analogs of Problems 2 and 3 for e^z instead of $1/(1 - z)$.
5. Apply the Parseval equality to $1/(1 - z)$.
6. A certain function $u(x,y)$ is harmonic for $|z| < 1$, assumes boundary values 1 on $z = e^{i\theta}$ for $|\theta| < \theta_0$, where $\theta_0 < \pi$, and assumes boundary values 0 for $|\theta| > \theta_0$. By (4.6), find the power series for an analytic function $f(z)$ of which u is the real part.
7. Let $f(z)$ be analytic for $|z| \le R$ and let $|\mathrm{Re}\, f(z)| \le M$ for $|z| = R$, where M is constant. Deduce from the Schwarz formula that

$$|f(z)| \le |f(0)| + \frac{2RM}{R - r}, \qquad |z| = r < R.$$

8. Let $f(z)$ be analytic for $|z| < R$, let $f'(z) \neq 0$, and let $g(z)$ be an analytic square root of $f'(z)$. Thus, $f'(z) = [g(z)]^2$ for $|z| < R$. If $L(r)$ is the length of the image of the circle $|z| = r < R$ under the mapping $w = f(z)$, show that

$$L(r) = 2\pi r \sum_{n=0}^{\infty} |b_n|^2 r^{2n} \quad \text{where} \quad g(z) = \sum_{n=0}^{\infty} b_n z^n.$$

(Use the formula for $L(r)$ given in Chapter 3, Section 5, Problem 6.)

9. Let $f(z)$ be analytic for $|z| < R$ and let $A(r)$ be the area of the image of the disk $|z| < r < R$ under the mapping $w = f(z)$, where areas covered more than once are counted a corresponding number of times. By Chapter 5, Problem 1.6,

$$A(r) = \int_0^r \left(\int_0^{2\pi} |f'(re^{i\theta})|^2 \, d\theta \right) r \, dr$$

and this is taken for granted here. Evaluate the inner integral by the Parseval equality for $f'(z)$ and then evaluate the outer integral by (2.5). Thus show that

$$A(r) = \pi r^2 \sum_{n=1}^{\infty} n|a_n|^2 r^{2n-2} \quad \text{where} \quad f(z) = \sum_{n=0}^{\infty} a_n z^n.$$

In particular, $A(r)/(\pi r^2) \geq |f'(0)|^2$ with strict inequality, unless $f(z)$ is a linear function, $a_0 + a_1 z$.

10. Apply Problems 8 and 9 with $f(z) = 1/(1 - z)$ and with $f(z) = e^z$.

5. Functions defined by integrals. Many functions arising in pure and applied mathematics can be represented by means of integrals. For example, the gamma function is defined by

$$\Gamma(z) = \int_0^\infty e^{-t} t^{z-1}\, dt \tag{5.1}$$

for $\operatorname{Re} z > 0$; the Bessel function by

$$J_n(z) = \frac{1}{\pi} \int_0^\pi \cos(n\theta - z \sin \theta)\, d\theta \tag{5.2}$$

for $|z| < \infty$; and for $\operatorname{Re} s > 1$, the Riemann zeta function satisfies

$$\zeta(s)\Gamma(s) = \int_0^\infty \frac{1}{e^u - 1} u^{s-1}\, du. \tag{5.3}$$

Among integral formulas which contain an arbitrary function are the Laplace transform,

$$F(s) = \int_0^\infty e^{-st} f(t)\, dt, \tag{5.4}$$

and the Schwarz formula

$$f(z) = \frac{1}{2\pi i} \int_C \frac{F(\zeta)}{\zeta} \frac{\zeta + z}{\zeta - z}\, d\zeta, \tag{5.5}$$

where C is $|\zeta| = R$.

All of these integrals have the general form

$$f(z) = \int_a^b F(z,t)\, dt, \tag{5.6}$$

where $F(z,t)$ is analytic for each z in a domain D and for t on (a,b) or

$[a,b]$.[1] The integral (5.5), for example, can be expressed as an integral over $[0,2\pi]$ by setting $\zeta = Re^{it}$.

THEOREM 5.1. *For t on $[a,b]$ let*

$$F(z,t) = \sum_{j=0}^{\infty} a_j(t)(z-\alpha)^j \tag{5.7}$$

where each $a_j(t)$ is continuous. Suppose $|a_j(t)| \leq A_j$ where A_j are constants such that the M series

$$\sum_{j=0}^{\infty} M_j \doteq \sum_{j=0}^{\infty} A_j r^j$$

has radius of convergence $R > 0$. Then $f(z)$ in (5.6) is analytic for $|z-\alpha| < R$. It has the power series expansion

$$f(z) = \sum_{j=0}^{\infty} (z-\alpha)^j \int_a^b a_j(t)\,dt. \tag{5.8}$$

Proof. Let z be fixed and $|z-\alpha| = r < R$. By the M test, the series (5.7) converges uniformly for t on $[a,b]$ and can be integrated term by term, as indicated below (2.5). The result is (5.8). Since z could be any number satisfying $|z-\alpha| < R$, the power series (5.8) must converge for $|z-\alpha| < R$ and hence defines an analytic function. This completes the proof.

The proof is unchanged if the $a_j(t)$ are only piecewise continuous rather than continuous, provided $F(z,t)$ is integrable as a function of t.

Example 5.1. If $F(\zeta)$ is piecewise continuous, show that the Schwarz formula defines an analytic function in $|z| < R$. Expansion in geometric series for $|z| < |\zeta| = R$ gives

$$F(\zeta)\frac{\zeta+z}{\zeta-z} = F(\zeta)\left(1 + 2\frac{z}{\zeta} + 2\frac{z^2}{\zeta^2} + 2\frac{z^3}{\zeta^3} + \cdots\right).$$

Since $\zeta = Re^{it}$ gives $d\zeta/\zeta = i\,dt$, we can take A_n to be any number such that $2|F(\zeta)/\zeta^n| \leq A_n$ on C. Evidently $A_n = 2M/R^n$ will do, where M is

[1] We use (a,b) to denote an open interval $a < t < b$ and $[a,b]$ for $a \leq t \leq b$. Similarly, $[a,b)$ denotes $a \leq t < b$.

bound for $|F(\zeta)|$ on C, and in this case the radius of convergence of the M series is R. Theorem 5.1 gives the conclusion.

THEOREM 5.2. *Under the hypothesis of Theorem 5.1*

$$f'(z) = \int_a^b \frac{\partial F(z,t)}{\partial z}\, dt, \qquad |z - \alpha| < R$$

and similarly for higher derivatives.

Proof. Since (5.7) is a power series, it can be differentiated to give

$$\frac{\partial F}{\partial z}(z,t) = \sum_{j=0}^{\infty} j a_j(t)(z - \alpha)^{j-1}.$$

Since $|ja_j(t)| \leq jA_j$, we can form a new M series with terms $jA_j r^{j-1}$. This series is obtained by differentiating the original M series and therefore has the same radius of convergence R. Accordingly, Theorem 5.1 applies to $\partial F/\partial z$ and gives

$$\int_a^b \frac{\partial F(z,t)}{\partial z}\, dt = \sum_{j=0}^{\infty} j(z - \alpha)^{j-1} \int_a^b a_j(t)\, dt.$$

This equals $f'(z)$ by (5.8). Clearly the process can be repeated, and Theorem 5.2 follows.

Here too, the extension to functions with discontinuities is immediate. However, neither Theorem 5.1 nor 5.2 can be applied directly to integrals like (5.1) or (5.3), for which the interval of integration is infinite. The extension to such integrals is discussed next.

Let $F(z,t)$ be piecewise continuous for $a \leq t < \infty$. It is said that the equation

$$f(z) = \int_a^{\infty} F(z,t)\, dt$$

holds uniformly in a region G, or that the integral satisfies the uniform Cauchy criterion in a region G, if the following is true: For every $\epsilon > 0$ there is an N such that, for z in G

$$p > N \text{ and } q > N \quad \text{implies} \quad \left| \int_p^q F(z,t)\,dt \right| < \epsilon.$$

The value of N depends on ϵ but not on z, so long as z is in G.

The uniform Cauchy criterion ensures that the sequence of functions defined by

$$f_n(z) = \int_a^n F(z,t)\,dt \tag{5.9}$$

is a uniform Cauchy sequence in G and hence has a limit $f(z)$ as $n \to \infty$. Indeed, if the integer n and the real number b are large enough, the Cauchy criterion gives

$$\left| \int_a^b F(z,t)\,dt - \int_a^n F(z,t)\,dt \right| = \left| \int_b^n F(z,t)\,dt \right| < \epsilon. \tag{5.10}$$

The choice $b = m$, an integer, shows that $|f_n(z) - f_m(z)| < \epsilon$ and hence that $\{f_n(z)\}$ is a uniform Cauchy sequence.

By letting $n \to \infty$ through integral values in (5.10), we get an inequality which shows that the same limit $f(z)$ is obtained when $b \to \infty$ without restriction. That is,

$$f(z) = \int_a^\infty F(z,t)\,dt \tag{5.11}$$

where the improper integral is as in Chapter 4, Section 4.

THEOREM 5.3. *Theorem 5.2 retains its validity for $b = \infty$, provided the hypothesis holds in $[a,b]$ for each $b > a$ and provided (5.11) holds uniformly in $|z - \alpha| < R$.*

Proof. The foregoing discussion shows that the restriction of b to integral values n, as $b \to \infty$, is not a serious impediment, and we give the proof for that case. (See also Problem 5.6.) By Theorem 5.2 the function $f_n(z)$ in (5.9) is analytic and satisfies

$$f_n'(z) = \int_a^n \frac{\partial F}{\partial z}(z,t)\,dt, \qquad |z - \alpha| < R$$

as well as similar equations for higher derivatives. Since the sequence

$\{f_n(z)\}$ converges uniformly, Theorem 1.3 shows that the limit function $f(z)$ in (5.11) satisfies

$$f'(z) = \lim_{n\to\infty} f_n'(z), \qquad f''(z) = \lim_{n\to\infty} f_n''(z), \tag{5.12}$$

and so on. The first of these equations gives

$$\frac{d}{dz}\int_a^\infty F(z,t)\,dt = \int_a^\infty \frac{\partial F}{\partial z}(z,t)\,dt$$

and the remaining equations give corresponding results for higher derivatives. This completes the proof.

It should be emphasized that different sequences $\{A_j\}$ can be used for different values of b in Theorem 5.3, so that the theorem by no means implies that the power series expansion of Theorem 5.1 remains valid for $b = \infty$.

In using Theorem 5.3 it is helpful to have a test for uniform convergence. Let $F(z,t)$ be piecewise continuous on $a \le t \le b$ for each $b > a$ and let

$$|F(z,t)| \le M(t) \quad \text{where} \quad \int_a^\infty M(t)\,dt < \infty.$$

Then, as we shall show, the integral (5.11) satisfies the uniform Cauchy criterion. This is an analog of the Weierstrass M test. For proof, note that $q > p$ gives

$$\left|\int_p^q F(z,t)\,dt\right| \le \int_p^q |F(z,t)|\,dt \le \int_p^q M(t)\,dt \le \int_p^\infty M(t)\,dt.$$

For large p, the right-hand side is $<\epsilon$, and the result follows.

Example 5.2. Let $f(t)$ be piecewise continuous in $0 \le t \le b$ for every $b > 0$ and suppose $|f(t)| \le Me^{ct}$ for some constants M and c. Show that the Laplace transform (5.4) defines an analytic function in $\text{Re}\, s > c$ and that its derivatives are given by

$$F^{(n)}(s) = \int_0^\infty (-t)^n e^{-st} f(t)\,dt, \qquad \text{Re}\, s > c.$$

Note that the roles of f and F are reversed here.

If the upper limit ∞ is replaced by any number $b > 0$, the resulting function $F_b(s)$ is entire and satisfies the hypothesis of Theorem 5.1 in $|s| < R$ for every R. (This was shown in Example 3.1.) Let $\operatorname{Re} s \geq c + \delta$ where $\delta > 0$ is constant. Then

$$|e^{-st} f(t)| \leq e^{-(c+\delta)t} M e^{ct} = M e^{-\delta t}.$$

The choice $M(t) = Me^{-\delta t}$ in the M test establishes uniform convergence, and the conclusion follows from Theorem 5.3.

THEOREM 5.4. *Let $F(z,t)$ be a continuous function of (z,t) for z in a domain D and for t on $[a,b]$. For each t on $[a,b]$ let $F(z,t)$ be an analytic function of z in D. Then the function $f(z)$ defined by the first of the following formulas is analytic in D, and its derivative is given by the second formula:*

$$f(z) = \int_a^b F(z,t)\, dt, \qquad f'(z) = \int_a^b \frac{\partial F}{\partial z}(z,t)\, dt.$$

It will be seen that higher derivatives are also given by differentiating under the integral sign.

The statement that $F(z,t)$ is continuous means the following: Given any point t_0 on $[a,b]$, any point z_0 in D, and any $\epsilon > 0$, there is a $\delta > 0$ such that

$$|t - t_0| + |z - z_0| < \delta \quad \text{implies} \quad |F(z,t) - F(z_0,t_0)| < \epsilon$$

provided t is on $[a,b]$ and z is in D. The value of δ can depend on z_0 and t_0, as well as on ϵ. However, if $|z - \alpha| \leq R$ is any closed disk in D, it is known from real analysis that $F(z,t)$ is bounded and uniformly continuous for z in this disk and t on $[a,b]$. This is taken for granted here.

Proof. Let α be a point of D and let C be a circle $|z - \alpha| = R$ so small that C and its interior are in D. Then for $|z - \alpha| < R$, the Taylor series of $F(z,t)$ is (5.7) with

$$a_j(t) = \frac{1}{2\pi i} \int_C \frac{F(\zeta,t)}{(\zeta - \alpha)^j} \frac{d\zeta}{\zeta}.$$

Clearly $|a_j(t)| \leq M/R^j$ where M is a bound for $|F(\zeta,t)|$ for ζ on C and t on $[a,b]$. To show that $a_j(t)$ is continuous, let t_0 and t be on $[a,b]$ and let $\epsilon > 0$. By uniform continuity, there is a $\delta > 0$, independent of ζ, such that

$$|t - t_0| < \delta \quad \text{implies} \quad |F(\zeta,t) - F(\zeta,t_0)| < \epsilon$$

for all ζ on C. Since $|\zeta - \alpha| = R$ on C, this gives

$$|a_j(t) - a_j(t_0)| \leq \epsilon R^{-j}$$

and shows that $a_j(t)$ is continuous. Hence Theorems 5.1 and 5.2 apply, and give the desired conclusion on $|z - \alpha| < R$. Since α is an arbitrary point of D, this completes the proof.

THEOREM 5.5. *Theorem 5.4 retains its validity for $b = \infty$ provided the hypothesis is given in $[a,b]$ for each $b > a$, and provided (5.11) holds uniformly in D.*

This follows from Theorem 5.4 just as Theorem 5.3 followed from Theorem 5.2.

Example 5.3. Show that the Bessel function $J_n(z)$ is entire and that its derivative can be calculated by differentiating under the integral sign. If $\theta = t$, the integrand in (5.2) is

$$\cos(nt - z \sin t). \tag{5.13}$$

This is an entire function of z and is a continuous function of (z,t). Hence, both conclusions follow from Theorem 5.4.

Continuity of (5.13) can be deduced from the fact that (5.13) has bounded partial derivatives with respect to z and t. However, it is simpler to note that $-z$, $\sin t$, nt, and $\cos \zeta$ are continuous, and to use familiar theorems pertaining to continuity of products, sums, and composite functions. The latter are proved for functions of (z,t) in the same way as for real-valued functions of a real variable t.

Problems

1. (a) With $\zeta = Re^{it}$, the Schwarz formula can be written

$$f(z) = \frac{1}{2\pi} \int_0^{2\pi} F(Re^{it}) \frac{Re^{it} + z}{Re^{it} - z}\, dt.$$

If $F(\zeta)$ is continuous for $|\zeta| = R$, deduce from Theorem 5.4 that $f(z)$ is analytic for $|z| < R$. (b) Extend to piecewise continuous functions by writing the integral as a sum of integrals over intervals $a_k < t < b_k$ on each of which $F(Re^{it})$ is continuous.

2. Show that the following are entire functions and find $f'(z)$:

$$f(z) = \int_{-\infty}^{\infty} e^{-t^2} \cos zt \, dt, \qquad f(z) = \int_1^{\infty} \frac{t^z}{\cosh t} \, dt.$$

3. (a) Show that the integral (5.1) represents a function analytic at least for $\operatorname{Re} z > 1$. (b) By partial integration,

$$\int_0^b e^{-t}t^x \, dt = -e^{-t}t^x \Big|_0^b + \int_0^b e^{-t}xt^{x-1} \, dt$$

for $x > 1$. Letting $b \to \infty$, get $\Gamma(x + 1) = x\Gamma(x)$. (c) Show that $\Gamma(1) = 1$ and deduce in succession $\Gamma(2) = 1$, $\Gamma(3) = 2 \cdot 1, \ldots, \Gamma(n + 1) = n!$.

4. Show that $\Gamma(z)$ in (5.1) differs only by an entire function from the integral considered in Example 3.2. Hence, $\Gamma(z)$ can be extended to be a meromorphic function with poles at $0, -1, -2, \ldots$ only. The extended function is also denoted by $\Gamma(z)$. Show that $\Gamma(z + 1) = z\Gamma(z)$ at all points where both sides of this equation are analytic. (By Problem 3, the function $f(z) = \Gamma(z + 1) - z\Gamma(z)$ vanishes for $z = x > 1$. Hence, it vanishes in its whole domain of analyticity.)

5. *A form of the M test.* Let $F(z,t)$ be piecewise continuous on $\epsilon \le t \le 1$ for z in a domain D and for each small $\epsilon > 0$. Suppose

$$|F(z,t)| \le M(t) \quad \text{where} \quad \int_0^1 M(t) \, dt < \infty.$$

Show that the sequence of functions defined by

$$f_n(z) = \int_{1/n}^1 F(z,t) \, dt, \qquad n = 1,2,3, \ldots$$

is a uniform Cauchy sequence for z in D and hence has a limit $f(z)$. Conclude that

$$f(z) = \int_0^1 F(z,t) \, dt$$

is analytic in D if each $f_n(z)$ is analytic in D.

6. By Problem 5, show that the following are analytic for $\operatorname{Re} z > 0$, $\operatorname{Re} z > -2$, and $-1 < \operatorname{Re} z < 2$, respectively:

$$\int_0^1 t^{z-1}e^{-t} \, dt, \quad \int_0^1 t^z \sin t \, dt, \quad \int_0^1 t^z(1 - t)^{1-z} \, dt.$$

The Poisson integral

7. Let $u(\phi)$ and $|u(\phi)|$ be integrable, $0 \le \phi \le 2\pi$, let $u(0) = u(2\pi)$, and let $\phi = \theta_0$ be a point at which $u(\phi)$ is continuous. If $u(r,\theta)$ is defined by the Poisson integral for $r < R$ and by $u(R,\theta) = u(\theta)$ for $r = R$, deduce from (4.12) that

$$u(r,\theta) - u(R,\theta_0) = \int_0^{2\pi} K(r,\theta - \phi)[u(\phi) - u(\theta_0)]\, d\phi$$

for $r < R$, where

$$K(r,\theta - \phi) = \frac{1}{2\pi} \frac{R^2 - r^2}{R^2 - 2Rr\cos(\theta - \phi) + r^2}.$$

8. Let $\theta_0 \ne 0$, $\theta_0 \ne 2\pi$. (If $\theta_0 = 0$ or 2π, the limits of integration $[0,2\pi]$ can be replaced by $[-\pi,\pi]$ since the integrand is periodic, and the following analysis is virtually unchanged.) Given $\epsilon > 0$, choose small $\delta_0 > 0$ so that

$$|u(\phi) - u(\theta_0)| < \epsilon/2$$

for $|\phi - \theta_0| < 2\delta_0$. Then show, for $r < R$, that

$$\left| \int_{\theta - 2\delta_0}^{\theta_0 + 2\delta_0} K(r,\theta - \phi)[u(\phi) - u(\theta_0)]\, d\phi \right| \le \frac{\epsilon}{2} \int_0^{2\pi} K(r,\theta - \phi)\, d\phi = \frac{\epsilon}{2}.$$

9. For $|\phi - \theta_0| \ge 2\delta_0$ and for $|\theta - \theta_0| \le \delta_0$, show that $|\theta - \phi| \ge \delta_0$ and hence

$$\left| \int_{|\phi - \theta_0| \ge 2\delta_0} K(r,\theta - \phi)[u(\phi) - u(\theta_0)]\, d\phi \right| \le K(r,\delta_0) \int_0^{2\pi} |u(\phi) - u(\theta_0)|\, d\phi.$$

10. Show that $K(r,\delta_0) \to 0$ as $r \to R-$ and hence there is a $\delta_1 > 0$ such that the result of Problem 9 does not exceed $\epsilon/2$ if $|r - R| < \delta_1$. For $r < R$ and

$$|\theta - \theta_0| + |r - R| < \delta = \min(\delta_0,\delta_1),$$

conclude that $|u(r,\theta) - u(R,\theta_0)| \le \epsilon/2 + \epsilon/2 = \epsilon$. By choice of δ_0 in Problem 8, this also holds for $r = R$ and hence $u(r,\theta)$ is continuous at (R,θ_0).

6. Asymptotic series. The integral

$$\int_x^{\infty} \frac{e^{-t}}{t}\, dt$$

arises often enough so that it is a tabulated function of x. Setting $t = x + u$ reduces it to the form

$$e^{-x} \int_0^{\infty} \frac{e^{-u}}{u + x}\, du.$$

This latter integral will now be studied and indeed for complex x.

Let

$$F(z) = \int_0^\infty \frac{e^{-u}}{u+z}\,du \tag{6.1}$$

in the cut plane $|\arg z| < \pi$. By Theorem 5.3 or 5.5, $F(z)$ is an analytic function in the cut plane. One can get an expansion for $F(z)$ by use of

$$\frac{1}{1+w} = \sum_{k=0}^{\infty}(-1)^k w^k, \qquad \int_0^\infty e^{-u}u^k\,du = k!$$

The first equation is the formula giving the sum of a geometric series and is valid for $|w| < 1$. The second equation above states that $\Gamma(k+1) = k!$ and was established for $k = 0,1,2,\ldots$ in Section 5, Problem 3.

If one now proceeds to compound several errors,

$$\begin{aligned} F(z) &= \int_0^\infty e^{-u}\frac{1}{z}\frac{1}{1+(u/z)}\,du = \int_0^\infty \frac{e^{-u}}{z}\sum_{k=0}^{\infty}(-1)^k\left(\frac{u}{z}\right)^k du \\ &= \sum_{k=0}^{\infty}\frac{(-1)^k}{z^{k+1}}\int_0^\infty e^{-u}u^k\,du = \sum_{k=0}^{\infty}\frac{(-1)^k k!}{z^{k+1}}. \end{aligned} \tag{6.2}$$

Since the series diverges for all z (by the ratio test), the above formula is false. However, it is by no means without use or meaning. The series represents $F(z)$ asymptotically.

The definition of an asymptotic series is best expressed by a notation which was introduced by Landau in connection with the analytic theory of numbers. The symbol $O(z^n)$, read "terms of the order of z^n," represents any function satisfying

$$|O(z^n)| \le M|z|^n \qquad \text{as } |z| \to \infty$$

in a specified region G, where M is constant. For example,

$$(z^2 + z^3)^{-4} = O(z^{-12}), \qquad e^z = O(1)$$

in $|z| < \infty$ and in $\operatorname{Re} z < 0$, respectively.

DEFINITION 6.1. *Let G be an unbounded region (usually a sector) in the z plane. Let $h(z)$ be analytic in the finite part of G. A formal power series, $\Sigma a_j/z^j$, not necessarily convergent, is said to be an asymptotic series for h in G if for every $n \ge 0$*

$$h(z) - \sum_{j=0}^{n} \frac{a_j}{z^j} = O\left(\frac{1}{z^{n+1}}\right) \tag{6.3}$$

as $|z| \to \infty$ in G. In this case, we write

$$h(z) \sim \sum_{j=0}^{\infty} a_j z^{-j} \quad \text{in } G.$$

In applications it is sometimes useful to have a series in powers $(z^\lambda)^{-j}$ rather than z^{-j}, where λ is a positive constant and where $z^\lambda = \exp(\lambda \operatorname{Log} z)$. The error term is then $O[(z^\lambda)^{-n-1}]$. Such an expansion is also called an asymptotic expansion.

It is impossible for two distinct series to represent h asymptotically in G. Indeed, (6.3) implies that as $|z| \to \infty$ in G,

$$h(z) \to a_0, \quad z[h(z) - a_0] \to a_1, \quad z^2[h(z) - a_0 - a_1 z] \to a_2,$$

and so on. However, distinct functions may have the same asymptotic series in a region. For example, in the sector $|\arg z| \leq c$ where $c < \pi/2$, the function e^{-z} has as its asymptotic series the series with all coefficients $a_j = 0$. This follows from the fact that for $|\arg z| \leq c$, $|z| = r$, and any given n,

$$|e^{-z}| \leq e^{-r\cos c} < |z|^{-n} \tag{6.4}$$

for large enough $|z|$. Hence the functions e^{-z} and 0 have the same asymptotic series in this sector.

To show that the divergent series (6.2) represents (6.1) asymptotically in $|\arg z| \leq \pi - \delta$, $\delta > 0$, observe that, by partial integration,

$$\begin{aligned} F(z) &= -\frac{e^{-u}}{u+z}\Big|_0^\infty - \int_0^\infty \frac{e^{-u}}{(u+z)^2}\,du \\ &= \frac{1}{z} + \frac{e^{-u}}{(u+z)^2}\Big|_0^\infty + 2!\int_0^\infty \frac{e^{-u}}{(u+z)^3}\,du \\ &= \frac{1}{z} - \frac{1!}{z^2} + \frac{2!}{z^3} - 3!\int_0^\infty \frac{e^{-u}}{(u+z)^4}\,du. \end{aligned}$$

In general,

$$F(z) = \sum_{j=0}^{n} \frac{(-1)^j j!}{z^{j+1}} + R_{n+1}$$

where

$$R_{n+1} = (-1)^{n+1}(n+1)!\int_0^\infty \frac{e^{-u}}{(u+z)^{n+2}}\,du.$$

If $z = re^{i\theta}$ where $r > 0$, $\delta > 0$, and $\pi/2 \leq |\theta| \leq \pi - \delta$, then

$$|u + z| \geq |\text{Im}(u + z)| = |\text{Im}\, z| \geq r|\sin\delta|.$$

If $|\theta| \leq \pi/2$, then $\text{Re}\, z \geq 0$, and hence $|u + z| \geq |z| = r$. Hence the above estimate holds with $\sin\delta$ replaced by 1. In either case

$$|R_{n+1}| \leq (n+1)!\int_0^\infty \frac{e^{-u}}{(r\sin\delta)^{n+2}}\,du = \frac{(n+1)!}{(r\sin\delta)^{n+2}}. \tag{6.5}$$

This shows that for $|\arg z| \leq \pi - \delta$,

$$F(z) - \sum_{j=0}^{n} \frac{(-1)^j j!}{z^{j+1}} = O\left(\frac{1}{z^{n+2}}\right).$$

Hence, $F(z)$ is represented asymptotically by the divergent series (6.2) in the sector $|\arg z| \leq \pi - \delta$.

An asymptotic series can be extremely useful if $|z|$ is large in spite of the fact that the series diverges. On the other hand, for a given value of $|z|$ it is pointless to use terms in the series once the magnitude of the terms begins to increase. For example, for $F(6)$ one can use

$$\frac{1}{6} - \frac{1!}{6^2} + \frac{2!}{6^3} - \frac{3!}{6^4} + \frac{4!}{6^5} - \frac{5!}{6^6}.$$

Here the first term neglected is exactly as large as the last term used. After that, the numerator of each term grows faster than the denominator and the terms begin to diverge. In the above series for $F(6)$, the error according to (6.5) with $\delta = \pi/2$ is at worst $6!/6^7$, which in this case happens to be the size of the last term used.

From (6.3) it is easy to see that if $h_1(z)$ and $h_2(z)$ have asymptotic series in G, then so do their sum and their product. The sum is obtained by simply adding the coefficients of the two series. The product is obtained by multiplying the two series and collecting the terms in decreasing powers of z.

If G is a sector, $\alpha < \arg z < \beta$, then asymptotic series exist for the integral of $h(z)$ and for its derivative. For the integral, (6.3) gives

$$\int_z^{\infty}\left[h(\zeta) - a_0 - \frac{a_1}{\zeta}\right] d\zeta = \sum_{j=2}^{n}\frac{a_j}{(j-1)z^{j-1}} + O\left(\frac{1}{z^n}\right). \tag{6.6}$$

The derivative must be treated more carefully. We shall show that an asymptotic series can be differentiated term by term, in a sense made precise by the following theorem:

THEOREM 6.1. *Let* $0 < \beta - \alpha \leq 2\pi$. *If* $h(z)$ *is analytic and has the asymptotic series*

$$h(z) \sim \sum_{j=0}^{\infty} a_j z^{-j} \qquad \text{in } \alpha < \arg z < \beta,$$

then for every small $\delta > 0$

$$h'(z) \sim \sum_{j=0}^{\infty} -ja_j z^{-j-1} \quad \text{in } \alpha + \delta < \arg z < \beta - \delta.$$

Proof. If $|z|$ is large enough and if C is the unit circle in the ζ plane with center at $\zeta = z$, then

$$h'(z) = \frac{1}{2\pi i}\int_C \frac{h(\zeta)}{(\zeta - z)^2}\, d\zeta.$$

By (6.3),

$$h'(z) = \sum_{j=0}^{n} a_j \frac{1}{2\pi i}\int_C \frac{\zeta^{-j}}{(\zeta - z)^2}\, d\zeta + O(z^{-(n+1)}).$$

Since Cauchy's formula for the derivative of the analytic function z^{-j} gives

$$\frac{1}{2\pi i}\int_C \frac{\zeta^{-j}}{(\zeta - z)^2}\, d\zeta = -jz^{-j-1},$$

it follows that

$$h'(z) = -\sum_{j=1}^{n} a_j j z^{-j-1} + O(z^{-(n+1)}) = -\sum_{j=1}^{n-1} a_j j z^{-j-1} + O(z^{-(n+1)}),$$

which proves the result.

There is another way of treating $F(z)$ defined in (6.1). Let $x > 0$. Then

$$F(x) = \int_0^\infty \frac{e^{-t}}{t + x}\, dt.$$

Replace the variable t by u where $t = ux$. Then

$$F(x) = \int_0^\infty \frac{e^{-ux}}{u + 1}\, du, \qquad F(z) = \int_0^\infty \frac{e^{-uz}}{u + 1}\, du \tag{6.7}$$

where the second expression is obtained from the first by analytic continuation for $\operatorname{Re} z > 0$. Thus $F(z)$ appears as the Laplace transform of $1/(u + 1)$.

There is a theorem due to Watson which shows that under quite general conditions Laplace transforms have asymptotic expansions. This theorem, which applies, in particular, to (6.7), reads as follows:

THEOREM 6.2 (*Watson's lemma*). *Let λ, δ, a, B and M be positive constants. Suppose the series*

$$g(u) = \sum_{k=1}^{\infty} a_k u^{k\lambda - 1}$$

converges for $0 < u < a + \delta$ to a function $g(u)$ which is piecewise continuous for $u > 0$. Suppose also that

$$|g(u)| \leq M e^{Bu}, \qquad u \geq a. \tag{6.8}$$

Then the Laplace transform of g satisfies

$$\int_0^\infty e^{-zu} g(u)\, du \sim \sum_{k=1}^{\infty} a_k \Gamma(k\lambda) z^{-k\lambda}$$

for $|z|$ large and for $|\arg z| \leq \tfrac{1}{2}\pi - \delta$.

In the course of the proof it is shown that the series which one gets by the (erroneous) formal procedure of term by term integration coincides with the above expansion.

Proof. Let $n \geq 2$ be an integer which remains fixed throughout the discussion. The remainder after $n - 1$ terms in the series for $g(u)$ is

$$r_n(u) = g(u) - \sum_{k=1}^{n-1} a_k u^{k\lambda-1}. \tag{6.9}$$

We want to show that there is a constant c such that, $b = B + 1$

$$|r_n(u)| \leq c u^{n\lambda-1} e^{bu}, \qquad 0 < u < \infty. \tag{6.10}$$

If $u \geq 1$ and $u \geq a$, then by (6.8) and (6.9)

$$|r_n(u)| \leq M e^{Bu} + \sum_{k=1}^{n-1} |a_k| u^{K\lambda-1}.$$

Each of the terms on the right admits an estimate of the form (6.10) for $u \geq 1$ and hence the same is true of the sum. Thus we can pick $c = c_1$ so that (6.10) holds for $u \geq \max(1,a)$.

If $u \leq a$, then the infinite series for $g(u)$ gives

$$|r_n(u)| = \left| \sum_{k=n}^{\infty} a_k u^{k\lambda-1} \right| = u^{n\lambda-1} \left| \sum_{k=n}^{\infty} a_k u^{(k-n)\lambda} \right|.$$

The series on the right is a power series in u^λ which converges for $u < a + \delta$ and hence is bounded for $u \leq a$. We can therefore choose $c = c_2$ so that (6.10) holds for $u \leq a$.

In case $a < 1$, note by (6.9) that $|r_n(u)|$ is bounded for $a \leq u \leq 1$ and hence $c = c_3$ can be chosen to make (6.10) hold for $a \leq u \leq 1$. The largest of the three constants c_1, c_2, c_3 gives the constant c for (6.10).

Since the Laplace transform of g exists for $\operatorname{Re} z > b$, by (6.8), we can use (6.9) and (6.10) to get

$$\left| \int_0^\infty e^{-zu} g(u)\, du - \sum_{k=1}^{n-1} \int_0^\infty a_k e^{-zu} u^{k\lambda-1}\, du \right| \leq c \int_0^\infty u^{n\lambda-1} e^{bu} e^{-xu}\, du.$$

By the change of variable $t = ux$ and $t = u(x - b)$, respectively,

$$\int_0^\infty e^{-xu} u^{k\lambda-1}\, du = \Gamma(k\lambda) x^{-k\lambda}, \quad \int_0^\infty e^{(b-x)u} u^{n\lambda-1}\, du = \frac{\Gamma(n\lambda)}{(x-b)^{n\lambda}}.$$

The first of these equations holds with x replaced by z, for $\operatorname{Re} z > 0$, as seen from the fact that the two sides are analytic. Substitution into the preceding inequality gives

$$\left| \int_0^\infty e^{-zu} g(u)\, du - \sum_{k=1}^{n-1} a_k \Gamma(k\lambda) z^{-k\lambda} \right| \le c \frac{\Gamma(n\lambda)}{(x-b)^{n\lambda}} \tag{6.11}$$

for $\operatorname{Re} z = x > b$. Since $|\arg z| \le \pi/2 - \delta$, it follows that $x \ge |z| \sin \delta$. Hence the right side of (6.11) is $O(|z|^{-n\lambda})$ as $|z| \to \infty$ and Theorem 6.2 follows.

Example 6.1. Stirling's formula. The equation

$$\frac{d}{dz} \frac{\Gamma'(z)}{\Gamma(z)} = \int_0^\infty e^{-uz} \frac{u}{1 - e^{-u}}\, du$$

is derived in Section 8, Problem 13, and is taken for granted here. For small $|u|$

$$\frac{u}{1 - e^{-u}} = 1 + \frac{u}{2} + \frac{u^2}{12} + \cdots$$

and hence Watson's lemma gives

$$\frac{d}{dz} \frac{\Gamma'(z)}{\Gamma(z)} \sim \frac{1}{z} + \frac{1}{2z^2} + \frac{1}{6z^3} + \cdots, \qquad |\arg z| < \frac{\pi}{2} - \delta.$$

According to (6.6) this can be integrated to give

$$\int_z^\infty \left(\frac{d}{dz} \frac{\Gamma'(z)}{\Gamma(z)} - \frac{1}{z} \right) dz \sim \frac{1}{2z} + \frac{1}{12z^2} + \cdots.$$

Since the derivative of $\Gamma'(z)/\Gamma(z) - \operatorname{Log} z$ coincides with minus the derivative of the left side above,

$$\frac{\Gamma'(z)}{\Gamma(z)} - \operatorname{Log} z + \alpha \sim -\frac{1}{2z} - \frac{1}{12z^2} - \cdots$$

for some constant α. Repeating this whole procedure gives

$$\operatorname{Log} \Gamma(z) - (z \operatorname{Log} z - z) + \alpha z + \frac{1}{2} \operatorname{Log} z + \beta \sim \frac{1}{12z} + \cdots$$

for some constant β. The limit of the right side is 0 as $|z| \to \infty$ and hence the limit of the exponential of the left side is 1. This gives

$$\lim_{|z|\to\infty} \Gamma(z)z^{-z}e^{z}z^{1/2}e^{\alpha z+\beta} = 1 \tag{6.12}$$

where the branches of the many-valued functions are defined by

$$z^{-z} = e^{-z \operatorname{Log} z}, \qquad z^{1/2} = e^{(1/2)\operatorname{Log} z}.$$

In Problems 7 and 8 it will be seen that $\alpha = 0$ and $e^{-\beta} = (2\pi)^{1/2}$. Hence

$$\lim_{|z|\to\infty} \Gamma(z)z^{-z}e^{z}z^{1/2} = (2\pi)^{1/2} \tag{6.13}$$

which is known as Stirling's formula. Although the formula was established here only for $|\arg z| < \pi/2 - \delta$, it can be extended to $|\arg z| < \pi - \delta$ by the method of Problems 10–13.

Problems

1. Find the smallest number n for which the following are $O(z^n)$ as $z \to \infty$; (a) in the whole plane, (b) in the half-plane $\operatorname{Re} z > 0$, and (c) in the sector $|\arg z| < \pi/4$:

$$3z, \quad \frac{5z^2}{1+z^4}, \quad z^2e^{-z}, \quad \frac{\sin z}{1+z^4}, \quad e^{z^2}.$$

2. By partial integration, show that $F(z)$ in (6.7) satisfies

$$F(z) = \frac{1}{z} - \frac{1}{z}\int_0^\infty \frac{e^{-uz}}{(1+u)^2}\,du = \frac{1}{z} - \frac{1}{z^2} + \frac{2}{z^2}\int_0^\infty \frac{e^{-uz}}{(1+u)^3}\,du$$

for $\operatorname{Re} z > 0$. If $|\arg z| < \theta_0 < \pi/2$, use $(1+u)^3 \geq 1$ to show that the integral on the right, in magnitude, does not exceed $|x|^{-1} = |z|^{-1}\sec\theta_0$. Hence $F(z) - z^{-1} + z^{-2} = O(z^{-3})$ in $|\arg z| < \theta_0$.

3. Continue the partial integration in Problem 2 and get an asymptotic expansion in $|\arg z| < \theta_0$.
4. Using $\Gamma(k+1) = k!$ and geometric series, verify that the asymptotic series in Problem 3 agrees with Watson's lemma.
5. Show that if $\delta > 0$ and $|\arg z| \leq \pi - \delta$,

$$\int_0^\infty \frac{\cos t}{t+z}\,dt \sim \frac{1!}{z^2} - \frac{3!}{z^4} + \frac{5!}{z^6} - \cdots, \qquad \int_0^\infty \frac{\sin t}{t+z}\,dt \sim \frac{0!}{z} - \frac{2!}{z^3} + \cdots.$$

6. At what index n does the magnitude of terms in the series above start to increase?
7. By Section 5, Problem 3, $\Gamma(n+1) = n\Gamma(n)$ for positive integers n. If (6.12) is

written for $z = n$ and for $z = n + 1$, show that division of the resulting equations gives

$$\lim_{n\to\infty}\left(1+\frac{1}{n}\right)^n\left(1+\frac{1}{n}\right)^{1/2} = e^{1+\alpha}.$$

Since $(1 + 1/n)^n \to e$, it follows that $\alpha = 0$.

8. If $\alpha = 0$ and $e^{-\beta} = c$ in (6.12), show that (6.12) gives

$$c^2 = \lim_{n\to\infty} [\Gamma(n)]^2 n^{-2n} e^{2n} n, \quad c = \lim_{n\to\infty} \Gamma(2n)(2n)^{-2n} e^{2n} (2n)^{1/2}.$$

Since $\Gamma(n) = (n - 1)!$, division of these equations gives

$$c = \lim_{n\to\infty} \frac{2\cdot 4\cdot 6\cdots(2n-2)}{1\cdot 3\cdot 5\cdots(2n-1)} 4\left(\frac{n}{2}\right)^{1/2}.$$

According to a well-known[1] formula of Wallis,

$$\frac{\pi}{2} = \frac{2}{1}\frac{2}{3}\frac{4}{3}\frac{4}{5}\frac{6}{5}\frac{6}{7}\cdots = \lim_{n\to\infty}\frac{2}{1}\frac{2}{3}\frac{4}{3}\cdots\frac{2n-2}{2n-1}\frac{2n}{2n-1}$$

where the second expression above is defined by the third expression. Using this, show that $c^2 = 2\pi$.

9. If $F(z) = a/z + O(1/z^2)$ where a is constant, show that

$$\exp F(z) = 1 + \frac{a}{z} + O\left(\frac{1}{z^2}\right).$$

Hence show that the result of Example 6.1 can be improved to

$$\Gamma(z) z^{-z} e^z z^{1/2} = \sqrt{2\pi}\left[1 + \frac{1}{12z} + O\left(\frac{1}{z^2}\right)\right].$$

Test this formula at $z = 2$, dropping the O term.

*Extension of the domain of an asymptotic expansion

10. The integral (6.7) is valid only for $x > 0$ whereas it was seen earlier that $F(z)$ is analytic for $|\arg z| < \pi$. Using contour integration, show that if β is real and $|\beta| < \pi/2$, then, for $x > 0$,

$$F(x) = \int_0^\infty \frac{\exp(-\rho x e^{-i\beta})}{\rho e^{-i\beta} + 1} e^{-i\beta}\, d\rho.$$

Show that if x is replaced by z, the resulting integral represents an analytic function for $|\arg z - \beta| < \pi/2$, hence gives an analytic continuation of $F(x)$ onto $|\arg z - \beta| < \pi/2$.

[1] A proof is given in Section 7, Problem 10. It is also possible to get c from the duplication formula (Section 8, Problem 14) and then use Problem 8, if desired, to evaluate Wallis' product.

11. By Watson's lemma, show that the function

$$F_1(\zeta) = \int_0^\infty \frac{e^{-\rho\zeta}}{\rho + e^{i\beta}}\,d\rho$$

has an asymptotic expansion for $|\arg \zeta| < \pi/2 - \delta$, $\delta > 0$. Let $z = \zeta e^{i\beta}$ where $|\beta|$ is real and $|\beta| < \pi/2 - \delta$. Show that $F(z)$ of Problem 10 has an asymptotic expansion for $|\arg z - \beta| \leq \pi/2 - \delta$. Since this sector overlaps with $|\arg z| < \pi/2 - \delta$, the asymptotic series must be the same as that found for $F(x)$ in the text.

12. Verify the last statement in Problem 11 by direct use of Watson's lemma to find the asymptotic series for (6.7) and for $F_1(\zeta)$.

13. Show that the procedure of Problem 10 can be applied to $F_1(\zeta)$ of Problem 11 to get finally

$$F(z) = \int_0^\infty \frac{\exp(-uze^{-2i\beta})}{u + e^{2i\beta}}\,du, \qquad |\arg z - 2\beta| < \frac{\pi}{2}.$$

Hence show that on the Riemann surface for $\log z$, $F(z)$ has an analytic continuation onto the sector $|\arg z| < 3\pi/2$. (If β is close to $\pi/2$ and the process is repeated, a new term enters because a pole is crossed.)

7. Infinite products.

If $\{a_k\}$ is a sequence of complex numbers, a product with n factors $1 + a_k$ is denoted by

$$p_n = \prod_{k=1}^{n} (1 + a_k) = (1 + a_1)(1 + a_2) \cdots (1 + a_n). \tag{7.1}$$

For the present, it will be assumed that $a_k \neq -1$ for all k. Under this hypothesis, the sequence of products converges if $p_n \to p \neq 0$. The condition $p \neq 0$ is required, in part, to ensure the existence of $\operatorname{Log} p$. The fact that $p_n \to p$ is also written

$$\prod_{k=1}^{\infty} (1 + a_k) = p,$$

and (7.1) is said to give the partial products of this infinite product.

Since (7.1) gives $p_n = (1 + a_n)p_{n-1}$, or $a_n p_{n-1} = p_n - p_{n-1}$, convergence implies

$$\lim_{n\to\infty} a_n = \lim_{n\to\infty} \frac{p_n - p_{n-1}}{p_{n-1}} = \frac{p - p}{p} = 0.$$

This corresponds to the condition $a_n \to 0$ for convergent series. As is the case for series, the condition $a_n \to 0$ is necessary but not sufficient for convergence. We shall establish the following:

THEOREM 7.1. *The infinite product with $a_k \neq -1$ converges if and only if*

$$\sum_{k=1}^{\infty} \operatorname{Log}(1 + a_k) \tag{7.2}$$

converges.

Proof of sufficiency. If s_n denotes the nth partial sum of the series (7.2), then

$$e^{s_n} = e^{\operatorname{Log}(1+a_1)} e^{\operatorname{Log}(1+a_2)} \cdots e^{\operatorname{Log}(1+a_n)} = p_n.$$

Hence $s_n \to s$ gives $p_n \to e^s \neq 0$ and shows the sufficiency of the condition. The necessity is less important in applications and is established below, in Problem 9.

The foregoing proof yields the following result which is often useful. For each j, let $\log_j$ denote a definite branch of the logarithm which, however, can be different for different j. Suppose the series

$$s = \sum_{j=1}^{\infty} \log_j(1 + a_j)$$

converges. Then, as above, $e^{s_n} = p_n$, and hence

$$\prod_{k=1}^{\infty} (1 + a_j) = e^s.$$

So far, we have excluded the case $a_j = -1$. When several of the $a_j = -1$, the product is said to converge if it converges (in the previous sense) when the several factors which are equal to 0 are deleted. Only a finite number of such factors will be allowed here.

These definitions extend in a natural way to products in which the index k ranges over some other set than $\{1,2,\ldots,n\}$ or $\{1,2,3,\ldots\}$. The definitions also extend to products in which the constants a_k are replaced by functions $a_k(z)$. In particular, if such a product of functions converges for each z in a region G, then it converges in G. The convergence is uniform if the partial products satisfy $p_n(z) \to p(z)$ uniformly in G.

THEOREM 7.2 (*Weierstrass M test for products*). *For* $k = 1,2,3, \ldots$ *let* $a_k(z)$ *be analytic in a domain D and let* $M_k \geq 0$ *be constants such that*

$$\sum_{k=1}^{\infty} M_k < \infty.$$

Suppose there is an N, independent of z, such that

$$|\text{Log}[1 + a_k(z)]| \leq M_k, \qquad k \geq N. \tag{7.3}$$

Then the product

$$p(z) = \prod_{k=1}^{\infty} [1 + a_k(z)]$$

converges to an analytic function $p(z)$ *in D, and* $p(z)$ *is 0 only where one or several of its factors* $1 + a_k(z)$ *are 0 in D.*

Proof. For $n \geq N$ let

$$P_n(z) = \prod_{k=N}^{n} [1 + a_k(z)], \quad S_n(z) = \sum_{k=N}^{n} \text{Log}[(1 + a_k(z)].$$

The hypothesis gives

$$|S_m(z)| \leq \sum_{k=N}^{m} M_k \leq \sum_{k=1}^{\infty} M_k = M \tag{7.4}$$

where $m \geq N$ and where M is defined by this equation. Similarly $n > m \geq N$ implies

$$|S_n(z) - S_m(z)| \leq \sum_{k=m+1}^{n} M_k \leq \sum_{k=m+1}^{\infty} M_k = \epsilon_m \tag{7.5}$$

where ϵ_m is defined by this equation.

From $P_n(z) = \exp S_n(z)$ and $P_m(z) = \exp S_m(z)$ follows

$$P_n(z) - P_m(z) = e^{S_m(z)}[e^{S_n(z)-S_m(z)} - 1] \tag{7.6}$$

for $n > m \geq N$. The Taylor series for e^w gives also

$$|e^w| \leq e^{|w|}, \qquad |e^w - 1| \leq e^{|w|} - 1$$

and therefore, by (7.6), (7.5) and (7.4),

$$|P_n(z) - P_m(z)| \le e^M(e^{\epsilon_m} - 1).$$

This shows that $\{P_n(z)\}$ is a uniform Cauchy sequence. Hence $P_n(z) \to P(z)$ uniformly on D, and $P(z)$ is analytic. From (7.4),

$$|P_m(z)| = \exp \operatorname{Re} S_m(z) \ge e^{-M}$$

and hence $|P(z)| \ge e^{-M}$. This shows $P(z) \ne 0$ in D.

The original infinite product satisfies $p(z) = Q(z)P(z)$ where $Q(z)$ is the product of the first $N - 1$ factors,

$$Q(z) = \prod_{k=1}^{N-1} [1 + a_k(z)]. \tag{7.7}$$

The zeros of $p(z)$ can arise only from the zeros of $Q(z)$, and this completes the proof.

The following supplement to Theorem 7.2 is often useful:

THEOREM 7.3. *The convergence of the product in Theorem 7.2 is uniform if the function $Q(z)$ in (7.7) is bounded in D. Also the conclusion remains valid if (7.3) is replaced by $|a_k(z)| \le M_k$.*

Proof. The first statement follows from

$$p(z) - p_n(z) = Q(z)[P(z) - P_n(z)]$$

and from the uniformity of the convergence $P_n(z) \to P(z)$. The second statement follows from[1]

$$|\operatorname{Log}(1 + w)| \le 2|w|, \qquad |w| \le \tfrac{1}{2}, \tag{7.8}$$

which in turn follows from the Taylor series for $\operatorname{Log}(1 + w)$. Since the general term of a convergent series tends to 0, there is an N such that $M_k < \frac{1}{2}$ for $k \ge N$. Then (7.8) gives

$$|\operatorname{Log}[1 + a_k(z)]| \le 2M_k, \qquad k \ge N.$$

Hence the hypothesis of Theorem 7.2 is satisfied with $2M_k$ instead of M_k and Theorem 7.3 follows.

[1] This and subsequent inequalities of the sort were proved in Section 3, Problem 5.

Example 7.1. Show that the product

$$F(z) = \prod_{k=1}^{\infty}\left(1 - \frac{z^2}{k^2}\right)$$

represents an entire function with simple zeros at $z = \pm 1, \pm 2, \pm 3, \ldots$ and with no other zeros. Show also that the convergence is uniform in $|z| < R$ for every $R > 0$.

In any disk $|z| < R$, clearly $|z^2/k^2| < R^2/k^2$, and Theorem 7.3 can be applied with $M_k = R^2/k^2$. This gives the uniformity of the convergence, and also shows that $F(z)$ has the desired properties in $|z| < R$. Since R is unrestricted, this completes the proof.

Example 7.2. Let it be required to construct a function with simple zeros at $z = 1, 2, 3, \ldots$ and no other zeros. The attempt to represent such a function by

$$\prod_{k=1}^{\infty}\left(1 - \frac{z}{k}\right)$$

fails because, as shown in Problem 2, this product diverges for $z \neq 0$. A satisfactory representation is given by

$$G(z) = \prod_{k=1}^{\infty}\left(1 - \frac{z}{k}\right)e^{z/k}. \tag{7.9}$$

The role of the exponential factor $e^{z/k}$ is that its logarithm produces a term z/k that just cancels the first term in the series for $\text{Log}(1 - z/k)$.

To see that (7.9) has the required properties, note that the Taylor series for $\text{Log}(1 - w)$ gives

$$|\text{Log}[(1 - w)e^{w}]| \leq |w|^2, \qquad |w| \leq \tfrac{1}{2}.$$

Let $|z| < R$, where R is fixed. Then for $k > 2R$ the above gives

$$\left|\text{Log}\left[\left(1 - \frac{z}{k}\right)e^{z/k}\right]\right| \leq \left|\frac{z}{k}\right|^2 \leq \frac{R^2}{k^2}.$$

The conclusion follows from Theorem 7.2 with

$$M_k = \frac{R^2}{k^2}, \qquad 1 + a_k(z) = \left(1 - \frac{z}{k}\right)e^{z/k}.$$

As we have seen, $G(z)$ in (7.9) represents an entire function with simple zeros at the points $z = 1,2,3,\ldots$, and with no other zeros. Another such function would be obtained if $G(z)$ were multiplied by e^z, or indeed by any nonvanishing entire function. This shows that an entire function is not determined by its zeros. However, we shall establish the following:

THEOREM 7.4. *Let $F(z)$ and $G(z)$ be two entire functions which have the same zeros with the same multiplicities. Then there exists an entire function $g(z)$ such that*

$$F(z) = e^{g(z)}G(z).$$

Proof. The function $H(z) = F(z)/G(z)$ is entire except for singularities at the common zeros α_k of F and G. By Chapter 3, Theorem 7.2, the singularity at α_k is removable and, if the singularity is removed, $H(\alpha_k) \neq 0$. The function $H(z)$ so obtained is nonvanishing and entire. By Chapter 4, Example 1.2, it has an analytic logarithm. That is, $H(z) = e^{g(z)}$ where $g(z)$ is entire. Multiplication by $G(z)$ gives Theorem 7.4.

Example 7.3. If the left side is defined to be 1 at $z = 0$, show that

$$\frac{\sin \pi z}{\pi z} = \prod_{k=1}^{\infty}\left(1 - \frac{z^2}{k^2}\right), \qquad |z| < \infty.$$

By Example 7.1, the product on the right is an entire function with the same zeros as the function on the left. Thus Theorem 7.4 gives

$$\frac{\sin \pi z}{\pi z} = e^{g(z)}G(z)$$

where G denotes the infinite product and g is entire. Since all three factors have the value 1 for $z = 0$, we can take $g(0) = 0$. Taking logarithms gives

$$\operatorname{Log}\frac{\sin \pi z}{\pi z} = g(z) + \operatorname{Log} G(z) = g(z) + \sum_{k=1}^{\infty} \operatorname{Log}\left(1 - \frac{z^2}{k^2}\right)$$

for small $|z|$. The inequality $|\operatorname{Log}(1 - w)| \leq 2|w|$ shows that the series converges uniformly for small $|z|$ and hence term by term differentiation is justified by Theorem 1.3. The result is

$$\pi \cot \pi z - \frac{1}{z} = g'(z) + \sum_{k=1}^{\infty} \frac{2z}{k^2 - z^2}.$$

Upon substituting the series for $\pi \cot \pi z$ as given by Example 2.3 (see also Example 2.2), we get $g'(z) = 0$ for small $|z|$. Since g is entire, the same holds for all z, and hence g is constant. Setting $z = 0$ shows that $g(0) = 0$ and the result follows.

A more satisfactory approach to calculations of this sort, which does not require any use of logarithms, is given by Theorem 8.4 of the next section. Hence, we do not pause to justify the use of the principal branch Log in each term of the above series.

Problems

1. Show that $a^{-1} \operatorname{Log}(1 + a) \to 1$ as $a \to 0$ and conclude that

$$\tfrac{1}{2}|a| \leq |\operatorname{Log}(1 + a)| \leq 2|a|$$

for small $|a|$. Hence show for $a_i > 0$ that if any one of the following converges, all converge:

$$\prod_{k=1}^{\infty}(1 - a_i), \qquad \sum_{k=1}^{\infty} a_i, \qquad \prod_{k=1}^{\infty}(1 + a_i).$$

2. If $s(n) = 1 + 1/2 + 1/3 + \cdots + 1/n$, verify that

$$P_n(z) = \prod_{k=1}^{n}\left(1 - \frac{z}{k}\right) = e^{-s(n)z} \prod_{k=1}^{n}\left(1 - \frac{z}{k}\right)e^{z/k}$$

where $P_n(z)$ is defined by this equation. Using results of the text, show that $|P_n(z)| \to 0$ as $n \to \infty$ if $\operatorname{Re} z > 0$, and $|P_n(z)| \to \infty$ as $n \to \infty$ if $\operatorname{Re} z < 0$. Also deduce from Problem 1 that the infinite product diverges for real $z \neq 0$.

3. If $k^z = e^{z \operatorname{Log} k}$, show that

$$\prod_{k=1}^{\infty}(1 - e^{kz}), \qquad \prod_{k=1}^{\infty}(1 - k^z)$$

represent functions which are analytic and nonvanishing in $\operatorname{Re} z < 0$ and $\operatorname{Re} z < -1$, respectively.

4. By grouping together the factors for $k = -n$ and $k = n$, show that

$$\prod_{k=-\infty}^{\infty}{}' \left(1 - \frac{z}{k}\right)e^{z/k} = \frac{\sin \pi z}{\pi z}$$

where the prime on the product means that the factor for $k = 0$ is omitted.

5. By grouping together the terms for n and $-n$ and using the result of Example 2.3, show that

$$\pi \cot \pi z = \frac{1}{z} + \sum_{k=-\infty}^{\infty}{}' \left(\frac{1}{z-k} + \frac{1}{k}\right)$$

where the prime on the sum means that the term for $k = 0$ is omitted.

6. By differentiating the result of Problem 5, get the expansions

$$\left(\frac{\pi}{\sin \pi z}\right)^2 = \sum_{k=-\infty}^{\infty} \frac{1}{(z-k)^2}, \quad \left(\frac{\pi}{\cos \pi z}\right)^2 = \sum_{k=-\infty}^{\infty} \frac{1}{(z-k-\frac{1}{2})^2}.$$

7. By integrating the expansion for $\pi^2 \sec^2 \pi z$ as given by Problem 6, obtain an expansion for $\pi \tan \pi z$. Group terms for $-n$ and n and show that this expansion is equivalent to

$$\pi \tan \pi z = -\sum_{k=1}^{\infty} \frac{2z}{z^2 - (k-\frac{1}{2})^2}.$$

8. Integrate the result of Problem 7 and then take exponentials to get

$$\cos \pi z = \prod_{k=1}^{\infty}\left(1 - \frac{z^2}{(k-\frac{1}{2})^2}\right).$$

9. If no $\alpha_k = -1$, show that the convergence of (7.2) is a necessary condition for $p_n \to p \neq 0$ in (7.1). (If $\operatorname{Arg} p \neq \pi$, then for large enough n

$$\operatorname{Log} p_n = \sum_{k=1}^{n} \operatorname{Log}(1 + a_k) + 2m_n\pi i,$$

where m_n is an integer which depends on n. Since $\operatorname{Log} p_n \to \operatorname{Log} p$, it follows that m_n cannot change as n increases, when n is large. If $\operatorname{Arg} p = \pi$, consider $\operatorname{Log}(-p_n)$ and proceed as before.)

10. Derive Wallis' product (Section 6, Problem 8) by setting $z = \frac{1}{2}$ in Example 7.3.

*8. Weierstrass and Mittag-Leffler expansions.

It was seen in Example 7.2 that introduction of exponential factors can cause a product to converge. With this device in mind, we define the Weierstrass primary factors by

$$E(w,m) = (1-w)\exp\left(w + \frac{w^2}{2} + \frac{w^3}{3} + \cdots + \frac{w^m}{m}\right)$$

for $m = 1,2,3,\ldots$, and also $E(w,0) = 1 - w$. By the Taylor series for

Log$(1 - w)$, it is easily verified that

$$|\text{Log}\, E(w,m)| \leq 2|w|^{m+1}, \qquad |w| \leq \tfrac{1}{2}. \tag{8.1}$$

An independent derivation is given later in this section.

THEOREM 8.1 (*Weierstrass*). *For* $k = 1,2,3, \ldots$ *let* $\{\alpha_k\}$ *be a sequence of complex numbers and let* $m \geq 0$ *be an integer such that*

$$\sum_{k=1}^{\infty} \frac{1}{|\alpha_k|^{m+1}} < \infty. \tag{8.2}$$

Then the function

$$P(z) = \prod_{k=1}^{\infty} E(z/\alpha_k, m) \tag{8.3}$$

is an entire function with zeros at α_k *and at these points only. The order of the zero at* α_n *is equal to the number of indices* j *such that* $\alpha_j = \alpha_n$.

Proof. Let $R > 0$ and choose N so that $|\alpha_k| > 2R$ for $k \geq N$. Such an N exists, because (8.2) makes the general term of the series tend to 0, and hence $|\alpha_k| \to \infty$. For $k > N$ and $|z| < R$, clearly $|z/\alpha_k| < \tfrac{1}{2}$. By (8.1),

$$|\text{Log}\, E(z/\alpha_k,\ m)| \leq 2\left|\frac{z}{\alpha_k}\right|^{m+1} \leq \frac{2R^{m+1}}{|\alpha_k|^{m+1}}.$$

Because of (8.2) we can apply Theorem 7.2. Hence $P(z)$ represents a function which is analytic in $|z| < R$ and has zeros only at those points α_k, $k < N$, which are in $|z| < R$. Since this is true for every R, Theorem 8.1 follows.

The product (8.3) is called a Weierstrass product of genus m. If there is no m such that the series (8.2) converges, for example, if $\alpha_k = \text{Log}(1 + k)$, then it may be possible to use a product of infinite genus as in the following theorem:

THEOREM 8.2 (*Weierstrass*). *For* $k = 1,2,3, \ldots$ *let* $\{\alpha_k\}$ *be a sequence of complex numbers such that* $|\alpha_k| \to \infty$. *Then the function*

$$P(z) = \prod_{k=1}^{\infty} E(z/\alpha_k, k)$$

is an entire function with zeros at α_k and at these points only. The multiplicity of the zero at α_n is equal to the number of indices j such that $\alpha_j = \alpha_n$.

Proof. Let $R > 0$ and choose N such that $|\alpha_k| > 2R$ for $k \geq N$. Then for $|z| < R$ and $k \geq N$, we have $|z/\alpha_k| \leq \frac{1}{2}$ and hence, by (8.1),

$$|\text{Log}\, E(z/\alpha_k,\ k)| \leq \left|\frac{z}{\alpha_k}\right|^{k+1} \leq \left(\frac{1}{2}\right)^{k+1}$$

Since the sum on k converges, Theorem 7.2 can be applied, and shows that $P(z)$ has the desired properties in $|z| < R$ for every R. This completes the proof.

If $F(z)$ is an entire function, not identically 0, then the zeros α_k of $F(z)$ are either finite in number, or else they satisfy the hypothesis $|\alpha_k| \to \infty$ of Theorem 8.2. To prove this, suppose there are infinitely many α_k in some disk $|z| \leq R$. By a theorem of real analysis, which is taken for granted here, there must be a subsequence of $\{\alpha_k\}$ which tends to some point α of $|z| \leq R$ as a limit. Continuity of F gives $F(\alpha) = 0$, but this zero is not isolated. Hence $F(z) \equiv 0$, contrary to hypothesis.

To allow for the fact that an entire function $F(z)$ might have no zeros, it is convenient to consider 1 to be a Weierstrass product. A finite product of expressions $E(z/\alpha,\ m)$ is also considered to be a Weierstrass product.

Theorem 8.3 (*Weierstrass product theorem*). *Every entire function $F(z) \not\equiv 0$ can be written in the form*

$$F(z) = z^n e^{g(z)} P(z)$$

where n is a nonnegative integer, $g(z)$ is entire, and $P(z)$ is a Weierstrass product.

Proof. Let n be the order of the zero at $z = 0$ and let the other zeros, if any, be listed in a sequence $\{\alpha_k\}$. (This can be accomplished by taking the zeros satisfying $|\alpha| < 1$, then those satisfying $1 \leq |\alpha| < 2$, and so on.) A multiple zero is repeated in the sequence $\{\alpha_k\}$ a number of times equal to the order of the zero. Let $P(z)$ be any convergent Weierstrass product formed from the α_k. The existence of $P(z)$ is trivial if there are only finitely many α_k. If there are infinitely many α_k, then $|\alpha_k| \to \infty$ and we can use $P(z)$ from Theorem 8.1 if (8.2) holds, or else we can use Theorem 8.2. Since

$z^n P(z)$ has the same zeros as $F(z)$, and since the orders of the zeros agree, Theorem 8.3 follows from Theorem 7.4.

So far, we have given no clue as to how a product expansion of a given function can be obtained. A systematic method can be based on logarithmic differentiation. If

$$F(z) = \prod_{k=1}^{\infty} [1 + a_k(z)] \tag{8.4}$$

then formal calculation suggests that

$$\log F(z) = \sum_{j=1}^{\infty} \log[1 + a_k(z)]$$

and hence that

$$\frac{F'(z)}{F(z)} = \sum_{k=1}^{\infty} \frac{a_k'(z)}{1 + a_k(z)}. \tag{8.5}$$

Conversely, from the latter series we might expect to be able to deduce the product expansion (8.4).

THEOREM 8.4. *Let $a_j(z)$ be analytic in a domain D and let (8.4) hold uniformly in D. Then (8.5) holds at all points of D where $F(z) \neq 0$.*

Proof. Let $F(z) = F_n(z)G_n(z)$ where $F_n(z)$ is the product of the first n factors and $G_n(z)$ is the product of the remaining factors. Uniformity of the convergence shows that $F(z)$ is analytic in D. Let z_0 be a point of D where $F(z) \neq 0$. Then $|F(z)| > \eta$ in some disk $|z - z_0| < \delta$, where η and δ are positive constants. The fact that $F_n(z) \to F(z)$ uniformly in D gives $G_n(z) \to 1$ uniformly in $|z - z_0| < \delta$, and hence, by Theorem 1.3, $G_n'(z) \to 0$ uniformly in $|z - z_0| < \delta/2$. From

$$F(z) = G_n(z) \prod_{k=1}^{n} [1 + a_k(z)], \qquad F(z) \neq 0,$$

it follows that

$$\frac{F'(z)}{F(z)} = \frac{G_n'(z)}{G_n(z)} + \sum_{k=1} \frac{a_k'(z)}{1 + a_k(z)}. \tag{8.6}$$

(This can be deduced by taking logarithms or, more easily, by Equation (1.4) of Chapter 2.) Since $G_n'(z) \to 0$ and $G_n(z) \to 1$ in $|z - z_0| < \delta/2$, clearly $G_n'(z)/G(z) \to 0$, and the conclusion follows by letting $n \to \infty$ in (8.6).

THEOREM 8.5. *If (8.5) holds uniformly in a domain D, then*

$$\frac{F(z)}{F(z_0)} = \prod_{k=1}^{\infty} \frac{1 + a_k(z)}{1 + a_k(z_0)}$$

holds for all z and z_0 in D.

Proof. By uniform convergence, the series (8.5) can be integrated from z_0 to z along any contour in D. The result is

$$\log_0 \frac{F(z)}{F(z_0)} = \sum_{k=1}^{\infty} \log_j \frac{1 + a_j(z)}{1 + a_j(z_0)}$$

where the branch $\log_j$ can depend on z as well as on j. Taking exponentials gives Theorem 8.5.

In the remainder of this section we discuss an important infinite series expansion, due to Mittag-Leffler, which is closely related to the Weierstrass product. The idea is to express a meromorphic function $F(z)$ by a series, each term of which contains the principle part of F at one of the singularities. The familiar partial fraction expansion for a rational function is an instance of a Mittag-Leffler expansion, just as the factorization theorem for polynomials is an instance of the Weierstrass expansion.

For example, let it be required to construct a function with simple poles of residue 1 at $z = 1,2,3,\ldots$. The attempt to represent such a function by

$$\sum_{k=1}^{\infty} \frac{1}{z - k}$$

is not satisfactory, because the series diverges for every value of z. A satisfactory representation is given by

$$\sum_{k=1}^{\infty}\left(\frac{1}{z - k} + \frac{1}{k}\right) = \sum_{k=1}^{\infty} \frac{z}{k(z - k)}. \tag{8.7}$$

In this case the terms tend to 0 almost as fast as R/k^2, if $|z| < R$, and the series converges except at points where a denominator vanishes. The role of the term $1/k$ is that it just cancels the first term in the expansion of $1/(z-k)$ as a power series in z/k.

Wih this device in mind, we define the Mittag-Leffler primary term for $m = 1,2,3,\ldots$ by

$$L(w,m) = \frac{1}{w-1} + 1 + w + w^2 + \cdots + w^{m-1} \tag{8.8}$$

and also $L(w,0) = 1/(w-1)$. Chapter 1, Example 1.5, gives

$$L(w,m) = \frac{w^m}{w-1}$$

and therefore

$$|L(w,m)| \leq 2|w|^m, \qquad |w| \leq \tfrac{1}{2}. \tag{8.9}$$

There is an important connection between $L(w,m)$ and $E(w,m)$. Since the first term of (8.8) is $-1/(1-w)$,

$$\int_0^w L(z,m)\,dz = \operatorname{Log} E(w,m), \qquad |w| < 1,$$

where the path of integration can be taken along the radius from 0 to w. Estimation of the integral by (8.9) gives

$$|\operatorname{Log} E(w,m)| \leq 2\frac{|w|^{m+1}}{m+1}, \qquad |w| \leq \tfrac{1}{2}$$

which is the basic inequality for the Weierstrass theory.

THEOREM 8.6 (*Mittag-Leffler*). *For $k = 1,2,3,\ldots$, let $\{\alpha_k\}$ and $\{\beta_k\}$ be sequences of numbers, with the α_k distinct, such that $|\alpha_k| \to \infty$ and*

$$\sum_{k=1}^{\infty} \frac{|\beta_k|}{|\alpha_k|^{m+1}} < \infty. \tag{8.10}$$

Then the function

$$G(z) = \sum_{k=1}^{\infty} \frac{\beta_k}{\alpha_k} L(z/\alpha_k,\ m)$$

represents a meromorphic function for which only finite singularities are simple poles at α_k with residue β_k.

Proof. Let $R > 0$ and choose N so that $|\alpha_k| > 2R$ for $k > N$. Then for $|z| < R$ and $k > N$, we have $|z/\alpha_k| < ½$ and hence, by (8.9),

$$\left|\frac{\beta_k}{\alpha_k} L(z/\alpha_k, m)\right| \le \frac{|\beta_k|}{|\alpha_k|} 2 \left|\frac{z}{\alpha_k}\right|^m \le 2 \frac{|\beta_k| R^m}{|\alpha_k|^{m+1}}$$

Since the sum on k converges by (8.10), the series

$$\sum_{k=N}^{\infty} \frac{\beta_k}{\alpha_k} L(z/\alpha_k, m)$$

converges uniformly in $|z| < R$ and represents an analytic function. The remaining terms of the series for $G(z)$ differ only by a polynomial from

$$\sum_{k=1}^{N-1} \frac{\beta_k}{\alpha_k} \frac{1}{z/\alpha_k - 1} = \sum_{k=1}^{N-1} \frac{\beta_k}{z - \alpha_k}$$

and hence have simple poles at $z = \alpha_k$ with residues β_k. This shows that $G(z)$ has the specified properties in $|z| < R$ for every R, and completes the proof.

Theorem 8.6 is not the most general form of the Mittag-Leffler theorem. In its general form, the theorem states that a meromorphic function $f(z)$ can be constructed with singularities at the α_k such that, at each singularity, $f(z)$ has an arbitrarily prescribed principal part. Furthermore, this can be done not only in $|z| < \infty$ but in any domain D, provided the α_k are isolated points of D; that is, provided no subsequence of $\{\alpha_k\}$ has a limit α in D. Proof of the general theorem is similar, in principle, to the proof of Theorem 8.6. The proof is not given here:

Example 8.1. Derive a Mittag-Leffler expansion for $\pi \cot \pi z$ by considering the integral

$$I = \frac{1}{2\pi i} \int_C (\pi \cot \pi\zeta)\left(\frac{1}{z - \zeta} + \frac{1}{\zeta}\right) d\zeta.$$

For C we use the contour of Figure 2-1, taking N so large that z is within

C. Then, by the residue theorem,

$$I = -\pi \cot \pi z + \sum_{n=-N}^{N}{}' \left(\frac{1}{z-n} + \frac{1}{n}\right) + \frac{1}{z}$$

where the prime on the sum means that the term for $n = 0$ is excluded. The term $-\pi \cot \pi z$ arises from the simple pole at $\zeta = z$, the sum arises from simple poles at $\zeta = n$ where $n \neq 0$ is an integer, and the term $1/z$ arises from the double pole at $\zeta = 0$. Since $|\cot \pi\zeta| \leq |\coth \pi|$ on C, as mentioned in Section 2, it is easily shown that $I \to 0$ as $N \to \infty$. This gives

$$\pi \cot \pi z = \frac{1}{z} + \sum_{n=-\infty}^{\infty}{}' \left(\frac{1}{z-n} + \frac{1}{n}\right),$$

which is the required Mittag-Leffler expansion.

Problems

1. By Theorem 8.1 with $m = 0$, 1 or 2, respectively, construct entire functions having zeros at $\{k^{3/2}\}$, $\{k^{3/4}\}$, $\{k^{1/2}\}$, and at these points only, $k = 1,2,3,\ldots$.
2. (a) Derive Mittag-Leffler expansions for the following:

   $$\coth \pi z, \quad \csc \pi z, \quad \csc^2 \pi z, \quad \cos \pi z \csc^2 \pi z.$$

   (b) Verify agreement with Section 2, Problems 7 and 10.
3. Prove that if $f(z)$ is meromorphic for $|z| < \infty$, then $f(z)$ is the quotient of two entire functions. (Construct a Weierstrass product g so that $fg = h$ is entire.)
4. Construct an entire function which has zeros of multiplicity n at $z = n$, $n = 1,2,3,\ldots$, and no other zeros.

The gamma function

5. Let $H(z)$ be defined for $z \neq 0, -1, -2, \ldots$ by

   $$\frac{1}{H(z)} = z e^{\gamma z} \prod_{k=1}^{\infty} \left(1 + \frac{z}{k}\right) e^{-z/k}$$

   where $\gamma \geq 0$ is constant. Show that

   $$\frac{H'(z)}{H(z)} = -\frac{1}{z} - \gamma - \sum_{k=1}^{\infty} \left(\frac{1}{z+k} - \frac{1}{k}\right), \quad \frac{d}{dz}\frac{H'(z)}{H(z)} = \sum_{k=0}^{\infty} \frac{1}{(z+k)^2}.$$

6. If $g(z) = H(z+1)/[zH(z)]$, show that

   $$\frac{g'(z)}{g(z)} = \frac{H'(z+1)}{H(z+1)} - \frac{H'(z)}{H(z)} - \frac{1}{z}.$$

By Problem 5 deduce $g'(z) = 0$ and hence $H(z + 1) = czH(z)$ where c is constant. Letting $z \to 0$, get $c = H(1)$.

7. The constant γ is now chosen so that $H(1) = 1$. Show that this holds if and only if

$$e^{-\gamma} = \prod_{k=1}^{\infty}\left(1 + \frac{1}{k}\right)e^{-1/k} = \lim_{n\to\infty} \frac{2}{1}\frac{3}{2}\frac{4}{3}\cdots\frac{n+1}{n}e^{-s(n)}$$

where $s(n) = 1 + ½ + ⅓ + \cdots + 1/n$. Since the product on the right telescopes to give the single factor $n + 1$, deduce

$$\gamma = \lim_{n\to\infty}\left[1 + \frac{1}{2} + \frac{1}{3} + \cdots + \frac{1}{n} - \text{Log}(n + 1)\right].$$

(The limit exists and is not 0 because the product converges by Theorem 8.1.) The constant γ is called Euler's constant. In contrast to e and π, which are known to be transcendental, it is not even known that γ is irrational.

8. From Problems 6 and 7 deduce $H(z + 1) = zH(z)$. Since also $\Gamma(z + 1) = z\Gamma(z)$, show that the function

$$F(z) = \frac{1}{\sin \pi z}\left(1 - \frac{\Gamma(z)}{H(z)}\right)$$

satisfies $F(z + 1) = -F(z)$. Hence $F(z + 2) = F(z)$. Since $F(z)$ is analytic for $\text{Re}\, z > 0$, this shows that $F(z)$ is entire.

9. For $1 \le x \le 2$, show that

$$|\Gamma(z)| \le \int_0^1 e^{-u}\, du + \int_1^{\infty} e^{-u}u\, du \le \Gamma(1) + \Gamma(2) = 2.$$

10. If $z = x + iy$, show that

$$\frac{1}{|H(z)|^2} = |z|^2 e^{2\gamma x}\prod_{k=1}^{\infty} E_k, \quad \text{where} \quad E_k = \left[\left(1 + \frac{x}{k}\right)^2 + \left(\frac{y}{k}\right)^2\right]e^{-2x/k}.$$

For $x \ge 0$ show that $\partial E_k/\partial x \le 0$ and hence the value of E_k at $x = 0$ dominates the value at $x > 0$. Thus $E_k \le 1 + (y/k)^2$, and hence for $x \ge 0$

$$\frac{1}{|H(z)|^2} \le e^{2\gamma x}|z|^2\frac{\sin \pi iy}{\pi iy} = e^{2\gamma x}|z|^2\frac{\sinh \pi y}{\pi y}.$$

11. Using Problems 9 and 10 in Problem 8, show for $y \ge 1$, $1 \le x \le 2$ that

$$|F(z)| \le \frac{2}{e^{\pi y} - e^{-\pi y}}\left(1 + \frac{e^{\gamma x}|z|e^{\pi y/2}|\Gamma(z)|}{(\pi y)^{1/2}}\right) \le \frac{4}{e^{\pi y}}[1 + e^{2\gamma}(y + 2)e^{\pi y/2}2].$$

Thus $|F(z)|$ is bounded for $1 \le x \le 2, y \ge 1$. Being entire, $F(z)$ is also bounded for $1 \le x \le 2, 0 \le y \le 1$. Since $F(\bar{z}) = \overline{F(z)}$, $F(z)$ is bounded for $1 \le x \le 2$. Since $|F(z + 1)| = |F(z)|$ by Problem 8, $F(z)$ is bounded for $|z| < \infty$ and by

Liouville's theorem $F(z)$ is constant. Letting $y \to \infty$, show that the constant is 0, and hence $\Gamma(z) = H(z)$.

12. Show that $[1/H(z)][1/H(-z)] = -z^2(\sin \pi z)/(\pi z)$ in Problem 5. Since $H(z) = \Gamma(z)$ and $\Gamma(1 - z) = -z\Gamma(-z)$, deduce

$$\Gamma(z)\Gamma(1 - z) = \pi \csc \pi z.$$

Setting $z = ½$, get $\Gamma(½) = \sqrt{\pi}$ and hence

$$2\int_0^\infty e^{-x^2}\, dx = \int_0^\infty t^{-1/2}e^{-t}\, dt = \sqrt{\pi}.$$

13. If Re $z > 0$, prove the first formula by partial integration and deduce the second from the first:

$$\int_0^\infty e^{-uz}u\, du = \frac{1}{z^2}, \quad \sum_{k=0}^{N-1} \frac{1}{(z + k)^2} = \int_0^\infty e^{-uz}\frac{1 - e^{-uN}}{1 - e^{-u}} u\, du.$$

Letting $N \to \infty$ and using $\Gamma(z) = H(z)$ in Problem 5, deduce

$$\frac{d}{dz}\frac{\Gamma'(z)}{\Gamma(z)} = \sum_{k=0}^{\infty} \frac{1}{(z + k)^2} = \int_0^\infty e^{-uz}\frac{u}{1 - e^{-u}}\, du.$$

14. Separate even and odd k in the product for $\Gamma(2z)$; to get

$$\frac{1}{\Gamma(2z)} = 2ze^{2\gamma z}\prod_{k=1}^{\infty}\left(1 + \frac{z}{k}\right)e^{-z/k}\prod_{k=0}^{\infty}\left(1 + \frac{z}{k + ½}\right)e^{-z/(k+1/2)}.$$

Deduce that the function $R(z) = 2\Gamma(2z)/[\Gamma(z)\Gamma(z + ½)]$ satisfies $R(z) = e^{cz}/\Gamma(½)$ where c is constant. Setting $z = 1$, get $e^c = 4$ and hence, since $\Gamma(½) = \sqrt{\pi}$,

$$\Gamma(2z) = \frac{1}{\sqrt{\pi}} 2^{2z-1}\Gamma(z)\Gamma(z + ½).$$

This is known as the *duplication formula.*

***9. Analytic continuation.** Let two domains D_1 and D_2 each contain a third domain D, as shown in Figure 9-1. Let $f_1(z)$ be analytic in D_1. Then there can be at most one function $f_2(z)$ which is analytic in D_2 and agrees with f_1 in the common domain D. Indeed, if $\tilde{f}_2$ is analytic in D_2 and $\tilde{f}_2 = f$ in D, then $f_2 - \tilde{f}_2 = 0$ in D. Hence the same holds in D_2 by Chapter 3, Theorem 7.1. Thus, given f_1, D_1 and D_2, either there is a unique function f_2 as described above or else there is no such function.

When f_2 exists, f_1 is said to have been continued analytically from D_1

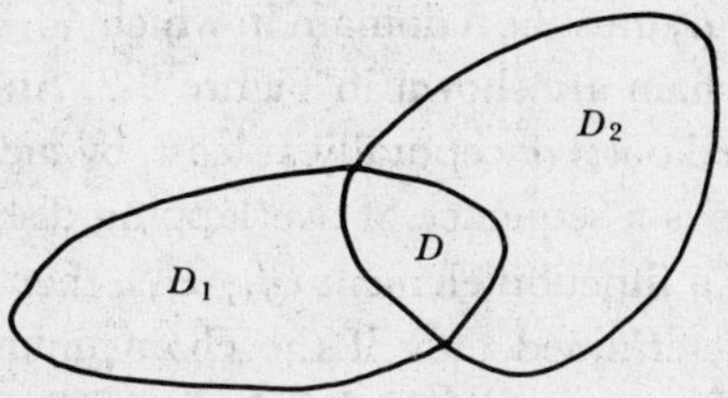

Figure 9-1

to D_2, and f_2 is an analytic continuation of f_1. Similarly, f_1 is an analytic continuation of f_2.

One of the main uses of analytic continuation is to extend functional relations, initially established on a small domain D_1, to a larger domain D_2. The small domain is chosen so that the various series and integrals involved in the calculation converge uniformly or so that branch points are avoided. Thus, the methods of this chapter apply in the small domain.

If the functions involved are known to be analytic it may not even be necessary to have a domain D common to D_1 and D_2; it may be sufficient to have just a short arc of a curve or a line. This also follows from results of Chapter 3, and was used in Chapter 2 to extend relations such as $\sin 2x = 2 \sin x \cos x$ from the real axis to the complex plane. As a more sophisticated example, the function $\Gamma(x)$ satisfies $\Gamma(x + 1) = x\Gamma(x)$ for $x > 1$, as seen by partial integration in (5.1). This derivation fails for $x < 0$, since the integral diverges. However, it follows from Section 5, Problem 4 that $\Gamma(x)$ can be continued as a meromorphic function in the whole plane. Hence, $\Gamma(x + 1) = x\Gamma(x)$ for $x > 1$ gives $\Gamma(z + 1) = z\Gamma(z)$ in the entire domain of analyticity.

The method of analytic continuation requires a more elaborate notation for application to more complicated functions than the meromorphic ones considered in previous examples. An analytic function f_1 defined on a domain D_1 is called a function element and is designated by (f_1, D_1). Of special importance in analytic continuation is the case of a sequence of function elements $\{(f_j, D_j)\}, j = 1,2, \ldots, n$, where each element is an analytic continuation of the one that preceded it. Such a sequence is called a chain. In a chain, D_j must overlap D_{j-1} for $j = 2,3, \ldots, n$ and the intersection of D_{j-1} with D_j must be a domain in which f_{j-1} and f_j are identical. The domains of a chain are shown in Figure 9-2. Analytic continuation can always be carried out, conceptually at least, by means of power series. A chain in this case is a sequence of overlapping disks.

If one starts with a function element (f_1, D_1), then in general the element (f_n, D_n) is determined only if the chain linking this element to (f_1, D_1) is given. For example, if $f(z) = \text{Log } z$ in the domain D_1 of Figure 9-3, the value in D_4 obtained by the chain D_1, D_2, D_3, D_4 differs (by $2\pi i$) from the value obtained in D_4 by means of the chain D_1, D_6, D_5, D_4. Another way of saying essentially the same thing is to say that the chain

$$D_1, D_2, D_3, D_4, D_5, D_6, D_7 = D_1$$

leads back to a different function element from that with which the process started. The original element and the new one are, respectively, $(\text{Log } z, D_1)$ and $(\text{Log } z + 2\pi i, D_1)$.

If distinct function elements in some neighborhood of z_0 can be obtained by analytic continuation of a given function element (f_1, D_1), the distinct function elements are called branches of the general analytic function defined by the given function element (f_1, D_1). For instance, $\text{Log } z + 2\pi i$ is a branch in the domain D_1 of Figure 9-3 obtained by continuation of the original function element $(\text{Log } z, D_1)$, along the chain shown in the figure. Other branches would be obtained by continuation along other chains.

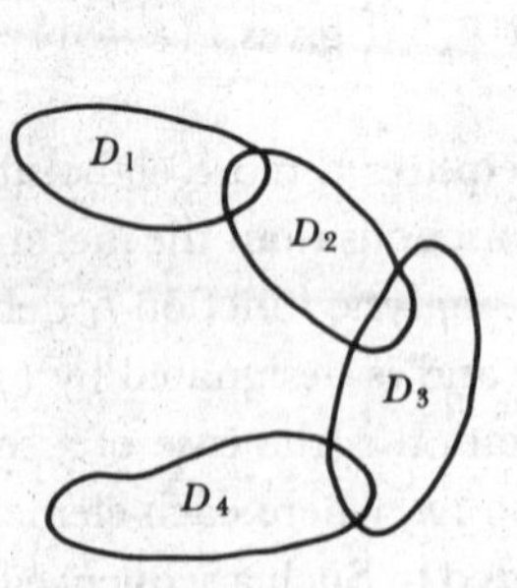

Figure 9-2

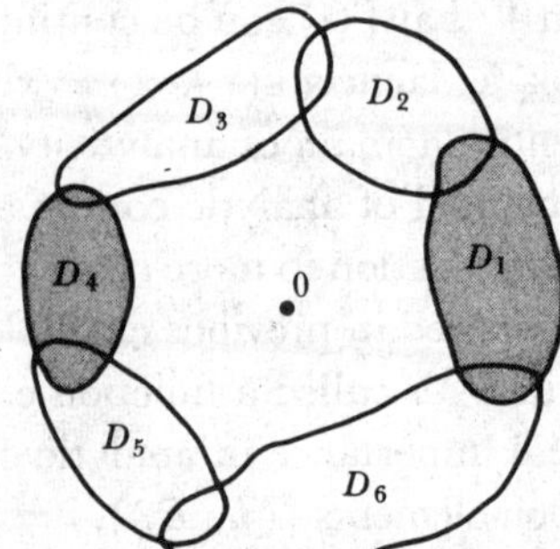

Figure 9-3

The notion of a general analytic function is that of the set of function elements obtainable by chains from a given element. To specify each function element, it is necessary to know the chain that joins it to the given element.

The domain formed by the union of the domains D_i of Figure 9-3 surrounds the origin and is not simply connected. In the case of a simply connected domain, an important theorem known as the monodromy theorem asserts that the result of the continuation will be unique, no matter what chain is used.

To formulate the monodromy theorem, we require the notion of continuation along a polygonal line. As in Chapter 4, Section 1, a polygonal line is a curve $z = \zeta(t)$, $a \le t \le b$, where the interval $[a,b]$ can be divided by points

$$a = t_0 < t_1 < t_2 < \cdots < t_n = b$$

such that $\zeta(t)$ is linear on each interval $[t_k,t_{k+1}]$. The trace of $z = \zeta(t)$ for t on $[t_k,t_{k+1}]$ is called an edge of the polygonal line and is denoted by E_k. It is important that $\zeta(t)$ assigns a definite order to the edges, giving a sequence $E_1,E_2,\ldots,E_n$. A polygonal line with its sequence of edges is illustrated for $n = 8$ in Figure 9-4.

The function element $(f_0,\ D_0)$ is said to be continued analytically along the polygonal line from $z_0 = \zeta(a)$ to $z_1 = \zeta(b)$ if there is a chain

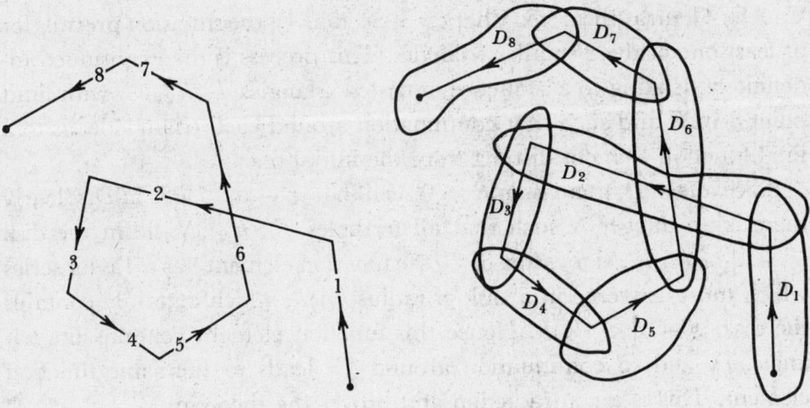

Figure 9-4 **Figure 9-5**

$$(f_0,\ D_0),\ (f_1,\ D_1),\ (f_2,\ D_2), \ldots, (f_n,\ D_n)$$

such that D_i contains the edge E_i for $i = 0,1,2, \ldots, n$. An example of analytic continuation along a polygonal line is illustrated in Figure 9-5 where $n = 8$.

THEOREM 9.1 (*Monodromy theorem*). *Let D be a simply connected domain and let $f_0(z)$ be analytic in some disk D_0 with center at a point z_0 of D. If the function element $(f_0,\ D_0)$ can be continued analytically along every polygonal line in D, then the continuation leads to a single-valued analytic function in D.*

Outline of proof. The theorem is false if and only if the continuation of (f_0, D_0) from z_0 to some point z_1 in D along each of two polygonal lines P_1 and P_2 in D leads to two distinct function elements at z_1. Suppose this occurs. Then continuation from z_0 to z_1 along P_1 followed by continuation from z_1 to z_0 along $-P_2$ must lead to a function element at z_0 distinct from $f_0(z)$. Much as in that part of the proof of Cauchy's theorem (Theorem 1.2 of Chapter 4) which uses Figure 1-5 of Chapter 4, this leads to the existence of a simple, closed polygon such that the continuation of the element at one vertex of the polygon around the polygon back to the starting point results in an element distinct from the initial one. If the polygon is triangulated, it then follows that at least along one of the triangles the same situation must prevail. If such a triangle is replaced by four triangles by dividing each edge in half as in the proof of the Cauchy-Goursat theorem (Chapter 3, Section 4), the situation prevails for at least one of these smaller triangles. This process is then continued indefinitely, leading to a sequence of nested triangles $T_1, T_2, \ldots$ with limit point α in D and such that continuation around each triangle leads to a final function element distinct from the initial one.

Since α is in D, there is an $R > 0$ such that $|z - \alpha| < R$ is in D. Clearly there is an integer N such that all triangles T_n, $n \geq N$, lie in the disk $|z - \alpha| < (1/4)R$. At a vertex of T_N the function element has a Taylor series which must converge in a disk of radius $(3/4)R$, which certainly contains the disk $|z - \alpha| < (1/4)R$. Hence this function element contains the triangle T_N and so continuation around T_N leads to the same function element. This is a contradiction and proves the theorem.

Additional problems on Chapter 6

1.1. *A theorem of Hurwitz.* Let each function of the sequence $\{f_n(z)\}$ be univalent in a domain D, and let $f_n(z) \to f(z)$ uniformly in every closed disk $|z - z_0| \leq R$ contained in D. Then $f(z)$ is either constant or univalent in D. Outline of solution: Suppose f is not constant and that $f(\alpha) = f(\beta)$ where α and β are in D. Then there is a $\delta > 0$ such that $f(z) \neq f(\alpha)$ for $0 < |z - \alpha| \leq \delta$ and $f(z) \neq f(\beta)$ for $0 < |z - \beta| \leq \delta$. Let $m > 0$ be chosen so that

$$|f(z) - f(\alpha)| \geq m \quad \text{and} \quad |f(z) - f(\beta)| \geq m$$

on $|z - \alpha| = \delta$ and $|z - \beta| = \delta$, respectively. If n is large enough, then

$$|f_n(z) - f(z)| < m$$

on these circles. By Rouché's theorem (Chapter 4, Section 6), $f_n(z) - f(\alpha)$ has at least one zero in $|z - \alpha| < \delta$ and $f_n(z) - f(\beta)$ has at least one zero in $|z - \beta| < \delta$. Since $f(\alpha) = f(\beta)$, this contradicts the fact that $f_n(z)$ is simple.

1.2. *Compact subsets.* Let D be a domain and let G be a compact subset of D. (This means that G is closed and bounded and is contained in D.) By considering $|\zeta - z|$ for z on G and ζ on the boundary of D, it is possible to show that the minimum distance from G to the boundary of D is a positive number δ. The number δ also equals the distance from G to the complement of D; that is, to the set of points not in D. These results follow from real analysis, since $|\zeta - z|$ is continuous, and are taken for granted here.

Let $s_n(z)$ be analytic in a domain D and suppose $s_n(z) \to s(z)$ uniformly in every compact subset of D. Show that $s(z)$ is analytic in D, and $s_n'(z) \to s'(z)$ uniformly in every compact subset of D. Outline of solution: The fact that $s(z)$ is analytic follows from Morera's theorem as in Theorem 1.2. To prove uniformity of convergence, let G be a compact subset of D and let the distance from G to the boundary be 4δ. (The distance is denoted by 4δ, rather than δ, to avoid fractions.) The coordinate lines

$$x = \ldots -2\delta, -\delta, 0, \delta, 2\delta, \ldots$$
$$y = \ldots -2\delta, -\delta, 0, \delta, 2\delta, \ldots$$

intersect in points $(j\delta, k\delta)$ where j and k are integers (see Figure 9-6). If z_j is such a point, let D_j be the disk centered at z_j and having radius δ. The disks D_j together cover the whole plane. Let

$$D_1, D_2, \ldots, D_m$$

be the disks of this kind that have a point in common with G, so that together these m disks cover G. The disk D_1 is contained in $|z - z_1| \leq 2\delta$, and the latter disk is a compact subset of D. Hence $s_n(z) \to s(z)$ uniformly in $|z - z_1| \leq 2\delta$, and Theorem 1.2 shows $s_n'(z) \to s'(z)$ uniformly in D_1. Similarly, $s_n'(z) \to s'(z)$

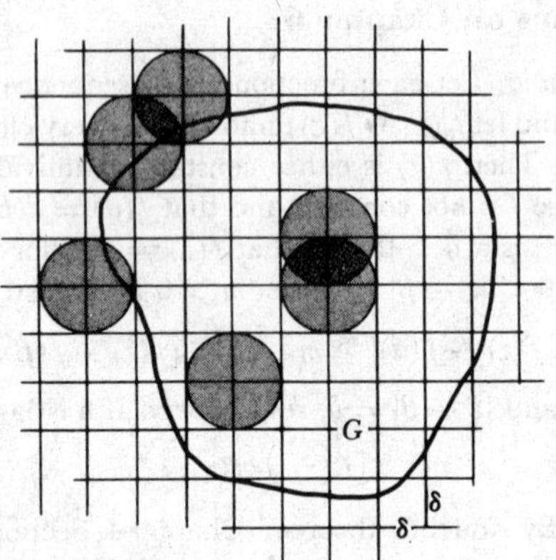

Figure 9-6

uniformly in D_2, and so on. Therefore the same holds in the union of the D_j, and hence it holds in G.

2.1. *Review.* Let $P(z)$ be a polynomial of degree $n \geq 1$. Prove that the equation $P(z) = \cot \pi z$ has $2N + n + 1$ roots within the contour C of Figure 2-1, if N is large. (Let N_1 and N_2 denote the number of zeros of $P(z)$ and $P(z) - \cot \pi z$ within C, respectively, and let P_1 and P_2 denote the respective numbers of poles. Problem 6 of Section 2 gives $|P(z)| > |\cot \pi z|$ on C if N is large and hence, as in the proof of Rouché's theorem, $N_1 - P_1 = N_2 - P_2$.)

2.2. By integrating $(2z + 1)^{-3}\, \pi \csc \pi z$, show that

$$\sum_{n=0}^{\infty} \frac{(-1)^n}{(2n+1)^3} = \frac{1}{2} \sum_{n=-\infty}^{\infty} \frac{(-1)^n}{(2n+1)^3} = \frac{\pi^3}{32}.$$

3.1. Show that a power series can converge at all, some, or no points of its circle of convergence. (Consider $\Sigma z^n/n^2$, $\Sigma z^n/n$, Σz^n.)

3.2. This problem requires familiarity with the concept of upper limit. Prove that the radius of convergence of a power series with coefficients a_n satisfies $1/R = \limsup |a_n|^{1/n}$. With suitable conventions for $R = 0$ or $R = \infty$, this holds in all cases. Thus deduce Theorem 3.2 and also deduce that the radius is unchanged by differentiation or integration of the power series.

3.3. The Fibonacci numbers 1,1,2,3,5,8, . . . satisfy $a_0 = a_1 = 1$, $a_n = a_{n-1} + a_{n-2}$ for $n \geq 2$. If

$$f(z) = a_0 + a_1 z + a_2 z^2 + a_3 z^3 + \cdots + a_n z^n + \cdots$$

show that $f(z) = 1 + zf(z) + z^2 f(z)$. Express $f(z) = (1 - z - z^2)^{-1}$ in partial fractions, expand each partial fraction in geometric series, and thus get

$$a_n = \frac{1}{\sqrt{5}}\left[\left(\frac{1+\sqrt{5}}{2}\right)^{n+1} - \left(\frac{1-\sqrt{5}}{2}\right)^{n+1}\right].$$

The function $f(z)$ is called the *generating function* of the sequence $\{a_n\}$.

3.4. The Hermite, Chebyshev and Legendre differential equations are, respectively,

$$Y'' + 2mY = 2zY', \quad (1-z^2)Y'' + m^2Y = zY',$$
$$(1-z^2)Y'' + m(m+1)Y = 2zY'$$

where m is a complex constant. (a) Find power series solutions satisfying $Y(0) = 1$, $Y'(0) = 0$. (b) Find solutions satisfying $Y(0) = 0$, $Y'(0) = 1$. (c) Show that the Laguerre equation $zY'' + mY = (z-1)Y'$ has no solution satisfying the above conditions if $m \neq 0$, but has a power-series solution satisfying $Y(0) = 1$, $Y'(0) = -m$.

3.5. Show how to get the coefficients of the series

$$\pi z \cot \pi z = a_0 + a_2z^2 + a_4z^4 + \cdots$$

by recursion. (The equation can be written

$$\cos \pi z = \frac{\sin \pi z}{\pi z}(a_0 + a_1z + a_2z^2 + \cdots)$$

or, upon using the Taylor series for $\sin \pi z$ and $\cos \pi z$,

$$1 - \frac{\pi^2z^2}{2!} + \cdots = \left(1 - \frac{\pi^2z^2}{3!} + \frac{\pi^4z^4}{5!} - \cdots\right)(a_0 + a_2z^2 + a_4z^4 + \cdots).$$

Equate coefficients of $1, z^2, z^4, \ldots$ on each side; see Chapter 3, Section 6, Problem 12.)

3.6. By integrating $z^{-2}\pi \cot \pi z$ or $z^{-4}\pi \cot \pi z$, show that

$$\sum_{n=1}^{\infty} \frac{1}{n^2} = \frac{\pi^2}{6}, \qquad \sum_{n=1}^{\infty} \frac{1}{n^4} = \frac{\pi^4}{90}$$

Note that the residues at $z = 0$ can be found by Problem 3.5.

3.7. Without using integration, prove that a power series can be differentiated term by term within its circle of convergence. Solution: Let $\alpha = 0$ and let z_0 be some point satisfying $|z_0| < R$. Let $|z_0| < R_0 < R_1 < R$. If $z \to z_0$, it can be assumed also that $|z| < R_0$. Then, for $z \neq z_0$,

$$\frac{f(z) - f(z_0)}{z - z_0} = \sum_{n=1}^{\infty} a_n \frac{z^n - z_0^n}{z - z_0} = \sum_{n=1}^{\infty} a_n(z^{n-1}z_0 + \cdots + zz_0^{n-1}).$$

Since $|z_0| < R_0$ and $|z| < R_0$, the general term satisfies

$$|a_n(z^{n-1}z_0 + \cdots + zz_0^{n-1})| \le |a_n|(R_0^n + \cdots + R_0^n) = n|a_n|R_0^n.$$

Convergence of the series at R_1 gives $|a_nR_1^n| < 1$ for large n, and hence

$$n|a_n|R_0^n = n|a_n|R_1^n\left(\frac{R_0}{R_1}\right)^n \le n\rho^n$$

where $\rho = R_0/R_1 < 1$. The series with $M_j = j\rho^j$ converges by the ratio test, and this establishes uniform convergence of the series above. Since the terms are continuous, Theorem 1.1 gives

$$f'(z_0) = \lim_{z\to z_0} \frac{f(z) - f(z_0)}{z - z_0} = \sum_{n=0}^{\infty} na_nz_0^{n-1}.$$

3.8. *Simple functions.* Let $f(z)$ be analytic for $|z| < R$, let $|f'(0)| = a_1 \neq 0$, and let $0 < R_0 < R$. As in Problem 3.7,

$$\left|\frac{f(z) - f(z_0)}{z - z_0}\right| \ge |a_1| - \sum_{n=2}^{\infty} n|a_n|R_0^n$$

for $|z| < R_0$, $|z_0| < R_0$, $z \neq z_0$. Show that the right-hand side is positive if $R_0 < \epsilon$, where ϵ is sufficiently small, and hence the mapping $w = f(z)$ is simple in $|z| < \epsilon$. Another proof was given in Chapter 5, Section 6.

4.1. *Gauss mean-value theorem.* Let $u(r,\theta)$ be continuous for $r \le R$ and harmonic for $r < R$. Show by the Poisson formula that

$$u(0,\theta) = \frac{1}{2\pi}\int_0^{2\pi} u(R,\theta)\, d\theta.$$

In other words, the value of a harmonic function at the center of a circle is the average of the values on the circumference.

4.2. *Harnack inequality.* If the function $u(r,\theta)$ of Problem 4.1 satisfies $u(R,\phi) = u(\phi) \ge 0$, and if $u(r,\theta) \not\equiv 0$, show that

$$\frac{R-r}{R+r} \le \frac{u(r,\theta)}{u(0,\theta)} \le \frac{R+r}{R-r}, \qquad 0 \le r < R.$$

(Note that the integrand in (4.11) lies between

$$\frac{R^2 - r^2}{(R+r)^2}u(\phi) \quad\text{and}\quad \frac{R^2 - r^2}{(R-r)^2}u(\phi).$$

Multiply by $1/2\pi$, integrate from 0 to 2π, and use Problem 4.1.)

4.3. *The Schwarz*[1] *inequality.* Let $a_i \ge 0$, $b_i \ge 0$ and $a > 0$, $b > 0$ where

$$a^2 = a_1^2 + a_2^2 + \cdots + a_n^2, \qquad b^2 = b_1^2 + b_2^2 + \cdots + b_n^2.$$

By summing $(ba_i - ab_i)^2 = b^2a_i^2 - 2aba_ib_i + a^2b_i^2$ from $i = 1$ to n, deduce

[1] The inequality is due to Cauchy.

$$b^2a^2 - 2ab\sum_{i=1}^{n} a_ib_i + a^2b^2 \geq 0$$

and hence, dividing by $2ab$,

$$\sum_{i=1}^{n} a_ib_i \leq \left(\sum_{i=1}^{n} a_i^2\right)^{1/2}\left(\sum_{i=1}^{n} b_i^2\right)^{1/2}.$$

(This also holds, trivially, for $a = 0$ or $b = 0$.) Taking $a_i = |\alpha_i|$ and $b_i = |\beta_i|$, get an extension to complex α_i and β_i.

4.4. Let $f(z)$ be analytic for $|z| < 1$ and satisfy $|f(z) \leq 1$. Show by the Parseval equality that $|f(0)|^2 + |f'(0)|^2 \leq 1$ and deduce

$$|f(0)| \cos t + |f'(0)| \sin t \leq 1$$

for all real t. (Use the Schwarz inequality with $n = 2$.)

4.5. Let $|f(0)| = 0$ and $|f'(0)| = a > 0$. If $f(z)$ is analytic for $|z| < 1$ and satisfies $|f(z)| \leq 1$, show that the coefficients of its power series expansion satisfy $|a_2|^2 + |a_3|^2 + \cdots \leq 1 - a^2$ and hence, by the Schwarz inequality,

$$\left|\sum_{n=2}^{\infty} a_nz^n\right|^2 \leq \left(\sum_{n=2}^{\infty}|a_n|^2\right)\left(\sum_{n=2}^{\infty} r^{2n}\right) \leq (1 - a^2)\frac{r^4}{1 - r^2}$$

for $|z| = r < 1$. Thus deduce that

$$|f(z)| \geq ar - \left(\frac{1 - a^2}{1 - r^2}\right)^{1/2} r^2.$$

4.6. *Landau's theorem.* Let $f(z)$ be as in Problem 4.5 and let $A = (16 - 12a^2)^{1/2}$, so that $2 \leq A < 4$. If $|w| < a^2/A$, show that the equation $f(z) = w$ has one and only one solution z such that $|z| < 2a/A$. (Apply Rouché's theorem to $f(z)$ and $f(z) - w$ in $|z| < r_0$, where r_0 is chosen so that

$$\left(\frac{1 - a^2}{1 - r_0^2}\right)^{1/2} r_0^2 = \frac{1}{2} ar_0.$$

Problem 4.5 gives $|f(z)| \geq ar_0/2$ on $|z| = r_0$.)

5.1. Show that the following integrals represent analytic functions in $|\text{Re } z| < 1$, $\text{Re } z > 1$, and $0 < \text{Re } z < 1$, respectively:

$$\int_0^{\pi/2} (\tan t)^z \, dt, \quad \int_0^{\infty} \frac{t^{z-1}}{e^t - 1} \, dt, \quad \int_0^{\infty} t^{z-1} \cos t \, dt.$$

(In the third case integrate by parts on $[1,b]$ before letting $b \to \infty$.)

5.2. (a) Let $\text{Re } s > 0$ and $n = 1,2,3,\ldots$. Obtain the first of the following formulas by setting $nu = t$ and deduce the second from the first:

$$\int_0^{\infty} e^{-nu}u^{s-1} \, du = \frac{\Gamma(s)}{n^s}, \quad \Gamma(s)\sum_{n=1}^{N} \frac{1}{n^s} = \int_0^{\infty} \frac{1 - e^{-Nu}}{1 - e^{-u}} e^{-u}u^{s-1} \, du.$$

(b) For $\operatorname{Re} s = \sigma > 1$ use $e^u - 1 \geq u$ and $|u^{s-1}| = u^{\sigma-1}$ to get

$$\left|\int_0^\infty \frac{e^{-Nu}}{1-e^{-u}} e^{-u} u^{s-1}\, du\right| \leq \int_0^\infty \frac{e^{-Nu}}{u} u^{\sigma-1}\, du = \frac{1}{N^{\sigma-1}} \int_0^\infty e^{-v} v^{\sigma-2}\, dv.$$

As $N \to \infty$ this tends to 0. Thus derive (5.3).

5.3. For $n = 1,2,3, \ldots$ let $u_n(r,\theta)$ be harmonic in $|z| < R_0$ and suppose that $u_n(r,\theta) \to u(r,\theta)$ uniformly in every disk $|z| \leq R$, where $R < R_0$. Prove that $u(r,\theta)$ is harmonic in $|z| < R_0$. Outline of solution: Since $u_n(r,\theta)$ is harmonic for $|z| \leq R$, uniqueness shows that $u_n(r,\theta)$ is given by the Poisson formula,

$$u_n(r,\theta) = \int_0^{2\pi} K(r,\theta - \phi) u_n(R,\phi)\, d\phi.$$

If $r \leq R_1 < R$, where R_1 is fixed, $K(r,\theta - \phi)$ is bounded and the integrands form a uniformly convergent sequence. By Theorem 1.1 the limit of the right side as $n \to \infty$ is the Poisson integral of $u(R,\phi)$, and by hypothesis the limit of the left side as $n \to \infty$ is $u(r,\theta)$. This shows that $u(r,\theta)$ is given by the Poisson formula for $r < R_1$, hence is harmonic for $r < R_1$.

5.4. Using Problem 5.3 prove: If a sequence of harmonic functions converges uniformly on a domain D, then the limit function is harmonic on D.

5.5. This problem requires knowledge of double and repeated integrals. Let C be as in the proof of Theorem 5.4, so that by the Cauchy integral formula

$$F(z,t) = \frac{1}{2\pi i} \int_C \frac{F(\zeta,t)}{\zeta - z}\, d\zeta, \qquad |z - \alpha| < R.$$

Thus get a formula for $f(z)$ in Theorem 5.4 as a double integral. By considering $[f(z+h) - f(z)]/h$ as in the proof of Chapter 3, Theorem 5.2, establish Theorem 5.4.

5.6. Let $F(z,t)$ be continuous for $a \leq t < \infty$ and let

$$f_n(z) = \int_a^{b_n} F(z,t)\, dt, \qquad f(z) = \int_a^\infty F(z,t)\, dt.$$

If the improper integral exists in the sense of Chapter 4, show that $f_n(z) \to f(z)$ for every real sequence $\{b_n\}$, $b_n \to \infty$, $b_n > a$. Conversely, if $f_n(z) \to f(z)$ for every such sequence $\{b_n\}$, then the improper integral exists and equals $f(z)$.

6.1. (a) If $x > 0$ and $1 > \operatorname{Re} \alpha > 0$, integrate around an indented quadrant and show that

$$\int_0^\infty e^{itx} t^{\alpha-1}\, dt = x^{-\alpha} e^{\pi i\alpha/2}\, \Gamma(\alpha).$$

(b) If $x > 0$ and $1 > \operatorname{Re} \alpha > 0$ as above, let

$$f(x) = \int_1^\infty e^{itx} t^{\alpha-1}\,dt, \qquad g(x) = \int_0^1 e^{itx} t^{\alpha-1}\,dt.$$

Investigate the behavior of $f(x)$ for large x by partial integration, and thus find the behavior of $g(x)$ for large x.

7.1. Let $f(z) \not\equiv 0$ be analytic for $|z| < 1$ and have zeros $\alpha_k \neq 0$. If $|f(z)| < 1$ in $|z| < 1$, use Jensen's inequality (Chapter 4, Problem 4.2) to show that the product $|\alpha_1||\alpha_2||\alpha_3|\cdots$ converges. Deduce from this that

$$\sum_{k=1}^{\infty} (1 - |\alpha_k|) < \infty. \tag{*}$$

7.2. *Blaschke products.* Let $\{\alpha_k\}$ be a sequence of complex numbers such that $0 < |\alpha_k| < 1$ and let (*) hold. Prove that

$$f(z) = \prod_{k=1}^{\infty} \frac{z - \alpha_k}{z - 1/\bar{\alpha}_k}$$

represents an analytic function for $|z| < 1$ which has its zeros at the points α_k. Show also that $|f(z)| < 1$.

8.1. By applying Theorem 8.5 to the result of Example 8.1, get

$$\frac{\sin \pi z}{\sin \pi z_0} = \frac{z}{z_0} \prod_{n=-\infty}^{\infty}{}' \frac{z - n}{z_0 - n} e^{(z-z_0)/n}$$

where the prime means that the factor for $n = 0$ is omitted. Multiply by $\sin \pi z_0$ and let $z_0 \to 0$ to get a product for $\sin \pi z$.

8.2. Let $g(z)$ be an entire function with simple zeros at the points α_k. Let $h(z)$ be a meromorphic function which has simple poles with residues ρ_k at the α_k and no other finite singularities. Show that $f(z) = g(z)h(z)$ has removable singularities at the α_k and that if $f(\alpha_k) = g'(\alpha_k)\rho_k$, then $f(z)$ is entire.

8.3. *Interpolation.* Let $\{\alpha_k\}$ be any sequence of complex numbers such that $|\alpha_k| \to \infty$ and let $\{\beta_k\}$ be any sequence of complex numbers. Prove that there exists an entire function $f(z)$ such that $f(\alpha_k) = \beta_k$. (See Problem 8.2.)

8.4. The decomposition in Theorem 8.3 is not unique. Explain how to make it unique by classifying Weierstrass products according to their simplicity and requiring that $P(z)$ be the simplest convergent product for the α_k. The unique product so obtained is called the *canonical product.*

9.1. The following are analytic for $|z| < 1$ and are then continued as general analytic functions. Which equation determines one entire function; which, more than one entire function; and which, one or more many-valued analytic functions?

$$(z^{1/2})^2, \quad (z^2)^{1/2}, \quad \cos z^{1/2}, \quad (1 - z)^{1/3}, \quad (e^z)^{1/3}, \quad (\cos z)^{1/2}.$$

9.2. Let $f(z)$ be an entire function of period 2π such that $|f(z)| \leq Me^{c|z|}$ for constants M and c. Show that $f(z)$ has the form

$$f(z) = \sum_{k=-n}^{n} a_k e^{ikz}$$

where a_k are complex constants. (The function $g(w) = f(i \log w)$ is independent of the branch used for $\log w$, by the periodicity of f, and hence is analytic in $0 < |w| < \infty$. If $n > c$, show that $w^n g(w)$ has only a removable singularity at 0 and grows like $|w|^{2n}$ at ∞. Hence $w^n g(w) = P(w)$ where P is a polynomial.)

9.3. *Natural boundary.* If $f(z)$ is analytic in a domain D, it may happen that every boundary point of D is a singular point, so that f cannot be continued beyond D. In this case the boundary of D is said to be a natural boundary for f. Show that $|z| = 1$ is a natural boundary for the following functions:

$$f(z) = \sum_{n=1}^{\infty} z^{n!}, \qquad g(z) = \sum_{n=1}^{\infty} \frac{z^{n!}}{n!}.$$

Note that the series for g converges just as rapidly as that for e^z in $|z| \leq 1$, and hence $g(z)$ is not only analytic in $|z| < 1$, but is continuous in $|z| \leq 1$. (Since $zg'(z) = f(z)$ it suffices to show that $|z| = 1$ is a natural boundary for f. The equation $f(z) = zg'(z)$ or the ratio test shows that f is analytic in $|z| < 1$. Assume that there is a point α on $|z| = 1$ which is not a singular point for $f(z)$. Then $f(z)$ can be continued analytically onto a neighborhood of α and hence there is some arc of $|z| = 1$ of length $\delta > 0$ on which $f(z)$ is analytic. Choose a positive integer $q > 2\pi/\delta$. By considering $0,\ q,\ 2q,\ 3q, \ldots$, find a point $\beta = \exp(2\pi i p/q)$, with integral p, which is on the arc of analyticity. Then $\beta^q = 1$ and hence $\beta^{n!} = 1$ for $n \geq q$. Thus, for $r < 1$,

$$f(r\beta) = \sum_{n=1}^{q-1} r^{n!}\beta^{n} + \sum_{n=q}^{\infty} r^{n!}.$$

This shows that

$$|f(r\beta)| \geq \sum_{n=q}^{\infty} r^{n!} - (q-1)$$

and $|f(r\beta)| \to \infty$ as $r \to 1-$. Thus f cannot be analytic at β.)

9.4. *Reflection principle for harmonic functions.* Let $v(x,y)$ be harmonic in the half-disk $|z| < R,\ y > 0$ and continuous in the closed half-disk (see Figure 9-7). Let $v(x,0) = 0$, and let v be extended to $y < 0$ by reflection; that is,

$$v(x,y) = -v(x,-y)$$

for $y \leq 0$, $|z| \leq R$. If $\tilde{v}(x,y)$ is defined by the Poisson formula with the same boundary values as those of v, show that $\tilde{v} = v$ on the boundary of both half-disks. Hence $\tilde{v} = v$ in $|z| < R$, and hence v is harmonic in $|z| < R$.

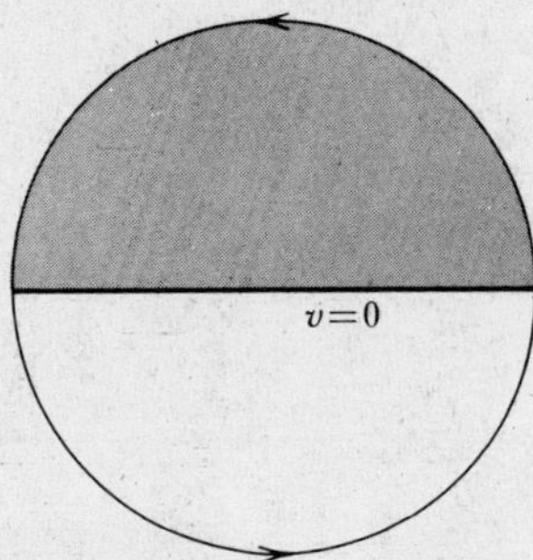

Figure 9-7

This shows that the requirement of continuity of $f = u + iv$ in the Schwarz method of analytic continuation (Chapter 5, Section 7) is more severe than necessary. It suffices for continuation across $y = 0$, for example, only to have v satisfy the above conditions, and no continuity hypothesis on u is needed. Once v is known to be harmonic there is no difficulty in constructing a harmonic conjugate u and reconstructing f from v.

Notes

(1) Chapter 1, Section 1. In analysis, two polynomials P and Q are considered to be equal if $P(z) = Q(z)$ for all complex z. One often writes just $P(z) = Q(z)$, the phrase "for all z" being understood. It is not hard to show that $P = Q$ in this sense only if corresponding coefficients agree, so that the analytic definition is consistent with the algebraic definition of polynomial equality (see Section 5, Problem 6). The equivalence of the two definitions hinges on the fact that the complex field has characteristic 0.

(2) Chapter 2, Section 3, Problem 1. The definition of α^z given here is not used when $\alpha = e$. That is, e^z denotes the function introduced in Section 2, and does not denote the many-valued function $\exp(z \log e)$.

(3) Chapter 2, Section 6. A more satisfactory basis for the theory of fluid flow is given by the formula

$$\int_C f'(z)\,dz = \int_C \mathbf{v} \cdot \mathbf{t}\,ds + i \int_C \mathbf{v} \cdot \mathbf{n}\,ds$$

which is stated in Problem 2.4 of Chapter 3. The discussion of the text is conditioned, in part, by the fact that line integrals and Cauchy's theorem are not available in Chapter 2.

(4) Chapter 2, Section 6. For additional applications see Lamb, *Hydrodynamics,* Dover Publications, New York, 1945, and Milne-Thompson, *Theoretical Hydrodynamics,* Macmillan and Company, London, 1955. A uniqueness theorem for fluid flow is proved in Pennisi, *Elements of Complex Variables,* Holt, Rinehart and Winston, New York, 1966.

(5) Chapter 3, Section 1. The "set of points occupied by the curve" $z = \zeta(t)$ is, by definition, $\{\zeta(t) \mid a \le t \le b\}$.

(6) Chapter 3, Section 2. Both of us have taught complex analysis for many years, using a variety of excellent texts. This experience has naturally influenced our own exposition of the subject. For example, for the derivation of (2.1) we are indebted to Ahlfors, *Complex Analysis,* McGraw-Hill Book Company, New York, 1953 (rev. ed. 1966).

(7) Chapter 3, Section 5, Problem 8. A function $\zeta(t)$ is *piecewise continuous* for $a \le t \le b$ if there is a sequence of points $a = t_0 < t_1 < \cdots < t_n = b$ such that $\zeta(t)$ is continuous on each interval $t_k < t < t_{k+1}$. It is required further that $\zeta(t)$ have a limit as $t \to t_k$ and $t \to t_{k+1}$ through values on this interval. Thus, piecewise continuous functions are integrable and bounded.

(8) Chapter 3, Section 8. A start toward the proof of Picard's theorem is given by Landau's theorem, Chapter 5, Section 7. Completion of the proof along these lines can be found in Titchmarsh, *The Theory of Functions,* Oxford University Press, Oxford, 1939 (reprint 1960) and also in Hille, *Analytic Function Theory,* Volume II, Blaisdell Publishing Company, New York, 1962.

(9) Chapter 3, Section 10, Problem 1.3. Derivation of Kepler's laws and their converses by use of complex-valued functions can be found in Sokolnikoff and Redheffer, *Mathematics of Physics and Modern Engineering,* McGraw-Hill Book Company, New York, 2nd Edition, 1966.

(10) Chapter 4, Section 1. Proof of the Jordan curve theorem for polygons and of the possibility of triangulation is in Hille, *Analytic Function Theory,* Volume I, Blaisdell Publishing Company, New York, 1959.

(11) Chapter 4, Section 2. The Jordan curve theorem in sufficient generality for this book is established in Pederson, "The Jordan Curve Theorem for Piecewise Smooth Curves," *The American Mathematical Monthly,* **76** (June–July 1969). The general case is dealt with in Newman, *Elements of the Topology of Plane Point Sets,* Cambridge University Press, Cambridge, 1954.

(12) Chapter 4, Section 7. An account of the Bode equations can be found in Guillemin, *The Mathematics of Circuit Analysis,* John Wiley & Sons, New York, 1949. The Bromwich inversion formula and a uniqueness theorem are proved in Widder, *The Laplace Transform,* Princeton University Press, Princeton, 1946.

(13) Chapter 4, Section 7. The Hilbert transform is an example of a singular integral, because the integrand becomes infinite at a point on the path of integration. Singular integrals can be formed when the path of integration is any simple contour, not necessarily the y axis. For an elementary discussion of singular integrals see Levinson, "Simplified Treatment of Integrals of Cauchy Type, the Hilbert Problem, and Singular Integral Equations," *SIAM Review,* **7** (Oct. 1965).

(14) Chapter 4, Section 9. The study of integrals over closed curves which bound regions belongs to a branch of topology known as homology theory. The proof of the strong form of Cauchy's theorem outlined in Section 9 is a homology argument, as is the proof of the Cauchy-Goursat theorem in Chapter 3. A general development of homology can be found in Ahlfors (see Note 6).

(15) Chapter 4, Section 10. The study of continuous deformation of curves belongs to a branch of topology known as homotopy theory. Proof of the principle of deformation of contours as sketched in Section 10 is a homotopy argument. It is possible to give a homotopy form of the monodromy theorem (Chapter 6, Section 9). For unified discussion see Redheffer, "The Homotopy Theorems of Function Theory," *The American Mathematical Monthly,* **76** (Aug.–Sept. 1969).

(16) Chapter 5, Sections 1–5 can be taken up before Chapters 3 and 4, from which we have borrowed only the concept of arc and the chain rule. In Chapter 5 an arc is a differentiable curve, as indicated in Section 1. The chain rule is proved in Chapter 3, Section 1, Problem 9.

(17) Chapter 5, Section 3. The use of conformal mapping in solving boundary value problems is briefly illustrated in Problems 3.1–3.6 and 4.1–4.4 at the end of this chapter. Further applications can be found in Churchill, *Complex Variables and Applications,* McGraw-Hill Book Company, New York, 1960.

(18) Chapter 5, Section 3, Problem 5. The discussion in Chapter 2, Problem 6.1 is in some respects more complete, but does not show the connection with conformal mapping.

(19) Chapter 5, Section 4. Proof of the Schwarz-Christoffel formula along lines suggested here is in Hille, Volume II (see Note 8). A somewhat different proof is given in Ahlfors (see Note 6).

(20) Chapter 5, Section 5. The paraconjugate is discussed in Belevitch, *Classical Network Theory,* Holden-Day, San Francisco, 1968. Positive real functions are discussed in Belevitch and in Guillemin (see Note 12).

(21) Chapter 5, Section 8. Some progress toward proof of the Riemann mapping theorem is given by this section together with Problems 8.1 and 8.2. Analysis needed to complete the proof can be found in most advanced texts on complex analysis, including those cited here.

(22) Chapter 6, Section 5. The gamma function and other special functions are discussed from the point of view of complex analysis in Copson, *Theory of Functions of a Complex Variable,* Clarendon Press, Oxford, 1935 (reprint Oxford University Press, 1960).

(23) Chapter 6, Section 6. An inequality such as $|\arg z| \leq \pi - \delta$ automatically specifies the intended branch of $\arg z$.

(24) Chapter 6, Section 5, Problems 7–10. It should be observed that this problem sequence does not just establish existence of the radial limit $u(r,\theta_0) \to u(\theta_0)$ as $r \to R-$, but allows (r,θ) to approach (R,θ_0) in any manner so long as $r \leq R$. This refinement is necessary for the assertion of continuity in $|z| \leq R$. Problems 6–9 of Chapter 4, Section 7 can be refined similarly, to allow $(x,y) \to (x_0,y_0)$ along any path in $\operatorname{Re} z \geq 0$.

(25) Chapter 6, Section 7. More specifically, $\log_j z = \operatorname{Log} z + 2\pi i N_j$ where N_j is an integer for each j. This notation does not conflict with that of Chapter 2, since there we used Log_j rather than $\log_j$.

(26) Chapter 6, Section 8. The Mittag-Leffler expansion is much more general than the Weierstrass product and leads to the latter when applied to $f'(z)/f(z)$ (see Theorem 8.5). The Mittag-Leffler theorem for a general domain D leads similarly to a Weierstrass product theorem for D.

(27) Chapter 6, Section 9. There is no loss of generality in associating each edge E_k with a single domain D_k, because if several domains are used for the continuation along E_k their union gives a single domain.

(28) Chapter 6, Section 9, Problem 9.3. Every domain D is the region of existence for some analytic function; that is, there exists a function which is analytic in D and which cannot be continued outside of D. This is proved by use of a Weierstrass product for D (see Note 26).

Answers

Chapter 1

Section 1, page 7

1. $1, -2, 3+2i, 4-i, 5+i$.
4. $2i, -32+i32, -1+i2, x^3-3xy^2+i(3x^2y-y^3), x^2+y^2,$

$$\frac{x^2-y^2}{x^2+y^2}+i\frac{-2xy}{x^2+y^2}, \frac{2x(1-y)}{x^2+(1-y)^2}+i\frac{x^2-(1-y)^2}{x^2+(1-y)^2}.$$

Section 2, page 15

1. $1, (x-1)^2+y^2, (x^2+y^2)^2, \left[\frac{(x+1)^2+y^2}{(x-1)^2+y^2}\right]^{1/2}, 1.$
3. Partial answer: (a) line, half-plane, circle, region outside a disk, line, half-plane, annulus; (b) circle, disk, line, boundary of square.
4. Partial answer: translation, translation, expansion, expansion and translation, reflection and translation.
5. 11^2+29^2. 6. $(\text{Re}\,\alpha)/(\text{Im}\,\alpha)$. 7. Circle or point. 11. (b) yes.

Section 3, page 22

1. (a) $1, (-1\pm i\sqrt{3})/2; -2, 1\pm i\sqrt{3}; -i, (i\pm\sqrt{3})/2$.
   (b) $z=\pm 2, \pm 2i, (1\pm i\sqrt{3})/2, -1$.
2. $(\pm 1\pm i)/\sqrt{2}; \pm\sqrt[4]{2}(\cos\pi/8+i\sin\pi/8)$; (same)$/\sqrt{2}$;
   $1, -1, (\pm 1\pm i\sqrt{3})/2; \pm 2\pm 2i; 2^{4/3}, 2^{1/3}(-1\pm i\sqrt{3})$.
3. Rotation about $z=0$ through an angle $-90°, -45°, 30°$.
5. (a) $\sin 3\theta = 3\cos^2\theta\sin\theta - \sin^3\theta$.

Section 4, page 32

1. Partial answer: $f[f(z)] = z+2, z^4, z, -1/z$.
2. (i), (iii), (iv), (v).
3. Partial answer: $\theta = 2\pi/3, \pi/2, \pi/3$.
4. Partial answer: $g(w) = -w, 1/w, (1-w)/(1+w), w^{1/2}, w^{1/3}$,
   $(w-i)^{1/4}+1$.

Section 5, page 39

1. (b), (c). Location of removable discontinuities: none; ∞; none; ∞; $-1, \infty$; ∞; $0, 1, \infty$. Appropriate definition: $-$; 0; $-$; 0; 1, 1; 1; 0, ¼, 0.
2. $1, 4z_0^3, -1/z_0^2, -2/z_0^3$.
4. Partial answer: The domains of ii, iii, iv are not connected sets, hence are not domains in the sense of Section 4.
7. $(z^2+1)(z+1)(z-1)(z^2+z\sqrt{2}+1)(z^2-z\sqrt{2}+1)$.

8. Partial answer: If $\omega_1, \omega_2, \ldots, \omega_n$ are the n nth roots of 1, then $\omega_1 + \omega_2 + \cdots + \omega_n = 0$, $\omega_1\omega_2\cdots\omega_n = (-1)^{n+1}$.

Chapter 2

Section 1, page 52

1. Partial answer: $10(2z+3)^4$, $2i/(z+i)^2$, $(2z+z^2)^3(1+z^3)(14z^4+20z^3+8z+8)$.
2. (a) neither; (b) both.

Section 2, page 61

1. -1, $\cos 1 + i\sin 1$, $i\sinh\pi$, $\sinh 1\cos 1 + i\cosh 1\sin 1$.
3. (b) Circle of radius R centered at z_0.
10. (a) $\pi/2 + k\pi$, $ik\pi$, $i\pi/2 + ik\pi$, $k\pi$, $ik\pi$; (b) $2k\pi$, $2ik\pi$, $2ik\pi$, $k\pi$, $ik\pi$.
14. $i\pi/2 + 2k\pi i$, $\operatorname{Log} 2 + i\pi/3 + 2k\pi i$, $\pi/2 + 2k\pi \pm i\operatorname{Log}(2+\sqrt{3})$, $\pi/2 + k\pi \pm i\operatorname{Log}(\sqrt{2}-1)$.

Section 3, page 69

3. 1st row: $-\pi i/2 + 2ik\pi$, $\tfrac{1}{2}\operatorname{Log} 2 + i\pi/4 + 2k\pi i$, $3\operatorname{Log} 2 + i\pi/2 + 6k\pi i$, $3\operatorname{Log} 2 + i\pi/2 + 2k\pi i$, $(\pi i/2)\operatorname{Log} 2 - \pi^2/4 - 2k\pi^2 + 2n\pi i$.
   2nd row: $e^{-\pi/2}e^{2k\pi}$, -1,
   $e^{\pi\operatorname{Log} 3}e^{2ik\pi^2}$, $e^{\pi i\operatorname{Log} 2}e^{-2k\pi^2}$, $e^{(1+i)[(\operatorname{Log} 2)/2+i\pi/4]}e^{-2k\pi}$, $e^{2k\pi}$.
4. $(1+i)/\sqrt{2}$; $1 + 2k\pi i$; $-i\operatorname{Log}(\sqrt{2}-1) + 2k\pi$, $\pi + i\operatorname{Log}(\sqrt{2}-1) + 2k\pi$; $\pi/4 + k\pi$; no solution.
9. Partial answer: $|z^i| = e^{-2k\pi-\theta}$, $z \neq 0$;
   $|i^z| = e^{-y(\pi/2+2k\pi)}$; $|z^z| = e^{x\operatorname{Log} r - y\theta + 2k\pi y}$, $z \neq 0$.

Section 5, page 82

1. i, iii, iv.
2. (b) Partial answer: $f(z) = z^2 - iz$, —, $-i(z^3 + z^2/2)$, $z^4(1 - i/4)$.
3. (b) See 2(b) above.
6. (a) See 2(b) above; also $f(z) = z^2/(z+i)$.
7. $i/z + ic$, c real.

Section 6, page 93

1. Partial answer: speed $= 1, 1, \sqrt{2}, 1/|z|^2, 1/|z|^2, 2|z|, 2|z|$. Some flow patterns relevant to this and subsequent problems are sketched in Chapter 5, Sections 3 and 4.
3. (b) Region exterior to circle through a, $ai/\sqrt{3}$, $-ai/\sqrt{3}$.

Section 7, page 94

1.2. Check by addition.

6.1. (a) For sketch see Chapter 5, Section 3.

Chapter 3

Section 1, page 105

1. $2\pi i$.

2. (b) $4z^3$: $-5, 0, 0, 0, -8 + 24i$. $\bar{z}$: $\frac{1}{2} + i, -\pi i, 18\pi i, 4\pi i, 2 + 2i/3$.
$1/z$: $(\text{Log } 2)/2 + i\pi/4, -\pi i, 2\pi i, 4\pi i, (\text{Log } 5)/2 + i \text{ Tan}^{-1}(\frac{1}{2})$.

Section 2, page 116

1. $2\pi i, 0, 0, 0, 0, 0$.
3. (a) $(\frac{20}{3})(2i - 1)$, $(\sqrt{5}/3)(12 + 16i)$; (b) $|I| \leq 20\sqrt{2}$.
5. $\text{Log}[(1 + i\sqrt{3})/(1 - i\sqrt{3})] = 2i\pi/3$.

Section 6, page 145

4. $\displaystyle\sum_{n=0}^{\infty}(-1)^n \frac{z^{2n+1}}{(2n+1)(2n+1)!}, \sum_{n=1}^{\infty}\frac{z^n}{nn!}, \sum_{n=1}^{\infty}(-1)^{n+1}\frac{z^n}{n^2}$.
6. $\displaystyle a_n = \frac{P(1)}{2} - \frac{P(2)}{10}\frac{1}{2^n} + \frac{P(i)}{2-6i}(-i)^n + \frac{P(-i)}{2+6i}(i)^n$.

Section 7, page 152

6. $1, \pi i/2, \infty$.

Section 8, page 161

1. (i) ess sing at ∞; (ii) pole ord 1 at 0, ess sing at ∞; (iii) pole ord 1 at 1, rem sing at 0, ess sing at ∞; (iv) pole ord 1 at i and $-i$, rem sing at ∞; (v) rem sing at 0, pole ord 1 at i, $-i$, pole ord 2 at ∞; (vi) ess sing at ∞; (vii) pole ord 1 at 0, rem sing at π, pole ord 2 at $n\pi \neq 0$ or π, ess sing at ∞ not isolated.
2. (i) ess sing at ∞; (ii) pole ord 1 at $z = \pi/2 + n\pi$, sing at ∞ not isolated; (iii) rem sing at 0, pole ord 1 at $z = 2\pi in \neq 0$, sing at ∞ not isolated; (iv) pole ord 1 at 1 and -1, rem sing at ∞; (v) pole ord 4 at 0, rem sing at ∞; (vi) ess sing at ∞; (vii) rem sing at 0 and π, ess sing at ∞.
6. $a_n = 1/(n+5)!$; $b_{2n-1} = (-1)^n/(2n+1)!$, $b_{2n} = 0$;
$c_k = -6!/[(3+k)!(3-k)!]$.
9. $1 + 2\Sigma_1^{\infty}(-1)^n(1/z)^n$.

11, 12. Check by addition.

Section 9, page 169

3. $\displaystyle n!\int_0^{\pi} e^{m\cos\theta}\cos(m\sin\theta - n\theta)\, d\theta = \pi m^n$ where $m = 1/c$.

Section 10, page 171

6.1. $\displaystyle B_n = \frac{n!}{2\pi}\int_0^{2\pi}\frac{e^{\cos\theta}\cos[(n-1)\theta + \sin\theta] - \cos(n-1)\theta}{e^{2\cos\theta} - 2e^{\cos\theta}\cos(\sin\theta) + 1}\, d\theta$.

6.5. $z - (\frac{1}{2})(\alpha + \beta) - (\frac{1}{8})(\alpha - \beta)^2 z^{-1}$.

8.1. $f(0) = e, f'(0) = -e/2$.

8.2. 10.

Chapter 4

Section 2, page 195

2. (singular point, residue) $= (0,-1)$, $(1,1)$; $(e^{(1+2k)\pi i/4}, (-1)^k/4i)$; $(0, 1/\pi)$; $(\pi, -1 - \pi i)$; $(1, \frac{3}{4})$, $(i, \frac{3}{4} + i/2)$, $(-i, \frac{3}{4} - i/2)$, $(-1, -\frac{9}{4})$.

3. (iii) changes to $\pi i e^{-3/2}$.
7. $I(a) = 0, 1, 3, 2, 3, 1, 0$ in intervals separated by the points $a = -\pi - 4, -4, -\pi, \pi - 4, 0, \pi$.

Section 3, page 202

4. Agrees.

Section 5, page 215

1. $DI_1 = \pi \sin \pi p \cosh \pi q$, $DI_2 = -\pi \cos \pi p \sinh \pi q$ where $D = \sin^2 \pi p + \sinh^2 \pi q$.
2. Note that $x^2 = t$ reduces the first integral to that in Example 5.1.

Section 7, page 231

3. (a) $\pi + 2\,\mathrm{Tan}^{-1}(y/x)$.
4. $\sin at$, $(1 - \cos at)/a$, te^{-at}, $t^7/7!$, $\cos at$.

Section 8, page 240

2. $-1, 0, -1, -1$.
6. $(\pi/4)(2 - \sqrt{2})$, $3\pi/8$, $(\pi/4)(\sqrt{2} - 1)$.

Section 10, page 254

1.3. Domain between two internally tangent circles.

Chapter 5

Section 1, page 266

5. $2z/(z + 1)$, $i(z - 1)/(z + 1)$, z, iz, $(z + i)/(z - 1)$.

Section 2, page 276

2. $w = (iz + \lambda)/(z + i\lambda)$; $w = 0$ for $z = i\lambda$; upper half-plane onto $|w| < 1$ for $\lambda > 0$ and onto $|w| > 1$ for $\lambda < 0$; only some.
3. $w = \beta b(z - \alpha a)/(\bar{\alpha} z - a)$.
4. $w = \gamma(\bar{\alpha} z - 1)/(z - \alpha)$, $|\gamma| = 1$, $|\alpha| < 1$.
5. $w = [z(\alpha - 2) + i(\alpha + 2)]/[z(2\alpha - 1) + i(2\alpha + 1)]$, $|\alpha| = 1$.
6. $w = i(\alpha + 1 - z)/(\alpha - 1 + z)$ where $|\alpha| = 2$.

Section 3, page 287

7. $R = 2 + \sqrt{3}$, $2h(z) = (z/R) + (R/z)$,
   $H(w) = [2w - \sqrt{3}(w^2 - 1)^{1/2}](2 + \sqrt{3})$.

Section 4, page 297

1. γ introduces a rotation of the half-plane and a change of scale.
6. $f'(z) = \gamma z^{-2/3}(z - 1)^{-2/3}$, $f'(z) = \gamma z^{\alpha - 1}(z - 1)^{\beta - 1}$ where two angles are $\pi\alpha$, $\pi\beta$.
7. (a) $f'(z) = \gamma(z + 1)^{-1/2} z^{1/2}$; (b) $f'(z) = \gamma(z + 1)^{1/2}(z - 1)^{-1/2}$. Both equations have elementary integrals $f(z)$.

Section 5, page 304

4. $0 < c < 2/3$, none, $0 < c < \sqrt{578} - 24 \doteq 0.00174$.

5. $c = a + ib$ satisfies $3a < 2$, $8b^2 < a(2 - 3a)^2$. As a check note that $8b^2 = a(2 - 3a)^2$ is the condition for an imaginary root, $z = iy$.
9. For instance, choose c suitably in Problem 4.

Section 8, page 323

1.2. $\dfrac{w-1}{w+1} = 2^n \dfrac{z-1}{z+1}$, hence $w = \dfrac{z(1+2^n) + (1-2^n)}{z(1-2^n) + (1+2^n)} = T^n z$;

$$\frac{w+1}{w-3} = \left(\frac{i-1}{i+3}\right)^n \frac{z+1}{z-3}, \frac{w-1+i}{w-1-i} = \left(\frac{1}{3}\right)^n \frac{z-1+i}{z-1-i}.$$

2.5. $\left|\dfrac{az+b}{cz+d}\right| \leq \dfrac{|bc - ad|\rho + |b\bar{d} - \rho^2 a\bar{c}|}{||d|^2 - \rho^2|c|^2|}$, $|z| = \rho$.

2.6. $|(az+b)/(cz+d)| \leq 1$ for $|z| \leq 1$ if and only if $|bc - ad| + |b\bar{d} - a\bar{c}| \leq |d|^2 - |c|^2$, $|c| < |d|$; $|a + bz/(1 - cz)| \leq 1$ for $|z| \leq 1$ if and only if $1 + 2\mathrm{Re}\, a\bar{b}c \leq (1 - |a|^2)(1 - |c|^2) + (1 - |b|)^2$, $|c|^2 + |b| \leq 1$. Deduce that also $|a|^2 + |b| \leq 1$, so that the final condition is symmetric in (a,c).

3.2. $(u_0 - u_1)\phi^* = A(u - u_1) + B(u_0 - u)$, $\phi^* \operatorname{Log}(r_0/r_1) = A \operatorname{Log}(r/r_1) + B \operatorname{Log}(r_0/r)$, $(\theta_0 - \theta_1)\phi^* = A(\theta - \theta_1) + B(\theta_0 - \theta)$ where $\theta = \operatorname{Arg} w$.

3.3. Parallel lines, concentric circles, radial lines.

3.4. $w = z/(1 - z)$ maps region onto an angle.

3.6. Equipotentials are circles through x_0 and x_1. Flow lines are circles with x_0 and x_1 as inverse points. Compare Figure 3-3.

4.1. $H_1(x,y) = (2/\pi)\operatorname{Arg} \sin z = (2/\pi)\operatorname{Tan}^{-1}(\cot x \tanh y)$;
$H_2(x,y) = H_1(\pi/2 - x, y) = (2/\pi)\operatorname{Tan}^{-1}(\tan x \tanh y)$;
$H_3(x,y) = 1 - H_1(x,y) - H_2(x,y)$. The solution for general A, B, C is $H = AH_1 + BH_2 + CH_3$, hence expressible in terms of H_1.

4.3. $A + (2/\pi)(B - A)\sin^{-1}x$, $0 \leq \sin^{-1}x \leq \pi/2$.

4.4. Let $\zeta = \sqrt{z} = \sin w$. This gives the above formula with $\sqrt{x}$ instead of x.

Chapter 6

Section 2, page 347

2. $\frac{1}{2}$, 1, $1/\sqrt{2}$, $\frac{2}{3}$, e.
5. Residue $(-1)^n/n!$ at $z = -n$; residue $\pm i/(2n)$ at $z = \mp ni$; residue 0 at $z = -n$; residue $(\sinh n^2)/(2n!n)$ at $z = \pm in$.

Section 4, page 362

1. (b) See Chapter 3, Problem 2.2.
2. Yes.
4. Partial answer: $e^{r\cos\theta}\cos(r\sin\theta) = \Sigma(r^n \cos n\theta)/n!$.
5. You get (4.12) with $R = 1$.
6. $f(z) = \dfrac{\theta_0}{\pi} + \dfrac{2}{\pi}\displaystyle\sum_{n=1}^{\infty} \frac{\sin n\theta_0}{n} z^n$.

10. $L(r) = 2\pi r/(1 - r^2)$, $2\pi r J_0(ir)$; $A(r) = \pi r^2/(1 - r^2)^2$, $\pi ir J_0'(2ir)$.

Section 5, page 370

2. $f'(z) = -\int_{-\infty}^{\infty} te^{-t^2} \sin zt.dt, \int_1^{\infty} t^z \operatorname{Log} t \operatorname{sech} t\, dt.$

Section 6, page 380

3. See Problem 4.
6. If $u_n = (2n-1)!/|z|^{2n}$ then $u_{n+1}/u_n > 1$ if and only if $4n^2 + 2n > |z|^2$. The second case is similar.

Section 8, page 396

1. $\Pi(1 - z/k^{3/2})$, $\Pi(1 - z/k^{3/4}) \exp(z/k^{3/4})$,
$\Pi(1 - z/k^{1/2}) \exp[z/k^{1/2} + z^2/(2k)]$.
4. $\Pi_{n=1}^{\infty}(1 - z/n)^n \exp[z + z^2/(2n)]$

Section 9, page 403

3.4. (a) $1 - \dfrac{2m}{2!}z^2 + \dfrac{2^2m(m-2)}{4!}z^4 - \dfrac{2^3m(m-2)(m-4)}{6!}z^6 + \cdots$

$$1 - \frac{m^2}{2!}z^2 + \frac{m^2(m^2-2^2)}{4!}z^4 - \frac{m^2(m^2-2^2)(m^2-4^2)}{6!}z^6 + \cdots$$

$$1 - \frac{m(m+1)}{2!}z^2 + \frac{m(m-2)(m+1)(m+3)}{4!}z^4 - \cdots.$$

(b) $z - \dfrac{2(m-1)}{3!}z^3 + \dfrac{2^2(m-1)(m-3)}{5!}z^5$

$$- \frac{2^3(m-1)(m-3)(m-5)}{7!}z^7 + \cdots$$

$$z - \frac{m^2-1^2}{3!}z^3 + \frac{(m^2-1^2)(m^2-3^2)}{5!}z^5$$

$$- \frac{(m^2-1^2)(m^2-3^2)(m^2-5^2)}{7!}z^7 + \cdots$$

$$z - \frac{(m-1)(m+2)}{3!}z^3 + \frac{(m-1)(m-3)(m+2)(m+4)}{5!}z^5 - \cdots.$$

(c) Setting $z = 0$ gives $mY(0) = -Y'(0)$ if $Y''(0)$ exists;

$$1 - mz - \frac{m(1-m)}{(2!)^2}z^2 - \frac{m(1-m)(2-m)}{(3!)^2}z^3 - \cdots.$$

3.5. $\{a_{2n}\} = \{1, -\pi^2/3, -\pi^4/45, -2\pi^6/945, -\pi^8/4725, \cdots\}$.

9.1. One entire, two entire, one entire, one three-valued, three entire, one two-valued.

Index